14
Springer Series on Chemical Sensors and Biosensors

Methods and Applications

Series Editor: G. Urban

For further volumes:
http://www.springer.com/series/5346

Springer Series on Chemical Sensors and Biosensors

Series Editor: G. Urban

Recently Published and Forthcoming Volumes

Applications of Nanomaterials in Sensors and Diagnostics
Volume Editor: A. Tuantranont
Vol. 14, 2013

Autonomous Sensor Networks: Collective Sensing Strategies for Analytical Purposes
Volume Editor: D. Filippini
Vol. 13, 2013

Designing Receptors for the Next Generation of Biosensors
Volume Editors: S.A. Piletsky, M.J. Whitcombe
Vol. 12, 2013

Solid State Gas Sensors – Industrial Application
Volume Editors: M. Fleischer, M. Lehmann
Vol. 11, 2012

Optical Nano- and Microsystems for Bioanalytics
Volume Editors: W. Fritzsche, J. Popp
Vol. 10, 2012

Mathematical Modeling of Biosensors
An Introduction for Chemists and Mathematicians
Volume Authors: R. Baronas, F. Ivanauskas, J. Kulys
Vol. 9, 2010

Optical Guided-wave Chemical and Biosensors II
Volume Editors: M. Zourob, A. Lakhtakia
Vol. 8, 2010

Optical Guided-wave Chemical and Biosensors I
Volume Editors: M. Zourob, A. Lakhtakia
Vol. 7, 2010

Hydrogel Sensors and Actuators
Volume Editors: Gerlach G., Arndt K. -F.
Vol. 6, 2009

Piezoelectric Sensors
Volume Editors: Steinem C., Janshoff A.
Vol. 5, 2006

Surface Plasmon Resonance Based Sensors
Volume Editor: Homola J.
Vol. 4, 2006

Frontiers in Chemical Sensors
Novel Principles and Techniques
Volume Editors: Orellana G., Moreno-Bondi M. C.
Vol. 3, 2005

Ultrathin Electrochemical Chemo- and Biosensors
Technology and Performance
Volume Editor: Mirsky V. M.
Vol. 2, 2004

Optical Sensors
Industrial, Environmental and Diagnostic Applications
Volume Editors: Narayanaswamy R., Wolfbeis O. S.
Vol. 1, 2003

Applications of Nanomaterials in Sensors and Diagnostics

Volume Editor:
Adisorn Tuantranont

With contributions by

U. Athikomrattanakul · A. Bhattacharya · A. Bonanni · B.H. Chung · B. Collins · D. Dechtrirat · Z. Dong · N. Gajovic-Eichelmann · D. Itoh · Y. Jang · H. Ju · J. Jung · T. Kerdcharoen · S. Kladsomboon · J. Lei · D. Pai · C. Renken · M. Saito · J. Sankar · F.W. Scheller · M. Schulz · M. Somasundrum · P. Sritongkham · H. Suzuki · E. Tamiya · A. Tuantranont · M. del Valle · M.C. Vestergaard · N. Watts · A. Wisitsoraat · M. Yokokawa · H. Yoshikawa · Y. Yun

Editor
Dr. Adisorn Tuantranont
Lab Director, Nanoelectronics
and MEMS Laboratory
National Electronics and Computer
Technology Center (NECTEC)
Pathumthani
Thailand

ISSN 1612-7617
ISBN 978-3-642-36024-4 ISBN 978-3-642-36025-1 (eBook)
DOI 10.1007/978-3-642-36025-1
Springer Heidelberg New York Dordrecht London

Library of Congress Control Number: 2013939353

Printed on acid-free paper

Springer is part of Springer Science+Business Media (www.springer.com)

Series Editor

Prof. Dr. Gerald Urban

IMTEK - Laboratory for Sensors
Institute for Microsystems Engineering
Albert-Ludwigs-University
Georges-Köhler-Allee 103
79110 Freiburg
Germany
urban@imtek.de

Aims and Scope

Chemical sensors and biosensors are becoming more and more indispensable tools in life science, medicine, chemistry and biotechnology. The series covers exciting sensor-related aspects of chemistry, biochemistry, thin film and interface techniques, physics, including opto-electronics, measurement sciences and signal processing. The single volumes of the series focus on selected topics and will be edited by selected volume editors. The *Springer Series on Chemical Sensors and Biosensors* aims to publish state-of-the-art articles that can serve as invaluable tools for both practitioners and researchers active in this highly interdisciplinary field. The carefully edited collection of papers in each volume will give continuous inspiration for new research and will point to existing new trends and brand new applications.

Preface

Nanomaterial is one of the hottest fields in nanotechnology that studies fabrication, characterization, and analysis of materials with morphological features on the nanoscale in at least one dimension. Recent progress in synthesis and fundamental understanding of properties of nanomaterial has led to significant advancement of nanomaterial-based gas/chemical/biological sensors. The most important aspect of nanomaterial is their special properties associated with nanoscale geometries. The most fundamental characteristic of nanomaterial is the high surface area to volume ratio, which results in a number of unusual physical and chemical properties such as high molecular adsorption, large surface tension force, enhanced chemical and biological activities, large catalytic effects, and extreme mechanical strength, but another unique property of nanomaterial and recently most studied is the quantum size effect that leads to their discrete electronic band structure like those of molecules. This quantum property of nanomaterial can lead to an extraordinary high sensitivity and selectivity of biosensors and can be benefit to the field of diagnostics.

In this book, we focus on a wide range of nanomaterials including nanoparticles, quantum dots, carbon nanotubes, molecularly imprinted nanostructures or plastibodies, nanometal, DNA-based structures, smart nanomaterials, nanoprobes, magnetic nanomaterials, organic molecules such as phthalocyanines and porphyrins, and the most amazing novel nanomaterial called graphene, for various gas/chemical/biological sensing applications. Moreover, perspectives of new sensing techniques such as nanoscaled electrochemical detection, functional nanomaterial-amplified optical assay, colorimetric fluorescence, and electrochemiluminescence are reviewed and extensively explained. This book includes recent progress of selected nanomaterials over a broad range of gas/chemical/biological sensing applications, and examples of nanomaterials in sensing and diagnostic application are given.

The use of biofunctional nanomaterials in signal amplification for ultrasensitive biosensing is extensively discussed. The biofunctional nanomaterials with the abilities of specific recognition and signal triggering can be employed as not only excellent carriers but also electronic and optical signal tags to amplify the detection signal. Nanomaterial-based electroanalytical biosensors are discussed to give some ideas and concepts of utilizing nanomaterials for cancer and bone disease diagnostics. Then, new nanomaterial-based electrochemical impedance biosensors applied in cancer and bone disease studies that can detect in real time without any pre-labeling specific biomolecules at previously unattainable ultra-low concentrations are specifically discussed. The hottest area of nanomaterial called "carbon nanomaterial" including carbon nanotube and graphene is up-to-date reviewed. Carbon nanotube-based chemical and biosensors and its integration to microfluidic systems are discussed. Carbon nanotube-based electrochemical sensors integrated into microfluidic systems are extensively surveyed and discussed. Moreover, a comprehensive review of graphene-based chemical and biosensors will help who interests to springboard to the new area of carbon nanomaterial-based sensors more easily. Graphene's synthesis methods, properties, and different types of chemical and biosensors including chemoresistive, electrochemical, and other sensing platforms are described. Newly invented organic nanomaterials such as molecularly imprinted polymers (MIPs) are expansively reviewed and analyzed for sensing and diagnostics of various biological species. Inorganic nanomaterials such as nanometal structures using in localized surface plasmon resonance (LSPR) biosensor platform are discussed including their biomedical diagnosis applications. Naturally derived nanomaterial-based sensors such as DNA sensors (genosensors) employing nanomaterials are extensively described. As quantum effect of nanomaterial is amazing, novel nanoprobes for in vivo cell tracking used for evaluating the therapeutic efficacy will show the potential of this quantum effect for diagnostics. Another organic nanomaterials made of metallo-porphyrin (MP) and metallo-phthalocyanine (MPc) which are optically active are used in optical-based gas sensors and electronic nose systems. Then, this book concludes with the uses of nanotechnology to attain highly sensitive detection in electrochemical microdevices. Issues relating to miniaturization of electrochemical electrode and system are discussed. Various techniques applicable to fabrication and integration of nanoelectrodes are included. With the extensive review of newly discovered nanomaterials used for sensors and diagnostics, this book will be interesting not only for scientists working in the field of nanomaterial-based sensor technology but also for students studying analytical chemistry, biochemistry, electrochemistry, material science, and micro- and nanotechnology.

Pathumthani, Thailand Adisorn Tuantranont

Contents

Nanomaterials for Sensing Applications: Introduction and Perspective

Adisorn Tuantranont

Abstract Recent progress in synthesis and fundamental understanding of properties of nanomaterials has led to significant advancement of nanomaterial-based gas/chemical/biological sensors. This book includes a wide range of nanomaterials including nanoparticles, quantum dots, carbon nanotubes, graphene, molecularly imprinted nanostructures, nanometal structures, DNA-based structures, smart nanomaterials, nanoprobes, magnetic-based nanomaterials, phthalocyanines, and porphyrins organic molecules for various gas/chemical/biological sensing applications. Perspectives of new sensing techniques such as nanoscaled electrochemical detection, functional nanomaterial-amplified optical assay, colorimetric, fluorescence, and electrochemiluminescense are explored.

Keywords Chemical and Biosensors, Gas, Nanomaterials

Contents

A. Tuantranont (✉)
Nanoelectronics and MEMS Laboratory, National Electronics and Computer Technology Center (NECTEC), National Sciences and Technology Development Agency (NSTDA), Pathum Thani, Thailand
e-mail: adisorn.tuantranont@nectec.or.th; adisorn.tuantranont@gmail.com

A. Tuantranont (ed.), *Applications of Nanomaterials in Sensors and Diagnostics*,
Springer Series on Chemical Sensors and Biosensors (2013) 14: 1–16
DOI 10.1007/5346_2012_41,
Published online: 8 December 2012

1 Introduction to Nanomaterials and Their Sensing Applications

Nanomaterial is one of the major fields in nanotechnology that studies fabrication, characterization, and analysis of materials with morphological features on the nanoscale in at least one dimension [1–5]. The nanoscale is usually defined as the size that is smaller than 100 nm. However, it is sometimes extended to a dimension smaller than 1 μm. Recently, the European Commission adopted the definition of a nanomaterial as a natural, incidental, or manufactured material containing particles, in an unbound state or as an aggregate or as an agglomerate and where, for 50% or more of the particles in the number size distribution, one or more external dimensions is in the size range 1–100 nm. In specific cases, the number size distribution threshold of 50% may be replaced by a threshold between 1% and 50%.

Nanomaterials may be classified based on dimensionality (D) of their features into 0D, 1D, 2D, and 3D nanostructures [6]. 0D nanostructures including nanoparticles, nanospheres, quantum dots, isolated molecules and atoms are point structures with nanoscale in all dimensions [7–9]. 1D nanostructures including nanotubes and nanowires are structures with non-nanoscale only in one dimension [8, 10–12]. 2D nanostructures such as nanosheet, nanoplates, nanobelts, and nanodisc are structures with nanoscale in one dimension [13–16]. Lastly, 3D nanostructures such as nanotetrapods, nanoflowers, and nanocombs are arbitrary structures, which contain nanoscale features in any of three dimensions [17, 18]. These nanomaterials can be made of large variety of functional materials, including metals, metal oxides, ionic compound, ceramics, semiconductors, insulators, organics, polymers, biological materials, bioorganisms, and so on. Each functional material can be made in many nanostructure forms. Carbon is one of the most notable examples that all dimensionalities of 0D fullerene (hollow bucky ball) [19], 1D carbon nanotubes (CNTs) [20–23], 2D graphene [13], and 3D graphite nanostructures are available. Apart from carbon, a wide range of nanomaterials with different dimensions of metal [24, 25], metal oxide [8, 15, 26–28], semiconductor [29–33], organic [11, 34, 35], polymers [36], biomaterials [37–40], and their composites [39, 41–44] have been widely reported.

Various forms of nanostructured materials can be synthesized or fabricated by many different methods. In general, nanomaterials can be made by three main approaches, including top-down, bottom-up, and the combination [45–47]. In the first approach, bulk starting materials will be broken down into nanoscale structures by various methods such as photolithographic patterning, wet etching, plasma etching, reactive-ion etching, laser processing, electrochemical etching, and grinding [30, 48–51]. The approach can be used for production of nanoparticles, nanorod, and nanowires of metal oxide, semiconductor, metal, and polymer materials. The main advantages of these methods include well-controlled parameters and large-scale manufacturability. However, they suffer from high material loss, relatively high cost, and slow production rate.

For the second scheme, nanostructures are formed by assembly of atoms or molecules controlled by suitable process parameters of each process [52]. Bottom-up methods are more widely used because they can be better controlled, faster, and more cost effective [53]. Bottom-up schemes can be mainly divided into vapor-phase and solution-route syntheses, in which nanostructures are built up from molecules or atoms in gas and liquid phases, respectively. Widely used vapor-phase methods include chemical vapor deposition (CVD) [54, 55], plasma-enhanced CVD [56, 57], atomic layer deposition [58, 59], thermal/e-beam evaporation [60–62], pulse laser deposition [63–65], sputtering [66], and flame-based synthesis [67]. These techniques have been widely applied for syntheses of metal oxide, semiconductor, metal, and composite nanostructures such as nanoparticles, nanowires, nanotetrapods, nanorods, nanobelts, nanosheets, and nanotubes made of carbon, SnO_2, TiO_2, ZnO, Si, GaAs, Ti, W, etc. They offer several advantages including well-controlled parameters, high-quality and aligned structure, very low contamination, and large-scale manufacturability. However, they normally involve expensive instrumentation, vacuum system, and high-temperature process.

Solution-phase methods including precipitation [68], sol–gel deposition [69], hydrothermal/solvothermal syntheses [70–72], electrochemical deposition [73], self-assembled monolayer [16], molecular self-assembly [74, 75], electrospinning [76, 77], electrospray [78], spray pyrolysis [79], and other chemical routes [80] are relatively simple, of low temperature, and of low cost. They are more suitable for syntheses of organic, polymer, and biological nanomaterials such as nanofibers, nanoparticles, nanosheets of phthalocyanines, porphyrins, polyanilene (PANI), poly (3,4 ethylenedioxythiophene):poly-styrene-sulfonic acid (PEDOT:PSS), polypyrol, polyvinlypyrolidone (PVP), polyacrylonitride (PAN), oxidase enzymes, and deoxyribonucleic acid (DNA) [81–89]. Nevertheless, these methods can also be used to synthesize some metal oxide, semiconductor, metal nanostructures such as nanowires of Au, Ni, Fe, and TiO_2, which often rely on self-assembly of polymer and biological materials such as cells and DNAs [90–97].

In the last approach, the bottom-up and top-down methodologies are combined to realize more sophisticated nanomaterials. First, initial nanomaterials in the form of film or nanostructures are synthesized by a top-down method. Next, initial nanomaterials are further broken down by bottom-down techniques such as wet etching and dry etching. The development of the approach is still in an early stage and there are not many examples of nanomaterial syntheses based on this concept. The first example is anodized alumina (AAO) nanoporous thin film fabricated by the deposition of aluminum thin film and electrochemical anodization in phosphoric acid. The nanopore structure can be used for subsequent bottom-up growth of nanowires [98–101]. Similarly, nanoporous silicon thin film can also be made by sputtering of amorphous silicon layer and electrochemical or plasma etching [102]. Another notable example is the fabrication of graphene sheet from CNTs. CNTs synthesized by CVD process were etched along their sidewall by photoresist masking and oxygen plasma etching [103]. Another interesting example is silver nanowire formed by laser shock on silver thin film [13].

The most important aspect of nanomaterials is their special properties associated with nanoscale dimensions. The most fundamental characteristic of nanomaterials is the high surface-area-to-volume ratio, which results in a number of unusual physical and chemical properties such as high molecular adsorption, large surface tension force, enhanced chemical and biological activities, large catalytic effects, and extreme mechanical strength [104–106]. Another unique property of nanomaterials is the quantum size effect that leads to their discrete electronic band structure like those of molecules. Unlike the increased surface-to-volume ratio that also occurs when going from macro to micro dimensions, quantum effect is only specific to deep nanoscale dimension of smaller than a few tens of nanometer [107, 108].

The nanomaterials are thus highly useful for a wide range of nanotechnology fields including nanoelectronics [108–110], optoelectronics [109], nanophotonics [111–114], nano-electromechanical systems (NEMS) [115], bioelectronics [116], nanobiotechnology [117, 118], nanochemistry [119], biochemistry [120, 121], biomedicine [122–124], electrochemistry [125], nanomechanics [126, 127], and so on. These lead to a large variety of applications such as quantum-effect lasers/solar cells/transistors [128, 129], photonic band gap devices [113, 114], catalyst [130, 131], photocatalyst [132, 133], molecular electronic device [8], surface-enhanced Raman spectroscopy (SERS) [134], nano fuel cells [135, 136], nano drug delivery systems [41, 137], nanosensors [20, 25, 138, 139], advanced energy storage devices [140–142], and nanoactuators. Among these, sensors are among the fastest-growing applications due to their huge demands in many real-world application fields such as automobiles, communication, consumer electronics, industrial, and biomedical. Sensors can be divided into several classes including mechanical, thermal, optical, magnetic, gas, chemical, and biological.

Among various kinds of sensors, gas/chemical/biological sensors can exploit the most benefits from high surface-to-volume-ratio property of nanomaterials [143, 144]. Gas/chemical/biological sensors generally comprise sensing material that responds to changes of gas/chemical/biological analytes and transducer that converts the changes into electrical signals. Gas sensor may be classified by sensing mechanisms into chemoresistive, surface acoustic wave (SAW), quartz crystal microbalance (QCM), chemiluminescent, optical absorption, and dielectric types [145–149]. Gas-sensing applications include toxic gases such as NO_2, CO, SO_2, NH_3, O_3, and H_2S; flammable gases such as H_2, CH_4, C_2H_2, and C_3H_8; and volatile organic compounds (VOCs) such as ethanol, acetone, methanol, and propanol [146–150]. Similarly, chemical sensors can be divided by sensing platforms into electrochemical, ion-sensitive field effect, chemiluminescent, optical, and mass spectroscopic ones [151–153]. Chemical sensing applications are much wider than gas-sensing ones as they include a large number of liquid-phase chemicals ranging from acids, bases, solvents, and inorganic substances to organic analytes [154]. Likewise, widely used biosensing platforms include electrochemical, fluorescent, surface plasmon resonance (SPR), QCM, and microcantilever [20, 155–158]. Biosensing applications also cover a very broad range of biologically relevant materials including bioanalytes found in living organisms such as glucose, cholesterol and uric acid, DNAs, RNAs, cells, proteins, organelles, and so on [12, 157–160].

The main and common requirement of these sensors is high sensitivity and specificity. The specific surface area of sensing material is one of the most important factors that dictate the sensitivity as it directly related to adsorption or reaction rate with target analytes [161]. Gas/chemical/biological sensors developed based on well-established microtechnology are now currently used in commercial applications. They provide good sensitivity and reproducibility along with low power consumption. However, their performances are still not satisfactory for many advanced applications that involve detection of very low concentration analytes. The use of nanomaterials in these sensors will provide substantial improvement of sensing performances due to several orders of magnitude increase of specific surface area and smaller size [162–164]. Well-controlled synthesis and fundamental understanding of properties of nanomaterials are very important for the advancement of nanomaterial-based gas/chemical/biological sensors.

Recently, there has been significant progress in development of nanomaterial-based sensors. A wide variety of nanostructured materials and composites have been devised on different sensing platforms by a number of preparation methods for various sensing applications. For instance, high-sensitivity chemoresistive gas sensors based on metal oxide nanostructures such as SnO_2 nanowires, ZnO nanotetrapods, and TiO_2 nanorods have been extensively explored [144]. In addition, highly sensitive electrochemical biosensors based on the combination of biofunctional materials such as enzymes, antibody and DNAs, and novel electrode materials such as carbon/metal/conductive polymer/metal-oxide nanostructures, and nanocomposites such as CNTs, graphene, gold nanoparticles, CNTs/polyaniline, CNTs/ZnO, CNTs/gold nanoparticles graphene/polythiophene and alike are of great interest [12, 165, 166]. This book includes recent progress of selected nanomaterials over a broad range of gas/chemical/biological sensing applications and it is organized as follows.

In Chap. II the use of biofunctional nanomaterials in signal amplification for ultrasensitive biosensing has been discussed. The biofunctional nanomaterials with the abilities of specific recognition and signal triggering can be employed as not only excellent carriers, but also electronic and optical signal tags to amplify the detection signal. Two approaches including noncovalent interaction and covalent route for the functionalization of nanomaterials with biomolecules are described. The performance in terms of sensitivity and specificity are also digested.

In Chap. III, nanomaterial-based electroanalytical biosensors are reported and emphasized for cancer and bone disease diagnostics. The existing biosensor technologies, the mechanisms and applications of two types of electroanalytical biosensors and advantages of nanomaterials in developing these biosensors are described. Then, new nanomaterial-based electrochemical impedance biosensors applied in cancer and bone disease studies that can detect in real time without any pre-labeling-specific biomolecules at previously unattainable ultra-low concentrations are specifically discussed.

Chapter IV deals with CNT-based chemical and biosensors and its integration to microfluidic systems. Different components necessary for the construction of a microfluidic system including micropump, microvalve, micromixer, and detection system utilizing CNT-based electrochemical sensors are extensively surveyed and discussed.

Chapter V covers a comprehensive review of graphene-based chemical and biosensors. These include graphene's synthesis methods, properties, and different types of chemical and biosensors including chemoresistive, electrochemical, and other sensing platforms. In addition, concluding remarks for further development of graphene-based chemical and biosensors are provided.

In Chap. VI, molecularly imprinted polymers (MIPs) for sensing and diagnostics of various biological species are expansively reviewed and analyzed. The design of novel artificial MIPs and the limitations of the classical non-covalent imprinting approach are discussed. Some novel strategies for the molecular imprinting of macromolecules such as the use of complementary functional monomers and a new electrochemical approach to the imprinting of peptides and proteins as well as new concepts for the integration with transducers and sensors are described.

Chapter VII reports design, synthesis, fabrication, properties, and biomedical diagnosis applications of nanometal structures including Au and Ag nanoparticles (NPs) based on localized surface plasmon resonance (LSPR) biosensor platform. The characteristics including enhanced sensitivity, label-free detection capability, specific changes in their absorbance responses upon binding with various molecules are demonstrated and discussed.

In Chap. VIII, DNA sensors (genosensors) employing nanomaterials for diagnostic applications are extensively described. These DNA sensors employ electrochemical impedance principle to detect hybridization of a target clinical diagnostic-related gene with the complementary probe genes with no labeling. The use of nanocomponents to improve sensor performance, mainly CNTs integrated in the sensor platform, or nanoparticles, for signal amplification and their diagnostic applications will be reviewed.

Chapter IX describes novel nanoprobes for in vivo cell tracking used for evaluating the therapeutic efficacy by measuring the changes in tumor volume and tumor markers after cell-based immunotherapy. Various molecular probes and imaging modalities including intrinsic or extrinsic therapeutic cells' modification with proper molecular probes and in vitro amplification as well as recent advances in molecular imaging probes are discussed. Their application in relation to in vivo tracking of dendritic cells (DCs), natural killer (NK) cells, and T cells are then addressed.

Chapter X includes optical chemical gas sensor and electronic nose based on optically active organic nanomaterials made of metallo-porphyrin (MP) and metallo-phthalocyanine (MPc). The gas-sensing mechanism, preparation methods of sensors, the optical absorption spectral measurement under ambient conditions, and application to electronic nose with principal component analysis (PCA) are described.

In Chap. XI, the uses of nanotechnology to attain highly sensitive detection in electrochemical microdevices are reviewed. Issues relating to miniaturization of electrochemical electrode and system are discussed. Various techniques applicable to fabrication and integration of nanoelectrodes are included.

2 Perspective of Nanomaterial Development for Sensing Applications

The development of nanomaterials for sensing applications is still in an early stage and there remains much more work to be done and some challenging issues to be overcome before nanomaterials can successfully be commercialized. Novel functional nanomaterials and new synthesis methods are still being further explored to achieve sensors with ultra-high sensitivity. Among novel nanomaterials, graphene and its composites are especially promising and their research in sensing applications has been growing tremendously [167–169]. Moreover, several nanomaterials have not yet been studied in many gas/chemical/biological applications due to application diversity, and these explorations are highly needed. Among these, new biological sensing applications such as virus-causing newly born infectious diseases and dangerous diseases such as cancer are of particular interest [170–172]. In addition, nanomaterials have not yet been applied in several sensing platforms to optimize their sensing capability. Thus, the integration of nanomaterials in novel sensing platforms such as plasmonic-based sensors is another important research direction [173, 174]. Furthermore, nanomaterial-based sensors should be integrated into processing systems such as microfluidics or lab-on-a-chip so that sample preprocessing and analysis can be automated. Presently, only some nanomaterial-based sensors have been successfully embedded in microfluidic devices [175–177]. Thus, fabrication of microfluidic devices with integrated nanomaterial-based sensors should be further developed.

One of the most important problems of nanomaterial-based sensors is their poor reproducibility because it is difficult to control the structure and arrangement of nanomaterial on sensor. Highly controlled synthesis and manipulation of nanomaterials are still major technological challenges [178]. Therefore, highly ordered nanomaterials and their implementation in sensing platforms are among the most important research topics in nanomaterials [179]. This leads to a new research field, namely *Nanoarchitectonics*, which is a conceptual paradigm for design and synthesis of dimension-controlled functional nanomaterials [180]. Self-assembled processes for various nanomaterials and structures are the most promising keys to achieve these nanostructures [46, 181–183]. However, effective methods and supporting instrumentation are still lacking and require significant technological development such as novel methods for arbitrary guiding assembly [184–187].

Another potential difficulty is high mass manufacturing cost due to sophisticated processing and instrumentation. Thus, development of fabrication process for low-cost and well-controlled large-scale nanostructure in sensing devices is another important future research topic. Chemical route syntheses [188–190] and printing techniques [191, 192] such as inkjet, gravure, and screen printing on low-cost, flexible substrates such as polymers and paper are among potential solutions to realize low-cost and disposable nanomaterial-based sensors, and research in this area should earn particular attention. Moreover, the integration of flexible and printed nanosensors with organic and printed electronics (OPE) for full functional sensing devices and systems will be a very active research field due to their important applications in smart textile, smart clothing, smart paper, and so on [193].

Acknowledgements Adisorn Tuantranont acknowledges Thailand Research Fund (TRF) for Researcher Career Development Funding (RSA5380005) and all financial support from NSTDA and NRCT.

References

1. Colvin VL (2003) The potential environmental impact of engineered nanomaterials. Nat Biotechnol 21(10):1166–1170
2. Martin CR (1994) Nanomaterials: a membrane-based synthetic approach. Science 266 (5193):1961–1966
3. Martin CR (1996) Membrane-based synthesis of nanomaterials. Chem Mater 8(8):1739–1746
4. Nel A, Xia T, Mudler L et al (2006) Toxic potential of materials at the nanolevel. Science 311 (5761):622–627
5. Oberdarster G, Oberdarster E, Oberdarster J (2005) Nanotoxicology: an emerging discipline evolving from studies of ultrafine particles. Environ Health Perspect 113(7):823–839
6. Pokropivny VV, Skorokhod VV (2008) New dimensionality classifications of nanostructures. Physica E 40(7):2521–2525
7. Chan WCW, Maxwell DJ, Gao X et al (2002) Luminescent quantum dots for multiplexed biological detection and imaging. Curr Opin Biotechnol 13(1):40–46
8. Pashchanka M, Hoffmann RC, Gurlo A et al (2010) Molecular based, chimie douce approach to 0d and 1d indium oxide nanostructures. Evaluation of their sensing properties towards co and h 2. J Mater Chem 20(38):8311–8319
9. Sun EY, Josephson L, Weissleder R (2006) "Clickable" nanoparticles for targeted imaging. Mol Imaging 5(2):122–128
10. Hillhouse HW (2011) Development of double-gyroid nanowire arrays for photovoltaics. In: Proceedings of the 2011 AIChE annual meeting. October 16–21, 2011, Minneapolis, MN
11. Zhao YS, Fu H, Peng A et al (2009) Construction and optoelectronic properties of organic one-dimensional nanostructures. Acc Chem Res 43(3):409–418
12. Chopra N, Gavalas VG, Hinds BJ et al (2007) Functional one-dimensional nanomaterials: applications in nanoscale biosensors. Anal Lett 40(11):2067–2096
13. Wang Y, Li Z, Wang J et al (2011) Graphene and graphene oxide: biofunctionalization and applications in biotechnology. Trends Biotechnol 29(5):205–212
14. Yi DK, Lee JH, Rogers JA et al (2009) Two-dimensional nanohybridization of gold nanorods and polystyrene colloids. Appl Phys Lett 94(8)
15. Chen JS, Archer LA, Wen Lou X (2011) SnO_2 hollow structures and TiO_2 nanosheets for lithium-ion batteries. J Mater Chem 21(27):9912–9924
16. Ciesielski A, Palma CA, Bonini M et al (2010) Towards supramolecular engineering of functional nanomaterials: pre-programming multi-component 2d self-assembly at solid–liquid interfaces. Adv Mater 22(32):3506–3520
17. Lu J, Dongning Y, Jie L et al (2008) Three dimensional single-walled carbon nanotubes. Nano Lett 8(10):3325–3329
18. Song HS, Zhang WJ, Cheng C et al (2011) Controllable fabrication of three-dimensional radial ZnO nanowire/silicon microrod hybrid architectures. Cryst Growth Des 11(1):147–153
19. Ivanovskii AL (2003) Fullerenes and related nanoparticles encapsulated in nanotubes: synthesis, properties, and design of new hybrid nanostructures. Russ J Inorg Chem 48 (6):846–860

20. Wang J (2005) Carbon-nanotube based electrochemical biosensors: a review. Electroanalysis 17(1):7–14
21. Chen RJ, Bangsaruntip S, Drouvalakis KA et al (2003) Noncovalent functionalization of carbon nanotubes for highly specific electronic biosensors. Proc Natl Acad Sci USA 100 (9):4984–4989
22. Lam CW, James JT, McCluskey R et al (2004) Pulmonary toxicity of single-wall carbon nanotubes in mice 7 and 90 days after intractracheal instillation. Toxicol Sci 77(1):126–134
23. Jeong S, Shim HC, Kim S et al (2009) Efficient electron transfer in functional assemblies of pyridine-modified nQDs on SWNTs. ACS Nano 4(1):324–330
24. Sondi I, Salopek-Sondi B (2004) Silver nanoparticles as antimicrobial agent: a case study on EColi as a model for gram-negative bacteria. J Colloid Interface Sci 275(1):177–182
25. Colaianni L, Kung SC, Taggart D et al (2009) Gold nanowires: deposition, characterization and application to the mass spectrometry detection of low-molecular weight analytes. In: 3rd international workshop on advances in sensors and interfaces (IWASI 2009), pp 20–24. June 25–26, 2009, Trani (Bari), Italy
26. Patzke GR, Krumeich F, Nesper R (2002) Oxidic nanotubes and nanorods – anisotropic modules for a future nanotechnology. Angew Chem Int Ed 41(14):2446–2461
27. Chen X, Mao SS (2007) Titanium dioxide nanomaterials: synthesis, properties, modifications and applications. Chem Rev 107(7):2891–2959
28. Sayle TXT, Maphanga RR, Ngoepe PE et al (2009) Predicting the electrochemical properties of MnO_2 nanomaterials used in rechargeable Li batteries: simulating nanostructure at the atomistic level. J Am Chem Soc 131(17):6161–6173
29. Pietryga JM, Zhuravlev KK, Whitehead M et al (2008) Evidence for Barrierless auger recombination in PbSe nanocrystals: a pressure-dependent study of transient optical absorption. Phys Rev Lett 101(21)
30. Chen Y, Xu Z, Gartia MR et al (2010) Ultrahigh throughput silicon nanomanufacturing by simultaneous reactive ion synthesis and etching. ACS Nano 5(10):8002–8012
31. Fang X, Zhang L (2006) One-dimensional (1d) ZnS nanomaterials and nanostructures. J Mater Sci Technol 22(6):721–736
32. Madsen M, Takei K, Kapadia R et al (2010) Nanoscale semiconductor "X" on substrate "Y" – processes, devices, and applications. Adv Mater 23(28):3115–3127
33. Okamoto H, Sugiyama Y, Nakano H (2011) Synthesis and modification of silicon nanosheets and other silicon nanomaterials. Chem Eur J 17(36):9864–9887
34. Forey C, Mellot-Draznieks C, Serre C et al (2005) Chemistry: a chromium terephthalate-based solid with unusually large pore volumes and surface area. Science 309 (5743):2040–2042
35. Zhao YS, Fu H, Peng A et al (2008) Low-dimensional nanomaterials based on small organic molecules: preparation and optoelectronic properties. Adv Mater 20(15):2859–2876
36. Barbero CA, Acevedo DF, Yslas E et al (2010) Synthesis, properties and applications of conducting polymer nano-objects. Mol Cryst Liq Cryst 521:214–228
37. Katz E, Willner I (2004) Integrated nanoparticle-biomolecule hybrid systems: synthesis, properties, and applications. Angew Chem Int Ed 43(45):6042–6108
38. Liu L, Busuttil K, Zhang S et al (2011) The role of self-assembling polypeptides in building nanomaterials. Phys Chem Chem Phys 13(39):17435–17444
39. George J, Ramana KV, Bawa AS et al (2011) Bacterial cellulose nanocrystals exhibiting high thermal stability and their polymer nanocomposites. Int J Biol Macromol 48(1):50–57
40. Lo PK, Karam P, Aldaye FA et al (2010) Loading and selective release of cargo in DNA nanotubes with longitudinal variation. Nat Chem 2(4):319–328
41. Arami H, Stephen Z, Veiseh O et al (2011) Chitosan-coated iron oxide nanoparticles for molecular imaging and drug delivery. Adv Polym Sci 243:169–184
42. Bao C, Tian F, Estrada G (2010) Improved visualization of internalized carbon nanotubes by maximising cell spreading on nanostructured substrates. Nano Biomed Eng 2(4):201–207

43. Chen J, Cheng F (2009) Combination of lightweight elements and nanostructured materials for batteries. Acc Chem Res 42(6):713–723
44. Sotiriou GA, Teleki A, Camenzind A et al (2011) Nanosilver on nanostructured silica: antibacterial activity and ag surface area. Chem Eng J 170(2–3):547–554
45. Wang X, Zhuang J, Peng Q et al (2005) A general strategy for nanocrystal synthesis. Nature 437(7055):121–124
46. Zhang S (2003) Fabrication of novel biomaterials through molecular self-assembly. Nat Biotechnol 21(10):1171–1178
47. Gasparotto A, Barreca D, MacCato C et al (2012) Manufacturing of inorganic nanomaterials: concepts and perspectives. Nanoscale 4(9):2813–2825
48. Bang J, Bae J, Lwenhielm P et al (2007) Facile routes to patterned surface neutralization layers for block copolymer lithography. Adv Mater 19(24):4552–4557
49. Chomistek KJ, Panagiotou T (2011) Large scale nanomaterial production using microfluidizer high shear processing. In: Materials research society symposium proceedings, pp 85–94
50. Diegoli S, Hamlett CAE, Leigh SJ et al (2007) Engineering nanostructures at surfaces using nanolithography. Proc Inst Mech Eng G J Aerosp Eng 221(4):589–629
51. Nakata Y, Miyanaga N, Okada T (2007) Topdown femtosecond laser-interference technique for the generation of new nanostructures. J Phys Conf Ser 59(1):245–248
52. Dessert PE (2010) Additive manufacturing of nano-materials: cornerstone for the new manufacturing paradigm. In: Technical paper – Society of Manufacturing Engineers, pp 1–8
53. Zhang S (2003) Building from the bottom up. Mater Today 6(5):20–27
54. Reina A, Jia X, Ho J et al (2009) Large area, few-layer graphene films on arbitrary substrates by chemical vapor deposition. Nano Lett 9(1):30–35
55. Wei D, Liu Y, Wang Y et al (2009) Synthesis of n-doped graphene by chemical vapor deposition and its electrical properties. Nano Lett 9(5):1752–1758
56. Mariotti D, Sankaran RM (2010) Microplasmas for nanomaterials synthesis. J Phys D Appl Phys 43(32)
57. Shang NG, Papakonstantinou P, McMullan M et al (2008) Catalyst-free efficient growth, orientation and biosensing properties of multilayer graphene nanoflake films with sharp edge planes. Adv Funct Mater 18(21):3506–3514
58. Sun X, Dong Y, Li C et al (2010) Characterization and fabrication of rare-earth doped amplifying fibers based on atomic layer deposition. In: Proceedings of SPIE – The International Society for Optical Engineering
59. Xu XD, Wang YC, Liu ZF (2005) Large-scale fabrication of uniform gold nanoparticles in ultrahigh vacuum. J Cryst Growth 285(3):372–379
60. Kim HW, Lee JW, Shim SH et al (2007) Controlled growth of SnO_2 nanorods by thermal evaporation of Sn powders. J Korean Phys Soc 51(1):198–203
61. Liu WC, Cai W (2008) One-dimensional and quasi-one-dimensional ZnO nanostructures prepared by spray-pyrolysis-assisted thermal evaporation. Appl Surf Sci 254(10):3162–3166
62. Wang M, Fei GT (2009) Synthesis of tapered CdS nanobelts and CdSe nanowires with good optical property by hydrogen-assisted thermal evaporation. Nanoscale Res Lett 4 (10):1166–1170
63. Ageeva SA, Bobrinetskii II, Konov VI et al (2009) 3D nanotube-based composites produced by laser irradiation. Quantum Electron 39(4):337–341
64. Choopun S, Tabata H, Kawai T (2005) Self-assembly ZnO nanorods by pulsed laser deposition under argon atmosphere. J Cryst Growth 274(1–2):167–172
65. Park HK, Schriver KE, Haglund Jr RF (2011) Resonant infrared laser deposition of polymer-nanocomposite materials for optoelectronic applications. Appl Phys A Mater Sci Process 105 (3):583–592
66. Dolatshahi-Pirouz A, Jensen T, Vorup-Jensen T et al (2010) Synthesis of functional nanomaterials via colloidal mask templating and glancing angle deposition (glad). Adv Eng Mater 12(9):899–905

67. Merchan-Merchan W, Saveliev AV, Desai M (2009) Novel flame-gradient method for synthesis of metal oxide nanomaterials. In: Nanotechnology 2009: fabrication, particles, characterization, MEMS, electronics and photonics – technical proceedings of the 2009 NSTI nanotechnology conference and Expo, NSTI-Nanotech 2009, pp 76–79
68. Li X, Wang H, Robinson JT et al (2009) Simultaneous nitrogen doping and reduction of graphene oxide. J Am Chem Soc 131(43):15939–15944
69. Malfatti L, Innocenzi P (2011) Sol–gel chemistry: from self-assembly to complex materials. J Sol–Gel Sci Technol 60(3):226–235
70. Chen J, Li W (2011) Hydrothermal synthesis of high densified CdS polycrystalline microspheres under high gravity. Chem Eng J 168(2):903–908
71. Li H, Lu Z, Li Q et al (2011) Hydrothermal synthesis and properties of controlled α-Fe_2O_3 nanostructures in HEPES solution. Chem Asian J 6(9):2320–2331
72. Li Q, Kang Z, Mao B et al (2008) One-step polyoxometalate-assisted solvothermal synthesis of ZnO microspheres and their photoluminescence properties. Mater Lett 62(16):2531–2534
73. Lai Y, Lin Z, Chen Z et al (2010) Fabrication of patterned CdS/TiO_2 heterojunction by wettability template-assisted electrodeposition. Mater Lett 64(11):1309–1312
74. Houlton A, Pike AR, Angel Galindo M et al (2009) DNA-based routes to semiconducting nanomaterials. Chem Commun (Cambridge, England) 14:1797–1806
75. Samano EC, Pilo-Pais M, Goldberg S et al (2011) Self-assembling DNA templates for programmed artificial biomineralization. Soft Matter 7(7):3240–3245
76. Eid C, Brioude A, Salles V et al (2010)Iron based 1d nanostructures by electrospinning process. In: Nanotechnology 2010: electronics, devices, fabrication, MEMS, fluidics and computational – technical proceedings of the 2010 NSTI nanotechnology conference and Expo, NSTI-Nanotech 2010, pp 95–98
77. Miao J, Miyauchi M, Simmons TJ et al (2010) Electrospinning of nanomaterials and applications in electronic components and devices. J Nanosci Nanotechnol 10(9):5507–5519
78. Jayasinghe SN (2008) Electrospray self-assembly: an emerging jet-based route for directly forming nanoscaled structures. Physica E 40(9):2911–2915
79. Rider DA, Liu K, Eloi JC et al (2008) Nanostructured magnetic thin films from organometallic block copolymers: pyrolysis of self-assembled polystyrene-block-poly(ferrocenylethylmethylsilane). ACS Nano 2(2):263–270
80. Feng X, Hu G, Hu J (2011) Solution-phase synthesis of metal and/or semiconductor homojunction/heterojunction nanomaterials. Nanoscale 3(5):2099–2117
81. Cha JN, Hung AM, Noh H (2010) Biomolecular architectures and systems for nanoscience engineering. In: Proceedings of SPIE – The International Society for Optical Engineering
82. Darling SB (2007) Directing the self-assembly of block copolymers. Prog Polym Sci (Oxford) 32(10):1152–1204
83. Gerasopoulos K, McCarthy M, Banerjee P et al (2010) Biofabrication methods for the patterned assembly and synthesis of viral nanotemplates. Nanotechnology 21(5)
84. Giacomelli C, Schmidt V, Aissou K et al (2010) Block copolymer systems: from single chain to self-assembled nanostructures. Langmuir 26(20):15734–15744
85. Kumar CV, Deshapriya IK, Duff MR Jr et al (2010) Novel, simple, versatile and general synthesis of nanoparticles made from proteins, nucleic acids and other materials. J Nano Res 12:77–88
86. Li S, Ji Y, Chen P et al (2010) Surface-induced phase transitions in dense nanoparticle arrays of lamella-forming diblock copolymers. Polymer 51(21):4994–5001
87. Liu LH, Yan M (2010) Perfluorophenyl azides: new applications in surface functionalization and nanomaterial synthesis. Acc Chem Res 43(11):1434–1443
88. Lohuller T, Aydin D, Schwieder M et al (2011) Nanopatterning by block copolymer micelle nanolithography and bioinspired applications. Biointerphases 6(1):MR1–MR12
89. Zhang Z, Fu Y, Li B et al (2011) Self-assembly-based structural DNA nanotechnology. Curr Org Chem 15(4):534–547

90. Carter JD, Labean TH (2011) Organization of inorganic nanomaterials via programmable DNA self-assembly and peptide molecular recognition. ACS Nano 5(3):2200–2205
91. Hegmann T, Qi H, Marx VM (2007) Nanoparticles in liquid crystals: synthesis, self-assembly, defect formation and potential applications. J Inorg Organomet Polym Mater 17 (3):483–508
92. Holmes JD, Lyons DM, Ziegler KJ (2003) Supercritical fluid synthesis of metal and semiconductor nanomaterials. Chem Eur J 9(10):2144–2150
93. Hung AM, Micheel CM, Bozano LD et al (2010) Large-area spatially ordered arrays of gold nanoparticles directed by lithographically confined DNA origami. Nat Nanotechnol 5 (2):121–126
94. Hung AM, Noh H, Cha JN (2010) Recent advances in DNA-based directed assembly on surfaces. Nanoscale 2(12):2530–2537
95. Ikkala O, Ras RHA, Houbenov N et al (2009) Solid state nanofibers based on self-assemblies: from cleaving from self-assemblies to multilevel hierarchical constructs. Faraday Discuss 143:95–107
96. Jaswal VS, Banipal PK, Kaura A et al (2011) Bovine serum albumin driven interfacial growth of selenium-gold/silver hybrid nanomaterials. J Nanosci Nanotechnol 11(5):3824–3833
97. Liyao Z, Mi Z, Qiang Y et al (2009) Asymmetric modification and controlled assembly of nanoparticles. Prog Chem 21(7–8):1389–1397
98. Jagminasa A, Valsiūnas I, Šimkūnaitėa B, Vaitkus R (2008) Peculiarities of bi0 nanowire arrays growth within the alumina template pores by ac electrolysis. J Crys Growth 310 (19):4351–4357
99. Ma MG, Zhu JF (2009) A facile solvothermal route to synthesis of a-alumina with bundle-like and flower-like morphologies. Mater Lett 63(11):881–883
100. Tsai KT, Huang YR, Lai MY et al (2010) Identical-length nanowire arrays in anodic alumina templates. J Nanosci Nanotechnol 10(12):8293–8297
101. Wan L, Fu H, Shi K et al (2008) Simple synthesis of mesoporous alumina thin films. Mater Lett 62(10–11):1525–1527
102. Chifen AN, Knoll W, Forch R (2007) Fabrication of nano-porous silicon oxide layers by plasma polymerisation methods. Mater Lett 61(8–9):1722–1724
103. Jiao L, Zhang L, Wang X et al (2009) Narrow graphene nanoribbons from carbon nanotubes. Nature 458(7240):877–880
104. Barbieri O, Hahn M, Herzog A et al (2005) Capacitance limits of high surface area activated carbons for double layer capacitors. Carbon 43(6):1303–1310
105. Chae HK, Siberio-Perez DY, Kim J et al (2004) A route to high surface area, porosity and inclusion of large molecules in crystals. Nature 427(6974):523–527
106. McIlroy DN, Corti G, Cantrell T et al (2009) Engineering high surface area catalysts for clean tech applications. In: Technical proceedings of the 2009 NSTI nanotechnology conference and expo, NSTI-Nanotech 2009, pp 111–114
107. Acharya S, Sarma DD, Golan Y et al (2009) Shape-dependent confinement in ultrasmall zero-, one-, and two-dimensional pbs nanostructures. J Am Chem Soc 131(32):11282–11283
108. Tsai C, Tseng RJ, Yang Y et al (2008) Quantum dot functionalized one dimensional virus templates for nanoelectronics. J Nanoelectron Optoelectron 3(2):133–136
109. Kaul AB, Megerian K, Bagge L et al (2010) Carbon-based nanodevices for electronic and optical applications. In: Nanotechnology 2010: electronics, devices, fabrication, MEMS, Fluidics and computational – technical proceedings of the 2010 NSTI nanotechnology conference and expo, NSTI-Nanotech 2010, pp 304–307
110. Ravindiran, Shankar P (2011) Nanoelectronics approach based on nano structures & nanomaterial. In: IET conference publications, pp 721–726. July 20–22, 2011, Chennai, India
111. Botey M, Martorell J, Lozano G et al (2010) Anomalous group velocity at the high energy range of real 3d photonic nanostructures. In: Proceedings of SPIE – The International Society for Optical Engineering

112. Chung YW, Leu IC, Lee JH et al (2005) Fabrication and characterization of photonic crystals from colloidal processes. J Cryst Growth 275(1–2):e2389–e2394
113. Botey M, Lozano G, Marguez H et al (2011) Anomalous light propagation, finite size-effects and losses in real 3d photonic nanostructures. In: International conference on transparent optical networks. June 26–30, 2011, Stockholm, Sweden
114. Jiang P, Sun CH, Linn NC et al (2007) Self-assembled photonic crystals and templated nanomaterials. Curr Nanosci 3(4):296–305
115. Kratochvil BE, Dong L, Zhang L et al (2007) Automatic nanorobotic characterization of anomalously rolled-up sige/si helical nanobelts through vision-based force measurement. In: Proceedings of the 3rd IEEE international conference on automation science and engineering (IEEE CASE 2007), pp 57–62. September 22–25, 2007, Scottsdale, AZ, USA
116. Kim J, Kim BC, Lopez-Ferrer D et al (2010) Nanobiocatalysis for protein digestion in proteomic analysis. Proteomics 10(4):687–699
117. Euliss LE, DuPont JA, Gratton S et al (2006) Imparting size, shape, and composition control of materials for nanomedicine. Chem Soc Rev 35(11):1095–1104
118. Soto CM, Ratna BR (2010) Virus hybrids as nanomaterials for biotechnology. Curr Opin Biotechnol 21(4):426–438
119. Stender CL, Sekar P, Odom TW (2008) Solid-state chemistry on a surface and in a beaker: unconventional routes to transition metal chalcogenide nanomaterials. J Solid State Chem 181(7):1621–1627
120. Geng L, Jiang P, Xu J et al (2009) Applications of nanotechnology in capillary electrophoresis and microfluidic chip electrophoresis for biomolecular separations. Prog Chem 21 (9):1905–1921
121. Vicens J, Vicens Q (2011) Emergences of supramolecular chemistry: from supramolecular chemistry to supramolecular science. J Incl Phenom Macrocycl Chem 71(3–4):251–274
122. Petrovie ZL, Radmilovic-Radenovic M, Maguire P et al (2010) Application of non-equilibrium plasmas in top-down and bottom-up nanotechnologies and biomedicine. In: Proceedings of the 2010 27th international conference on microelectronics (MIEL 2010), pp 29–36. May 16-19, 2010, Aleksan Nis, Serbia
123. Pierstorff E, Ho D (2007) Monitoring, diagnostic, and therapeutic technologies for nanoscale medicine. J Nanosci Nanotechnol 7(9):2949–2968
124. Verma S, Domb AJ, Kumar N (2011) Nanomaterials for regenerative medicine. Nanomedicine 6(1):157–181
125. Chen D, Gao Y, Wang G et al (2007) Surface tailoring for controlled photoelectrochemical properties: effect of patterned tio2 microarrays. J Phys Chem C 111(35):13163–13169
126. Ji C, Park HS (2007) Characterizing the elasticity of hollow metal nanowires. Nanotechnology 18(11)
127. Park HS, Cai W, Espinosa HD et al (2009) Mechanics of crystalline nanowires. MRS Bull 34 (3):178–183
128. Yu M, Long YZ, Sun B et al (2012) Recent advances in solar cells based on one-dimensional nanostructure arrays. Nanoscale 4(9):2783–2796
129. Bernardi M, Giulianini M, Grossman JC (2010) Self-assembly and its impact on interfacial charge transfer in carbon nanotube/p3ht solar cells. ACS Nano 4(11):6599–6606
130. Jakhmola A, Bhandari R, Pacardo DB et al (2010) Peptide template effects for the synthesis and catalytic application of pd nanoparticle networks. J Mater Chem 20(8):1522–1531
131. Jiao H (2009) Recent developments and applications of iron oxide nanomaterials. Fenmo Yejin Cailiao Kexue yu Gongcheng/Mater Sci Eng Powder Metallurgy 14(3):131–137
132. Baikousi M, Bourlinos AB, Douvalis A et al (2012) Synthesis and characterization of γ-Fe_2O_3/carbon hybrids and their application in removal of hexavalent chromium ions from aqueous solutions. Langmuir 28(8):3918–3930
133. Banerjee AN (2011) The design, fabrication, and photocatalytic utility of nanostructured semiconductors: focus on TiO_2-based nanostructures. Nanotechnol Sci Appl 4(1):35–65

134. Guo S, Dong S (2011) Metal nanomaterial-based self-assembly: development, electrochemical sensing and sers applications. J Mater Chem 21(42):16704–16716
135. Antolini E (2009) Carbon supports for low-temperature fuel cell catalysts. Appl Catal Environ 88(1–2):1–24
136. Guo S, Wang E (2011) Functional micro/nanostructures: simple synthesis and application in sensors, fuel cells, and gene delivery. Acc Chem Res 44(7):491–500
137. Ochekpe NA, Olorunfemi PO, Ngwuluka NC (2009) Nanotechnology and drug delivery. Part 1: background and applications. Trop J Pharm Res 8(3):265–274
138. Di Francia G, Alfano B, La Ferrara V (2009) Conductometric gas nanosensors. Journal of Sensors 2009:18, Article ID 659275, doi:10.1155/2009/659275
139. Sugiyama S, Toriyama T, Nakamura K et al (2010) Evaluation and analysis of physical properties of nanomaterials for highly sensitive mechanical sensing devices. IEEJ Trans Sens Micromach 130(5):146–151
140. Simon P, Gogotsi Y (2008) Materials for electrochemical capacitors. Nat Mater 7 (11):845–854
141. Arici AS, Bruce P, Scrosati B et al (2005) Nanostructured materials for advanced energy conversion and storage devices. Nat Mater 4(5):366–377
142. Lee KT, Cho J (2011) Roles of nanosize in lithium reactive nanomaterials for lithium ion batteries. Nano Today 6(1):28–41
143. Basu S, Basu PK (2011) Nanomaterials and chemical sensors. Sens Transducers 134 (11):1–31
144. Jimenez-Cadena G, Riu J, Rius FX (2007) Gas sensors based on nanostructured materials. Analyst 132(11):1083–1099
145. Jung I, Dikin D, Park S et al (2008) Effect of water vapor on electrical properties of individual reduced graphene oxide sheets. J Phys Chem C 112(51):20264–20268
146. Lu G, Ocola LE, Chen J (2009) Gas detection using low-temperature reduced graphene oxide sheets. Appl Phys Lett 94(8)
147. Lu G, Ocola LE, Chen J (2009) Reduced graphene oxide for room-temperature gas sensors. Nanotechnology 20(44):445502–445510
148. Shafiei M, Spizzirri PG, Arsat R et al (2010) Platinum/graphene nanosheet/SiC contacts and their application for hydrogen gas sensing. J Phys Chem C 114(32):13796–13801
149. Arsat R, Breedon M, Shafiei M et al (2009) Graphene-like nano-sheets for surface acoustic wave gas sensor applications. Chem Phys Lett 467(4–6):344–347
150. Qin L, Xu J, Dong X et al (2008) The template-free synthesis of square-shaped SnO_2 nanowires: the temperature effect and acetone gas sensors. Nanotechnology 19(18)
151. Spencer MJS (2012) Gas sensing applications of 1d-nanostructured zinc oxide: insights from density functional theory calculations. Prog Mater Sci 57(3):437–486
152. Madamopoulos N, Siganakis G, Tsigara A et al (2005) Diffractive optical elements for photonic gas sensors. In: Proceedings of SPIE – The International Society for Optical Engineering
153. Gracheva IE, Moshnikov VA, Karpova SS et al (2011) Net-like structured materials for gas sensors. J Phys Conf Ser 291(1)
154. Fowler JD, Allen MJ, Tung VC et al (2009) Practical chemical sensors from chemically derived graphene. ACS Nano 3(2):301–306
155. Wang J, Yang S, Guo D et al (2009) Comparative studies on electrochemical activity of graphene nanosheets and carbon nanotubes. Electrochem Commun 11(10):1892–1895
156. Wang Y, Lu J, Tang L et al (2009) Graphene oxide amplified electrogenerated chemiluminescence of quantum dots and its selective sensing for glutathione from thiol-containing compounds. Anal Chem 81(23):9710–9715
157. Wang J (2008) Electrochemical glucose biosensors. Chem Rev 108(2):814–825
158. Drummond TG, Hill MG, Barton JK (2003) Electrochemical DNA sensors. Nat Biotechnol 21(10):1192–1199

159. Dong CK, Dae JK (2008) Molecular recognition and specific interactions for biosensing applications. Sensors 8(10):6605–6641
160. Lu CH, Yang HH, Zhu CL et al (2009) A graphene platform for sensing biomolecules. Angew Chem Int Ed 48(26):4785–4787
161. Hyun S, Park TH (2011) Integration of biomolecules and nanomaterials: towards highly selective and sensitive biosensors. Biotechnol J 6(11):1310–1316
162. Lin YM, Jenkins KA, Alberto VG et al (2009) Operation of graphene transistors at giqahertz frequencies. Nano Lett 9(1):422–426
163. Lu J, Drzal LT, Worden RM et al (2007) Simple fabrication of a highly sensitive glucose biosensor using enzymes immobilized in exfoliated graphite nanoplatelets nafion membrane. Chem Mater 19(25):6240–6246
164. Xu H, Zeng L, Xing S et al (2008) Ultrasensitive voltammetric detection of trace lead(ii) and cadmium(ii) using MWCNTs-nafion/bismuth composite electrodes. Electroanalysis 20 (24):2655–2662
165. Tamiya E (2006) Nanomaterials based optical and electrochemical biosensors. In: 2006 I.E. nanotechnology materials and devices conference (NMDC), pp 288–289. October 22–25, 2006, Gyeongju, Korea
166. Zheng G, Lieber CM (2011) Nanowire biosensors for label-free, real-time, ultrasensitive protein detection. Methods Mol Biol 790:223–237
167. Choi BG, Park H, Park TJ et al (2010) Solution chemistry of self-assembled graphene nanohybrids for high-performance flexible biosensors. ACS Nano 4(5):2910–2918
168. Li L, Du Z, Liu S et al (2010) A novel nonenzymatic hydrogen peroxide sensor based on MnO_2/graphene oxide nanocomposite. Talanta 82(5):1637–1641
169. Robinson JT, Perkins FK, Snow ES et al (2008) Reduced graphene oxide molecular sensors. Nano Lett 8(10):3137–3140
170. Bi S, Zhou H, Zhang S (2009) Multilayers enzyme-coated carbon nanotubes as biolabel for ultrasensitive chemiluminescence immunoassy of cancer biomarker. Biosens Bioelectron 24:2961–2966
171. Shao N, Lu S, Wickstrom E et al (2007) Integrated molecular targeting of IGF1R and HER2 surface receptors and destruction of breast cancer cells using single wall carbon nanotubes. Nanotechnology 18:315101
172. Teker K, Sirdeshmukh R, Sivakumar K et al (2005) Applications of carbon nanotubes for cancer research. Nanobiotechnology 1:171–182
173. Dasgupta A, Kumar GVP (2012) Palladium bridged gold nanocylinder dimer: plasmonic properties and hydrogen sensitivity. Appl Opt 51(11):1688–1693
174. Langhammer C, Larsson EM, Kasemo B et al (2010) Indirect nanoplasmonic sensing: ultrasensitive experimental platform for nanomaterials science and optical nanocalorimetry. Nano Lett 10(9):3529–3538
175. Karuwan C, Wisitoraat A, Maturos T et al (2009) Flow injection based microfluidic device with carbon nanotube electrode for rapid salbutamol detection. Talanta 79:995–1000
176. Panini N, Messina G, Salinas E et al (2008) Integrated microfluidic systems with an immunosensor modified with carbon nanotubes for detection of prostate specific antigen (PSA) in human serum samples. Biosens Bioelectron 23:1145–1151
177. Phokharatkul D, Karuwan C, Lomas T et al (2011) AAO-CNTs electrode on microfluidic flow injection system for rapid iodide sensing. Talanta 84:1390–1395
178. Yuan Q, Duan HH, Li LL et al (2009) Controlled synthesis and assembly of ceria-based nanomaterials. J Colloid Interface Sci 335(2):151–167
179. Joo SH, Choi SJ, Oh I et al (2001) Ordered nanoporous arrays of carbon supporting high dispersions of platinum nanoparticles. Nature 412(6843):169–172
180. Ariga K, Li M, Richards GJ et al (2011) Nanoarchitectonics: a conceptual paradigm for design and synthesis of dimension-controlled functional nanomaterials. J Nanosci Nanotechnol 11(1):1–13

181. Feng W, Sun LD, Zhang YW et al (2010) Synthesis and assembly of rare earth nanostructures directed by the principle of coordination chemistry in solution-based process. Coord Chem Rev 254(9–10):1038–1053
182. Anthony NR, Bisignano AJ, Mehta AK et al (2012) Structural heterogeneities of self-assembled peptide nanomaterials. In: Progress in biomedical optics and imaging – Proceedings of SPIE. February 4–9, 2012, California, USA
183. Lazzari M, Rodriguez-Abreu C, Rivas J et al (2006) Self-assembly: a minimalist route to the fabrication of nanomaterials. J Nanosci Nanotechnol 6(4):892–905
184. Ariga K, Hill JP, Lee MV et al (2008) Challenges and breakthroughs in recent research on self-assembly. Sci Technol Adv Mater 9(1)
185. Arning V, Leenen MAM, Steiger J et al (2009) New nanomaterials enable low cost flexible electronics. When circuits are printed, labels can talk. Vakuum in Forschung und Praxis 21(2): A18–A23
186. Jakubowska M, Sloma M, Mlozniak A (2009) Polymer composites based on carbon nanotubes for printed electronics. In: ISSE 2009: 32nd International spring seminar on electronics technology: hetero system integration, the path to new solutions in the modern electronics – conference proceedings. May 13–17, 2009, Brno, Czech Republic
187. Jubete E, Loaiza OA, Ochoteco E et al (2009) Nanotechnology: a tool for improved performance on electrochemical screen-printed (bio)sensors. J Sens
188. Gilje S, Han S, Wang M et al (2007) A chemical route to graphene for device applications. Nano Lett 7(11):3394–3398
189. Qu Y, Li X, Chen G et al (2008) Synthesis of Cu_2O nano-whiskers by a novel wet-chemical route. Mater Lett 62(6–7):886–888
190. Singh DP, Ojha AK, Srivastava ON (2009) Synthesis of different $Cu(OH)_2$ and CuO (nanowires, rectangles, seed-, belt-, and sheetlike) nanostructures by simple wet chemical route. J Phys Chem C 113(9):3409–3418
191. Dua V, Surwade SP, Ammu S et al (2010) All-organic vapor sensor using inkjet-printed reduced graphene oxide. Angew Chem Int Ed 49(12):2154–2157
192. Manga KK, Wang S, Jaiswal M et al (2010) High-gain graphene-titanium oxide photoconductor made from inkjet printable ionic solution. Adv Mater 22(46):5265–5270
193. Joshi M, Bhattacharyya A, Ali SW (2008) Characterization techniques for nanotechnology applications in textiles. Indian J Fibre Textile Res 33(3):304–317

Signal Amplification Using Nanomaterials for Biosensing

Jianping Lei and Huangxian Ju

Abstract Signal amplification based on biofunctional nanomaterials has recently attracted considerable attention due to the need for ultrasensitive bioassays. Especially, most nanoscaled materials are biocompatible, which permits them to act in direct contact with the environment as carriers of biological recognition elements for obtaining lower and lower detection limit. In order to achieve the good performance for biosensing, two approaches including noncovalent interaction and covalent route have been introduced for the functionalization of nanomaterials with biomolecules. The biofunctional nanomaterials with the abilities of specific recognition and signal triggering can be employed as not only excellent carriers, but also electronic and optical signal tags to amplify the detection signal. These advantages provide a new avenue to construct a sensitive and specific platform in nanobiosensing.

Keywords Biosensing, Functionalization, Nanomaterials, Signal amplification

Contents

J. Lei and H. Ju (✉)
State Key Laboratory of Analytical Chemistry for Life Science, School of Chemistry and Chemical Engineering, Nanjing University, Nanjing 210093, People's Republic of China
e-mail: hxju@nju.edu.cn

A. Tuantranont (ed.), *Applications of Nanomaterials in Sensors and Diagnostics*,
Springer Series on Chemical Sensors and Biosensors (2013) 14: 17–42
DOI 10.1007/5346_2012_46,
Published online: 15 February 2013

1 Introduction

The need for ultrasensitive bioassays and the trend towards miniaturized assays make the biofunctionalization of nanomaterials become one of the hottest fields [1, 2]. These biofunctionalized nanomaterials can be used as carriers or tracers to obtain the amplified detection signal and the stabilized recognition probes. Based on the unique properties of nanomaterials, the biofunctional nanoparticles can produce a synergic effect among catalytic activity, conductivity, and biocompatibility to result in significantly signal amplification for designing a new generation of nanobiosensing device.

A lot of nanomaterials, such as metal nanoparticles, carbon-based nanostructures, and magnetic nanoparticles have been introduced as carriers for the signal amplification. In particularly, carbon-based nanomaterials and metal nanoparticles show to promote the direct electron transfer between the biomolecules and electrode surface. For example, based on excellent conductivity, the single-walled carbon nanotubes (SWNTs) can act as a nanoconnector that electrically contacts the active site of the enzyme and the electrode with the interfacial electron transfer rate constant of 42 s^{-1}, which provides a significant potential for constructing an electrochemical biosensor [3]. Using superparamagnetic particle as carrier for signal amplification, surface plasmon resonance (SPR) immunoassay has been achieved for the detection of cancer biomarker prostate specific antigen (PSA) in serum at an ultralow detection limit of 10 fg mL^{-1} [4].

As a signal trace, the biofunctionalized nanomaterials have the abilities of specific recognition and signal amplification in optical, electrochemical, and photoelectrochemical assays [5, 6]. In optical assay, nanoparticle probes such as fluorescence energy transfer nanobeads and quantum dots (QDs) provide significant advantages of signal brightness, photostability, wide dynamic range, and multiplexing capabilities comparison with organic dyes and fluorescent proteins. Electrochemical assays based on nanoprobes are attractive because of their low cost, high sensitivity, simplicity, and easy miniaturization. The electrochemiluminescent (ECL) and photoelectrochemical assays hold the advantages of both optical and electrochemical detections are a promising perspective.

In this chapter, the recent significant advances in signal amplification based on biofunctional nanomaterials are highlighted including the efficient functionalization of nanomaterials with biomolecules as recognition elements, and the functions of nanomaterials as carrier and signal trace in ultrasensitive nanobiosensing.

2 Biofunctionalization Method of Nanomaterials

2.1 *Biofunctionalization by Noncovalent Assembly*

Nanomaterials hold much promise for biological applications, but they require appropriate functionalization to provide biocompatibility in biological environments. Two approaches including noncovalent interaction and covalent interaction are introduced for the functionalization of nanoparticles (NPs). The noncovalent approach such as electrostatic interaction and π–π stacking can avoid destruction of conjugated skeleton and loss of electronic properties of the NPs. A general and attractive approach via π–π interaction has been designed by Dai and coworkers for the noncovalent functionalization of SWNTs sidewalls and the subsequent immobilization of biomolecules onto SWNTs via *N*-succinimidyl-1-pyrenebutanoate [7]. A functional nanocomposite of reduced graphene oxide (RGO) with water-soluble picket-fence iron porphyrin has been prepared by means of π–π interactions. The resulting nanocomposite has good biocompatibility and excellent electrocatalytic activity toward the reduction of chlorite [8].

The electrostatic interaction is alternative method to assembly biomolecules on the surface of NPs, particularly for deposition of macromolecules such as proteins or enzymes. Typically, the carboxylate group decorated carbon nanotube (CNT) can be functionalized with antibody molecules at pH values that lie slightly above the isoelectric point of the citrate ligand. Further, the electrostatic layer-by-layer (LBL) self-assembly onto CNT carriers maximizes the ratio of enzyme tags per binding event to offer the great amplification factor, which allows detection of DNA and proteins down to 80 copies (5.4 aM) and 2,000 protein molecules (67 aM), respectively [9].

Another noncovalent method for immobilizing biomolecules on NPs is to entrap them in biocompatible films such as phospholipid, polymer, and DNA. A conjugate of phospholipid and dextran has been found to not only be as a stable coating for nanomaterials, but also provide brighter photoluminescence than carbon nanotubes suspended by poly(ethylene glycol) [10]. Moreover, the coating films can provide the abundant positions for functionalization with second biomolecules. Figure 1 shows the formation of polymer vesicles by mixing hydroxyl (1) or azide termini (2) with controlled densities of surface azide groups. Dendrons with focal point alkynes can subsequently be conjugated to the surface azides providing controlled densities of dendritic groups on the vesicle surface. The dendritic systems exhibit one to two orders of magnitude enhancement in binding affinity relative to the

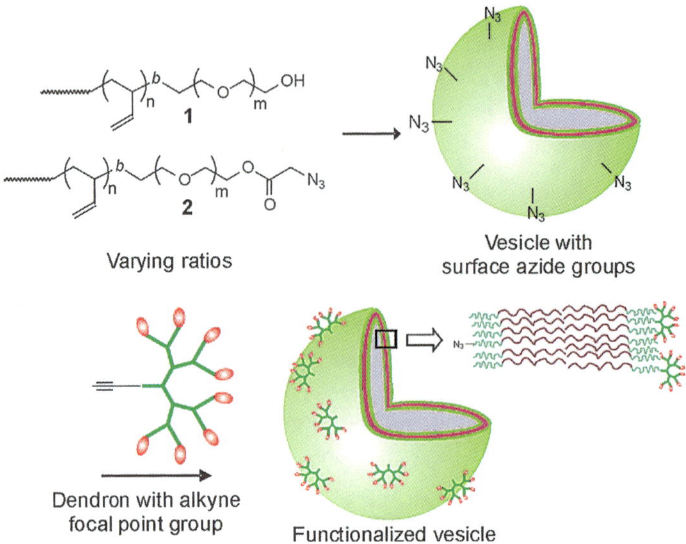

Fig. 1 General approach for functionalization of vesicle surfaces with dendritic groups. Reprinted with permission from Martin et al. [11]. © 2009, American Chemical Society

nondendritic displays, which is attributed to the ability of the dendritic groups to overcome steric inhibition by polymer chains at the material surface and also to the presentation of ligands in localized clusters [11]. A new strategy for the synthesis of metal–nanoparticle/CNT nanohybrids has been developed by ionic-liquid polymer (PIL)-functionalized CNTs. The PtRu/CNTs-PIL electrocatalyst shows better performance in the direct electrooxidation of methanol than the PtRu/CNTs electrocatalyst alone [12].

Affinity interactions, such as antigen–antibody, nucleic acid–DNA, lectin–glycan, streptavidin–biotin, and aptamer–protein, are highly stable and the strongest of all noncovalent linkages for bioconjugation of targeting ligands to NPs. Moreover, various biomolecules contain several binding sites, for example, streptavidin or concanavalin A, each displays four binding domains. This allows the multidirectional growth of NPs structures [13, 14]. In addition, barnase–barstar system is a new generic method for robust self-assembly of multifunctional particles to macroscopic superstructures [15].

2.2 Covalent Route

Biofunctionalization of nanomaterials employing covalent methods should be preferable to unspecific physisorption in terms of stability and reproducibility of the surface functionalization. In general, functional groups at the NP surfaces can be directly bound to reactive ligands by a linkage reaction facilitated with the aid of

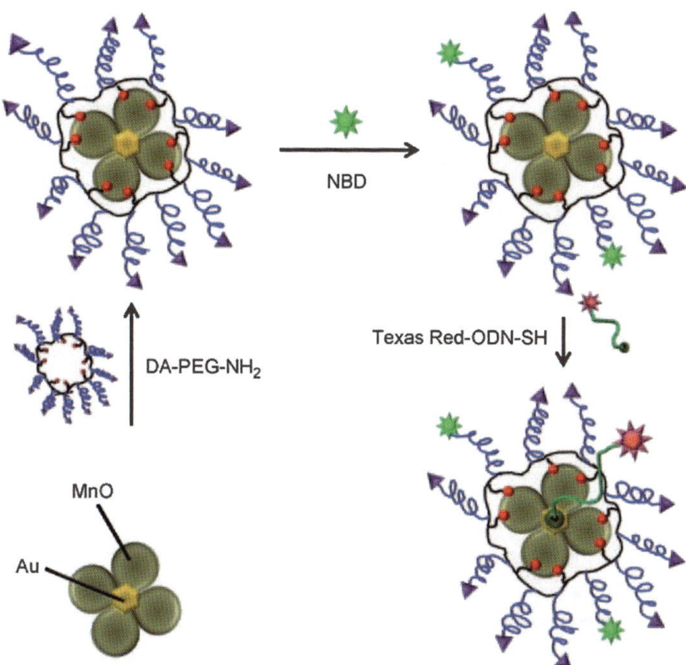

Fig. 2 Surface functionalization of Au@MnO nanoflowers with a multidentate copolymer and subsequent conjugation with NBD. The gold domain was selectively functionalized with a Texas-Red-tagged thiolated oligonucleotide. Reprinted with permission from Schladt et al. [20]. © 2010, Wiley

catalysts. Typically, CNTs can be firstly shortened by sonication in 3:1 H_2SO_4/HNO_3 for several hours refluxing to introduce hydrophilic carboxylic acid groups for functionalization. Then NPs decorated with carboxylic acid groups can be covalently bound to biomolecules bearing primary amines through *N*-hydroxysuccinimide linkers [16]. On the basis of the arginine–glycine–aspartic acid–serine (RGDS)-functionalized SWNTs, a novel electrochemical cytosensing strategy has been designed with detection limit down to 620 cells mL^{-1} and linear calibration range from 1.0×10^3 to 1.0×10^7 cells mL^{-1} of BGC-823 human gastric carcinoma cells [17]. Similarly, arginine–glycine–aspartic acid-labeled QDs have also been designed for in vivo targeting and imaging of tumor vasculature [18]. The tumor fluorescence intensity reaches maximum at 6 h postinjection with good contrast.

As to metal nanoparticles, the primary binding of thiolated molecules, such as thiolated oligopeptides, to gold nanoparticles (AuNPs) can provide a means for the covalent tethering of biomolecules to NPs [19]. Figure 2 depicts a functionalized Au@MnO nanoflower with selective attachment of catechol anchors to the metal oxide petals and thiol anchors to the gold core. Selective functionalization of the gold domain can be achieved by incubating an aqueous solution of the fluorescent dye 4-chloro-7-nitrobenzofurazan (NBD)-polymer-modified Au@MnO NPs with thiol-modified 24-mers customized oligonucleotide tagged with Texas red. The polymer-functionalized Au@MnO NPs are stable against aggregation and precipitation in various aqueous media, including deionized water and PBS buffer solution for several days. Viability assays of nanocomposite solutions with the renal cell carcinoma line Caki^{-1} show negligible toxicity of the nanoparticles even for concentrations as high as 140 mg mL^{-1} [20]. A DNA sensor based on a "sandwich" detection strategy has been designed, which involves capture probe DNA immobilized on gold electrodes and reporter probe DNA labeled with AuNPs via Au–S chemistry [21].

Some specific reactions under mild reaction conditions have been extensively used in the generation of covalent-tethered conjugates of biomolecules with various NPs. To design a modular and broadly applicable targeting platform, Weissleder and coworkers described a covalent bioorthogonal reaction between a 1,2,4,5-tetrazine and a trans-cyclooctene for small-molecule labeling. The [4+2] cycloaddition was fast, chemoselective, did not require a catalyst, which was adapted to targeting nanoparticle sensors in different configurations to improve binding efficiency and detection sensitivity [22]. On the other hand, "Click" chemistry, a Cu-catalyzed azide–alkyne cycloaddition, is a relatively new approach for easy and almost quantitative functionalization with high specificity, high stability, and extreme rigidity [23]. Via one-step Click reaction, the drug-loaded polymer nanoparticles can be functionalized with folate, biotin, and gold nanoparticles for drug delivery [24]. A general approach has been presented for functionalization of low-fouling, nanoengineered polymer capsules with antibodies by using click chemistry. Significantly, antibody-functionalized capsules can specifically bind to colorectal cancer cells even when the target cells constitute less than 0.1% of the total cell population [25].

3 Nanomaterials as Carriers for Signal Amplification

3.1 Metal Nanoparticles

Metal NPs have been extensively used in the detection of biologically important or toxic substances, usually by the change of spectral or SPR signal accompanying NP aggregation or dispersion in the presence of an analyte. Using AuNPs as carriers, a homogeneous colorimetric DNA biosensor has been developed by a novel nicking endonuclease-assisted AuNP amplification, resulting in a 103-fold improvement in amplification (ca. 10 pm) and the capability of recognizing long single-stranded oligonucleotides with single-base mismatch selectivity [26]. NP supracrystals and core–shell supracrystals stabilized by analyte-specific cross-linkers can enhance dramatically (by over two orders of magnitude compared to noncrystalline NP aggregates) the sensitivity of NPs-based detection [27].

A dual element amplification method based on AuNPs and RNA transcription is designed by using SPR imaging to detect single-stranded DNA (ssDNA) down to a concentration of 1 fM in a volume of 25 μL (25 zeptomoles) [28]. Surface plasmons (SP)-induced ECL enhancement has been applied for ultrasensitive detection of thrombin with the concentration of thrombin in a wide range from 100 aM to 100 fM. This system shows five fold enhancement of ECL intensity as compared to that without AuNPs, which might be attributed to the long-distance interaction between the semiconductor nanocrystal and SPR field of noble metal NPs [29].

Due to the unique electrochemical properties, AuNPs can significantly enhance the sensitivity and the selectivity in the electrochemical detection of DNA. DNA-functionalized AuNPs (DNA-AuNPs) have been used to enhance the sensitivity of the aptasensor because DNA-AuNPs-modified interface can load more $[Ru(NH_3)_6]^{3+}$ cations to produce electrochemical signal. The proposed aptasensor has a low detection limit (0.02 nM for adenosine and 0.01 μg mL^{-1} for lysozyme) [30]. A highly selective electrochemical biosensor for the ultrasensitive detection of Hg^{2+} in aqueous solution has been developed based on the strong and specific binding of Hg^{2+} by two DNA thymine bases (T–Hg^{2+}–T) and the use of AuNP-functionalized reporter DNA to achieve signal amplification [31, 32].

3.2 *Carbon-Based Nanomaterials*

Carbon-based NPs are excellent carriers to enhance the probe due to the good conductivity and biocompativity. Typically, SWNTs have been used to immobilize DNA probe for fabrication of DNA biosensor. Based on the direct current response of guanine, the biosensor can detect target DNA in the range of 40–110 nM with a detection limit of 20 nM [33]. Ready renewal for more than 3,000 times is the outstanding merit of this label-free biosensor. Moreover, CNTs play a dual amplification role in both the recognition and transduction events, namely as carriers for numerous enzyme tags and for accumulating the product of the enzymatic reaction. Using alkaline phosphatase (ALP)-CNTs as a tracer, a favorable response of DNA target indicates a remarkably low detection limit of around 1 fg mL^{-1} (54 aM), i.e., 820 copies or 1.3 zmol in the 25 μL sample [34]. Based on the specific recognitions of target DNA and streptavidin to biotin-labeled molecular beacon and the signal amplification of streptavidin–horseradish peroxidase (HRP)-functionalized CNTs, a biosensing strategy has been developed for selective electrochemical detection of DNA in five orders of magnitude with a detection limit of 2.8 aM [35].

Compared with SWNTs, single-walled carbon nanohorns (SWNHs) as immobilization matrixes show a better sensitizing effect. This material has been used as carrier to develop an immunosensor for microcystin-LR (MC-LR) ranging from 0.05 to 20 μg L^{-1} [36]. A nanoscaffold of nanohorns functionalized with RGDS has also been prepared on an electrode surface for cell capture and enhancing the electrical connectivity (Fig. 3). Combined with the AuNPs-Con A-HRP nanoprobe

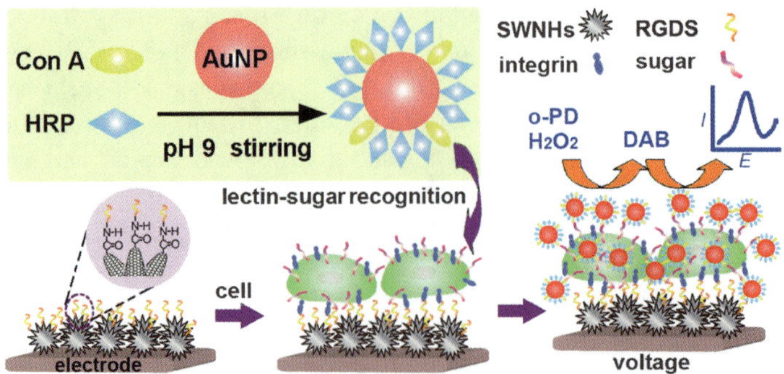

Fig. 3 Scheme of nanoprobe assembly and electrochemical strategy for in situ detection of mannose groups on living cells. Reprinted with permission from Ding et al. [37]. © 2010, American Chemical Society

and peptide-functionalized nanohorns, a highly sensitive electrochemical strategy is developed for cytosensing, which shows a detection limit down to 15 cells, broad dynamic range, acceptable rapidity, and low cost [37].

Nitrogen-doped carbon nanotubes (CNx-MWNTs) are suitable for loading biomolecules to construct biosensors due to lower cytotoxicity and better biocompatibility. An AuNPs/CNx-MWNTs nanocomposite has been used as an immobilization scaffold of antibodies for preparation of a sensitive immunosensor to detect MC-LR. The immunosensor exhibits a linear response to MC-LR ranging from 0.005 to 1 $\mu g\ L^{-1}$ with a detection limit of 0.002 $\mu g\ L^{-1}$ at a signal-to-noise of 3 [38].

The functionalized carbon nanospheres (CNSs) are often used for the biosensor platform to increase the surface area for capturing a large amount of primary antibodies, thus amplifying the detection response. For example, the AuNPs/CNSs hybrid material can be conjugated with HRP-labeled antibody (HRP-Ab_2) to fabricate HRP-Ab_2–AuNPs/CNSs bioconjugates, which can then be used as a label for the sensitive detection of human IgG (HIgG). This approach provides a linear response range between 0.01 and 250 ng mL^{-1} with a detection limit of 5.6 pg mL^{-1} [39]. On the basis of the dual signal amplification strategy of graphene sheets and the multienzyme labeling on CNSs, an immunosensor shows a seven fold increase in detection signal compared to the immunosensor without graphene modification and CNSs labeling. The proposed method can respond to 0.02 ng mL^{-1} α-fetoprotein (AFP) with a linear calibration range from 0.05 to 6 ng mL^{-1} [40].

Functionalized graphene oxide (GO) sheets coupled with the nanomaterial-promoted reduction of silver ions have been developed for the sensitive and selective detection of bacteria. Using an electrochemical technique, a linear relationship between the stripping response and the logarithm of the bacterial concentration is obtained for concentrations ranging from 1.8×10^2 to 1.8×10^8 cfu mL^{-1} [41]. Based on the supramolecular assembly of free-base cationic 5,10,15,20-tetrakis (1-methyl-4-pyridinio)porphyrin on reduced graphene, the resulting graphene-porphyrin hybrid as an optical probe has been constructed for rapid and selective sensing of Cd^{2+} ions in aqueous media [42].

3.3 *Magnetic Nanoparticles*

Magnetic NPs easily achieve concentration and purification of analysts, which is useful to enhance dramatically the sensitivity in biosensing. Using p19-functionalized magnetic beads, over 100,000-fold enrichment of the probe:miRNA duplex has been achieved from total RNA. This approach is validated by detecting picogram levels of a liver-specific miRNA (miR122a) from rat liver RNA [43]. The antiferromagnetic NPs have been used to enable magnetic detection of biomolecules at low analyte concentrations (10 pm) with better detection than conventional superparamagnetic materials [44].

Magnetic beads are the good candidates as supporters for largely loading signal trace in ultrasensitive detection. The sensitivity can greatly be amplified by synthesizing magnetic bioconjugates particles containing 7,500 HRP labels along with detection antibodies (Ab_2) attached to activated carboxyl groups on 1 μm diameter magnetic beads. The resulting sensor shows a sensitivity of 31.5 μA mL ng^{-1} and a detection limit of 0.5 pg mL^{-1} for PSA in 10 μL of undiluted serum [45]. When coupled to superparamagnetic beads massively loaded with about 500,000 HRP labels and Ab_2, an unprecedented detection limit has been obtained to be 1 fg mL^{-1} (100 am) for interleukin 8, which is lower than that of any method for direct biomarker protein detection in serum [46]. The near-single-protein sensor has great promise for extension to arrays for clinical cancer screening and therapy monitoring.

Iron oxide (Fe_3O_4) nanoparticles, as the well-known hard magnetic material, have been extensively applied in the nanoparticle-based assays due to good biocompatibility. With the employment of the AuNPs–Prussian blue (PB)–Fe_3O_4 nanohybrid, a signal amplification strategy has been developed based on bienzyme (HRP and glucose oxidase) functionalized Au–PB–Fe_3O_4 NPs for electrochemical immunosensing. The linear ranges span the concentrations of carcinoembryonic antigen (CEA) from 0.01 to 80.0 ng mL^{-1} with detection limit of 4 pg mL^{-1} and AFP from 0.014 to 142.0 ng mL^{-1} with detection limit of 7 pg mL^{-1}, respectively [47]. A sandwich-type electrochemical immunoassay has been designed for the detection of carbohydrate antigen 125 (CA125) using anti-CA125-coated magnetic beads for target capture and HRP-anti-CA125-coated silica beads containing HRP and thionine for signal enhancement. This immunoassay exhibits a range from 0.1 to 450 U mL^{-1} with a detection limit of 0.1 U mL^{-1} for CA125 [48].

A luminol–H_2O_2–HRP–bromophenol blue chemiluminescence (CL) system has been applied to a sandwich-type CL immunoassay based on the magnetic separation and the amplification feature of AuNPs as HRP labels. The linear range for AFP is from 0.1 to 5.0 ng mL^{-1} with the detection limit of 0.01 ng mL^{-1}, which is one order of magnitude lower than that obtained without using AuNPs [49]. A sensitive strategy, which integrates a DNA cycle device onto magnetic microbeads, amplifying the signal with GO and enhancing ECL intensity, has successfully been applied to thrombin detection [50].

Magnetic nanotags are a promising alternative as giant magnetoresistive sensors, such as spin valve sensors in biomolecular detection assays. With the addition of magnetic nanotag amplification, an inexpensive giant magnetoresistive sensor has

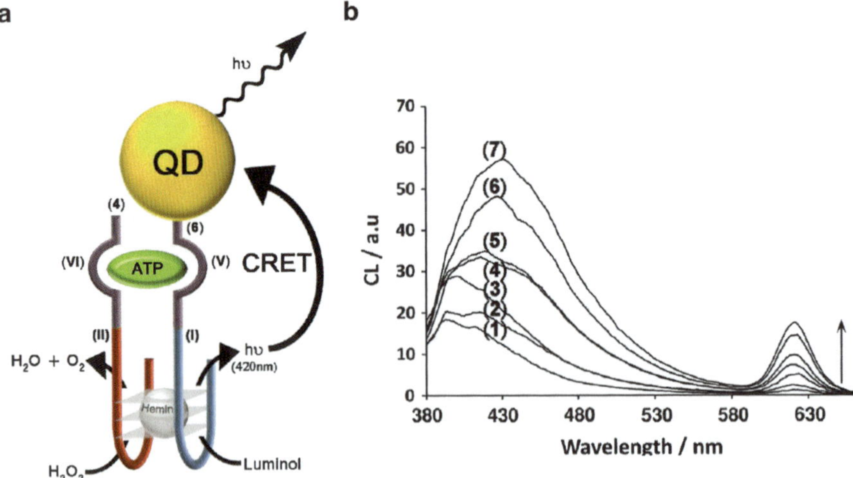

Fig. 4 (**a**) Analysis of ATP through the CRET from luminol, oxidized by the assembled hemin/G-quadruplex, to the QDs. (**b**) Luminescence spectrum corresponding to the CRET signal of the QDs at $\lambda = 612$ nm in the absence of ATP, curve (1), and in the presence of different concentrations of ATP: (2) 0.125, (3) 1.25, (4) 5, (5) 12.5, (6) 50, (7) 100 μM. Reprinted with permission from Freeman et al. [54]. © 2011, American Chemical Society

been constructed for multiplex protein detection of potential cancer markers at subpicomolar concentration levels and with a dynamic range of more than four decades [51]. In addition, combining with two-color photo-acoustic flow cytometry, a platform using targeted magnetic nanoparticles has been developed for in vivo magnetic enrichment and detection of rare circulating tumor cells from a large pool of blood with high spatial resolution [52].

3.4 *Other Nanomaterials*

QDs are most frequently used semiconductor nanoparticles for biological detection of DNA. By integrating CdTe QDs with different biomolecules, such as molecular beacon (MB) and aptamer, an effective sensing platform has been designed for DNA target based on fluorescence resonance energy transfer (FRET) between QDs and graphene oxide. The change in fluorescent intensity is used for the detection of the target with a detection limit down to 12 nM [53].

Most recently, Willner and coworkers implemented the DNAzyme-stimulated chemiluminescence resonance energy transfer (CRET) to CdSe/ZnS QDs for developing aptamer or DNA sensing platforms [54]. Figure 4 depicts the CRET-based analysis of adenosine-5′-triphosphate (ATP) by the hemin/G-quadruplex conjugated aptasensor. Glutathione (GSH)-capped CdSe/ZnS QDs ($\lambda_{em} = 620$ nm) are covalently tethered to the thiol-functionalized nucleic acid (6) (average loading ca. 10 units per particle), which consists of the anti-ATP aptamer subunit (V), and the HRP-mimicking DNAzyme subunit (I). Treatment of the (6)-functionalized QDs with ATP in the presence of the nucleic acid (4) that includes the complementary aptamer subunit, region (VI), and the second DNAzyme subunit (II) resulted in the formation of the ATP hemin/G-quadruplex-QDs complex. Figure 4b shows that, upon the addition of ATP, the resulting chemiluminescence stimulates a CRET process, which is intensified by the luminescence of the QDs, $\lambda_{em} = 620$ nm. Thus ATP can be detected with a sensitivity corresponding to 100 nM.

Due to the small size, high surface-to-volume ratio and good biocompatibility, silica NPs have become another normally used carrier for signal amplification. Labels based on mesoporous silica nanoparticles (MSN) loaded with mediator thionine (TH), HRP, and Ab_2 have been developed in order to improve the sensitivity of an amperometric immunosensor [55]. The sensitivity of the sandwich-type immunosensor using MSN–TH–HRP-Ab_2 as labels for HIgG detection is about 100 times higher than that using either MSN–TH–Ab_2 or MSN–HRP-Ab_2 as labels, indicating the high catalytic efficiency of HRP in the presence of mediator TH toward H_2O_2.

Based on dual signal amplification of poly-(guanine)-functionalized silica nanoparticles label and $Ru(bpy)_3^{2+}$-induced catalytic oxidation of guanine, an electrochemical immunosensor for the detection of tumor necrosis factor-alpha (TNF-α) is presented. The detection limit for TNF-α is found to be 5.0×10^{-11} g mL^{-1} (2.0 pM), which corresponds to 60 amol of TNF-α in 30 μL of sample [56]. The ECL of $Ru(bpy)_3^{2+}$ doped on silica nanoparticles is more than 1,000-fold increase than that of a single dye, suggesting that the use of this kind of nanostructures as luminescent labels represents a very promising system for ultrasensitive bioanalysis [57].

4 Functional Nanomaterial-Amplified Optical Assay

4.1 Colorimetric Detection of Biological Analytes

Colorimetric sensors are particularly important because they minimize or eliminate the necessity of using expensive and complicated instruments. Among the many colorimetric sensing strategies, metallic nanoparticle-based detection is desirable because of the high extinction coefficients and strong distance-dependent optical properties of the NPs. For example, colorimetric detection of DNA sequences based on electrostatic interactions with unmodified AuNPs can be completed within 5 min, and <100 fmol of target produces color change observable without instrumentation [58].

The colorimetric bio-barcode assay is a red-to-blue color change-based protein detection method with ultrahigh sensitivity. This assay is based on both the bio-barcode amplification method that allows for detecting miniscule amount of targets with attomolar sensitivity and AuNPs-based colorimetric DNA detection method that allows for a simple and straightforward detection of interleukin-2 [59]. Since AuNPs folded with aptamer are more stable toward salt-induced aggregation than those unfolded aptamers, colorimetric biosensors have been developed for the detection of adenosine, K^+, adenosine deaminase, and its inhibitors [60].

Based on detecting Cu^{II} released from copper monoxide nanoparticle-labeled antibodies as Ab_2 via click chemistry, a colorimetric immunoassay has been developed for the detection of human immunodeficiency virus with the naked eye [61]. This method is highly specific even in the presence of high concentrations of mixtures of other cations and interfering molecules.

The greater signal enhancement for colorimetric detection can be obtained by catalytic deposition of gold or silver NPs. For example, a convenient and label-free scanometric approach for DNA assay has been designed by integrating a metal ion-mediated conformational MB and silver-signal amplification regulated by AuNPs aggregation. By using scanometric detection, the concentration of the target DNA sequence can be conveniently read out within a linear range from 1.0 to 30 nM [62]. A PCR-free colorimetric assay has been developed for telomerase activity that relies on polyvalent oligonucleotide–nanoparticle conjugates as probes and the concept of elongated and unmodified oligonucleotides on one particle for amplification. The assay can detect telomerase activity with as few as 10 HeLa cells, with on-chip positive and negative controls [63]. An information transfer strategy has been developed for the visualization of carbohydrate expression by the competition of a primary cell-adhered solid surface with a carbohydrate assembled surface as an artificial secondary surface for one species. The strategy can be effectively utilized for in situ monitoring of dynamic carbohydrate expression on an adhesive cell surface [64]. Further, an ultrasensitive glycans array using iron oxide/gold core/shell nanoparticles conjugated with antibodies or proteins has been developed for ultrasensitive detection of carbohydrate–protein interactions [65].

Based on the catalytic activity of AuNPs covered with a self-assembled organic monolayer (Au-MPC) toward 2-hydroxypropyl-4-nitrophenylphosphate (HPNPP), a catalytic amplification process has been developed for the detection of proteases (Fig. 5). The strategy relies on a cascade of two catalytic events for signal generation. In the first event, an enzyme hydrolyzes a peptide substrate, which acts as an inhibitor for the catalytic monolayer. Upon hydrolysis, the catalytic activity of the monolayer is restored, and large quantities of a yellow reporter molecule are produced, leading to a sensitive colorimetric assay for the detection of enzyme activity [66].

In order to achieve an amplification of the optical signal, AuNPs have been used as carriers of the signaling HRP-anti-CA153 for the immunoassay of CA153 antigen. In the range up to 60 U mL^{-1}, the assay adopting AuNPs as an enhancer results in higher sensitivity and shorter assay time when compared to classical enzyme-linked immunosorbent assay [67].

4.2 Fluorescence Detection

Fluorescence detection is currently one of the most widely used methods in the areas of biotechnology, medical testing, and drug discovery. Based on the fluorescence signal recovery after digestion of RNA by RNase H, a strategy of

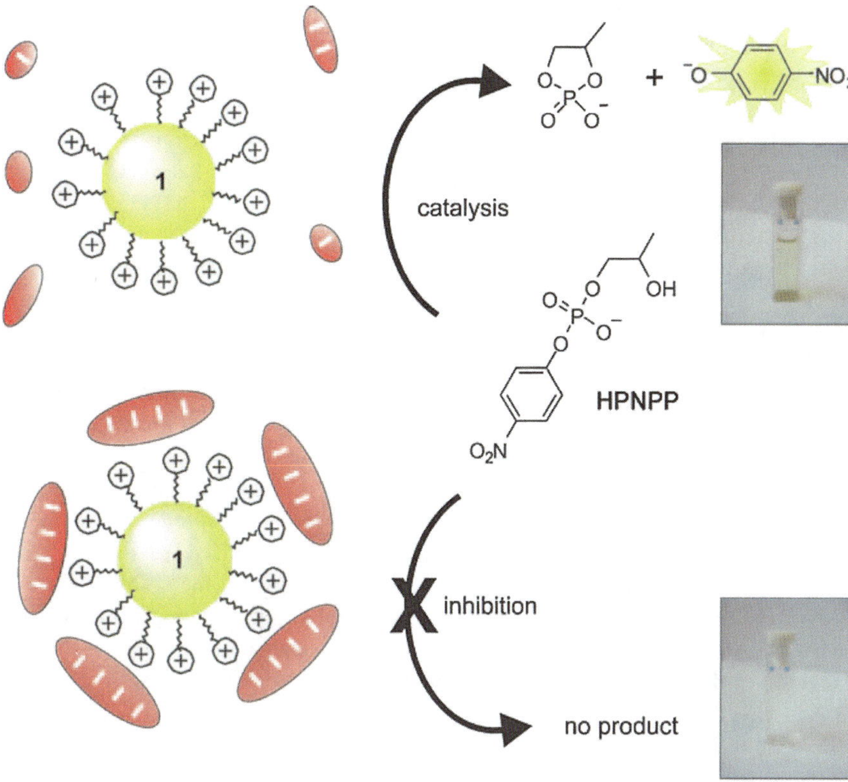

Fig. 5 Schematic representation of the catalytic assay. The presence of an enzyme able to hydrolyze a substrate enables catalysis of the transphosphorylation of HPNPP by Au-MPC resulting in the release of a yellow reporter molecule. In the absence of the enzyme, the catalytic activity of Au-MPC is suppressed because the enzyme substrate acts as an inhibitor for Au-MPC. Reprinted with permission from Bonomi et al. [66]. © 2011, Wiley

fluorescence signal amplification has been developed for highly sensitive and rapid protease assay at concentrations as low as 10 pM within 4 h [68]. A fluorescence-quenched peptide-based AuNP probes has been developed to visualize proteolytic activity in vivo. Optimal AuNP probes targeted to trypsin and urokinasetype plasminogen activator require the incorporation of a dark quencher to achieve five- to eight fold signal amplification [69]. In addition, a much simpler and milder strategy to amplify fluorescence signal has been proposed by the cation-exchange reaction with ionic nanocrystals. The Cd^{2+} released from CdSe QDs can trigger the fluorescence of dyes and lead to a 60-fold enhancement of the fluorescence signal and a limit of detection in protein detection 100 times lower than that of the organic fluorophore Alexa 488 [70].

Based on dual-color imaging and automated colocalization of bioconjugated nanoparticle probes, routine two-color super resolution imaging and single-molecule detection have been achieved at nanometer precision with standard fluorescence microscopes and inexpensive digital color cameras. This approach can apply in single-molecule studies in cell lysate samples with a detection dynamic range over three orders of magnitude [71].

FRET is a good opportunity to set up an ultrasensitive and reliable nanotechnology assay. This approach based on QDs FRET can detect as little as 15 pg of methylated DNA in the presence of a 10,000-fold excess of unmethylated alleles and allows for multiplexed analyses [72]. FRET-based probes incorporated with single-molecule fluorescence detection technologies can allow detection of DNA with low abundance without additional amplification. Unbound nanosensors produce near-zero background fluorescence, but binding to even a small amount of target DNA (~50 copies or less) can generate a very distinct FRET signal. The detection limit is 100-fold greater than conventional FRET probe-based assays as monitored by confocal fluorescence spectroscopy [13]. A bottom-up strategy has been developed to construct water-soluble fluorescent single-molecular nanoparticles based on polyhedral oligomeric silsesquioxanes (POSS) and cationic oligofluorene for fluorescence amplification in cellular imaging. The fluorescence of intercalated ethidium bromide is substantially amplified by 52-fold upon excitation of cationic oligofluorene substituted POSS in buffer, allowing naked-eye discrimination of dsDNA from ssDNA [73]. A plasmonic- and FRET-based DNA sensing scheme has also been designed based on core–shell multilayer dye-doped acceptor nanoparticles grafted with ssDNA probes and complexed with a cationic conjugated polymer, resulting in direct molecular detection of target nucleic acids at femtomolar concentrations [74].

An ECL resonance energy transfer (ECL-RET) system has been developed from CdS QDs to $Ru(bpy)_3^{2+}$. By the signal amplification of $Ru(bpy)_3^{2+}$ and the specific antibody–cell surface interactions, this ECL-RET system can sensitively respond down to 12.5 SMMC–7,721 cells mL^{-1} [75]. In the same group, an ultrasensitive DNA detection approach, which combines AuNPs enhanced ECL of the CdS nanocrystal film with isothermal circular amplification reaction of polymerase and nicking endonuclease, has been developed for the detection of DNA down to 5 aM [76].

4.3 Other Spectroscopic Measurements

Surface-enhanced Raman scattering (SERS)-based signal amplification and detection methods using plasmonic nanostructures have been widely investigated for imaging and sensing applications. A bifunctional adenosine-sensitive double-stranded DNA aptamer can create and control a SERS hot spot between a bulk Au surface and an AuNP attached to the aptamer via a biotin–avidin linkage. The AuNP is decorated with 4-aminobenzenethiol (4-ABT), a Raman reporter molecule. In the presence of adenosine, the target molecule, the SERS spectrum of 4-ABT increases in intensity by factors as large as ~4 [77]. In particular, SERS by molecules starts with an excitation, followed by inelastic coupling to internal vibrational levels of the molecule and a subsequent radiative decay, and therefore undergoing signal enhancement factors up to ten orders of magnitude by coupling

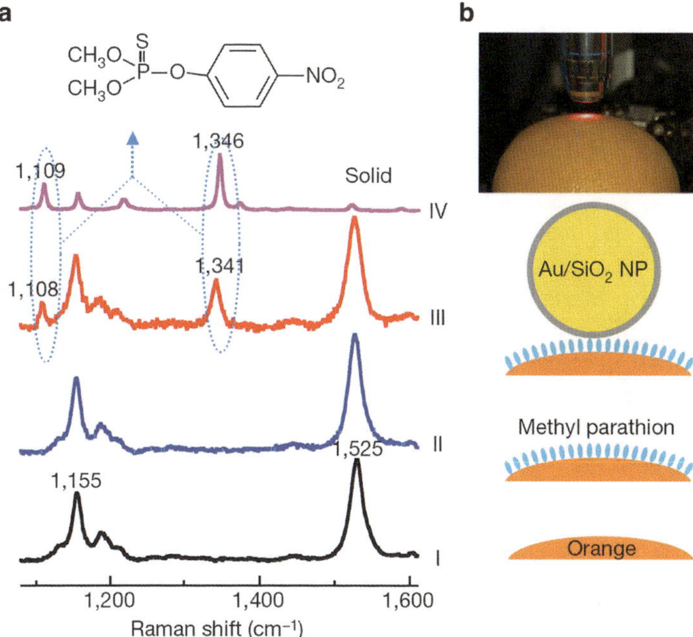

Fig. 6 In situ inspection of pesticide residues on food/fruit. (**a**) Normal Raman spectra on fresh citrus fruits. Curve I, with clean pericarps; curve II, contaminated by parathion. Curve III, SHINERS spectrum of contaminated orange modified by Au/SiO_2 nanoparticles. Curve IV, Raman spectrum of solid methyl parathion. Laser power on the sample was 0.5 mW, and the collected times were 30 s. (**b**) Schematic of the SHINERS experiment. Reprinted with permission from Li et al. [80]. © 2010, Nature

to plasmonic hot spots. Such a tremendous increase in SERS signal allows zeptomole detection [78]. A SERS-based single-molecule detection has also been reported by using gap-tailorable gold–silver core–shell nanodumbbells. Using a stoichiometric control over the number of tethering DNA molecules on the AuNPs surface and a subsequent magnetic-particle-based separation method, Au nanoparticle heterodimers are successfully synthesized in a relatively high yield by means of a single-target-DNA hybridization [79].

A shell-isolated nanoparticle-enhanced Raman spectroscopy has been proposed for inspecting pesticide residues on food and fruit via Raman signal amplification by AuNPs with an ultrathin silica or alumina shell. Figure 6 shows that normal Raman spectra recorded on fresh orange with clean pericarps (curve I) or contaminated by parathion (curve II), which shows only two bands at about 1,155 and 1,525 cm^{-1}, attributed to carotenoid molecules contained in citrus fruits.

By spreading shell-isolated NPs on the same surface, two bands can clearly be detected at 1,108 and 1,341 cm^{-1} (curve III) that are the characteristic bands of parathion residues. The shell-isolated nanoparticle-enhanced Raman spectroscopy demonstrates tremendous scope as a simple-to-use, field-portable, and cost-effective analyzer [80].

Controlled assembly of gold nanorods induced by Na_3PO_4 leads to a significant amplification of localized surface plasmon resonance (LSPR) signals. The strong affinity between Au and Hg alters the coupled LSPR signals due to the amalgamation of Hg and Au. This allows detection of Hg in aqueous solutions with ultrahigh sensitivity and excellent selectivity, without sample pretreatment [81]. A phase interrogation SPR system based on gold nanorod has led to a drastic sensitivity enhancement at a concentration as low as the femtomolar range for detecting antigen with more than 40-fold increase compared to the traditional SPR biosensing technique [82]. Two-photon Rayleigh scattering (TPRS) properties of gold nanorods can be used for rapid, highly sensitive, and selective detection of *Escherichia coli* bacteria from aqueous solution, without any amplification or enrichment in 50 colony forming units (cfu) mL^{-1} level with excellent discrimination against any other bacteria. TPRS intensity increases 40 times when anti-*E. coli* antibody-conjugated nanorods were mixed with various concentrations of *E. coli* O157:H7 bacterium [83].

5 Functional Nanomaterial-Amplified Electrochemical Detection

5.1 Enhanced Conductivity with Nanoparticles

Carbon-based nanomaterials show excellent conductivity to promote the direct electron transfer between the biomolecules and electrode surface [84]. Based on the excellent conductivity of SWNTs, a proof-of-principle of the terminal protection assay of small-molecule-linked DNA has been designed in quantitative analysis of the interaction of folate with a tumor biomarker folate receptor, and a detection limit of 3 pM folate receptor is achieved with desirable specificity and sensitivity [85]. Similarly, an electrochemical immunoassay strategy has been developed by using phospholipid-coated CNTs as the electrochemical labels. The quasilinear response is obtained in a logarithmic concentration scale within a four-order of magnitude concentration range from 5 pg mL^{-1} to 50 ng mL^{-1} with a readily achieved detection limit of 3 pg mL^{-1} [86].

Using SWNTs forest platforms with multi-label Ab_2-nanotube bioconjugates, a general amplification strategy has been designed for highly sensitive detection of a cancer biomarker in serum and tissue lysates. This approach provides a detection limit of 4 pg mL^{-1} (100 amol mL^{-1}), for PSA in 10 μL of undiluted calf serum, a mass detection limit of 40 fg [16]. A glucose oxidase-functionalized CNTs nanocomposite has been designed to label the signal antibodies for ultrasensitive multiplexed measurement of tumor markers using a disposable immunosensor array. The simultaneous multiplexed immunoassay method showed linear ranges of three orders of magnitude with the detection limits down to 1.4 and 2.2 pg mL^{-1} for CEA and AFP, respectively [87].

In cytosensing, RGDS-functionalized SWNTs are mainly used in two roles as nanoscaffolds for immobilization of cells and as nanoprobes to combine the specific recognition, signal transduction, and signal amplification abilities. The designed electrochemical cytosensor array has been used for simultaneous analyzing the dynamic change of the K562 cell-surface glycome during erythroid differentiation induced by sodium butyrate [88]. The result is consistent with the reference method.

AuNPs with quantum size effects and high electrical conductivity can accelerate electron transfer between redox enzymes and electrode surface. For example, when an apo-flavoenzyme, apo-glucose oxidase, is reconstituted on a 1.4-nanometer gold nanocrystal functionalized with the cofactor flavin adenine dinucleotide, the electron transfer turnover rate of the reconstituted bioelectrocatalyst (~5,000 s^{-1}) is much larger than that of the natural cosubstrate of the enzyme (~700 s^{-1}), providing an attractive route for electrochemical transduction of biorecognition events [89].

5.2 *Direct Electrochemistry of Nanoparticle Aggregations*

Utilizing the direct electrochemical signal of nanoparticle aggregations, many electrochemical assays have been developed for ultrasensitive detection. Typically, silver-enhanced labeling method is frequently employed in immunoassays for improving the sensitivity of detecting proteins. For example, an ultrasensitive multiplexed immunoassay has been developed by combining ALP-labeled antibody-functionalized AuNPs (ALP-Ab/AuNPs) catalyzed deposition of silver nanoparticles at a disposable immunosensor array. The deposited silver is then measured by anodic stripping analysis in KCl solution. This multiplexed immunoassay method shows wide linear ranges over four orders of magnitude with the detection limits down to 4.8 and 6.1 pg mL^{-1} for human and mouse IgG, respectively [90]. Subsequently, a streptavidin-functionalized silver–nanoparticle-enriched carbon nanotube (CNT/AgNP) is designed as trace tag for ultrasensitive multiplexed measurements of tumor markers (Fig. 7). The CNT/AgNP nanohybrid is prepared by one-pot in situ deposition of AgNPs on carboxylated CNTs. The nanohybrid is functionalized with streptavidin via the inherent interaction between the protein and AgNPs for further linkage of biotinylated signal antibodies to obtain tagged antibodies. Through a sandwich-type immunoreaction on the immunosensor array, numerous AgNPs are captured onto every single immunocomplex and are further amplified by a subsequent AgNP-promoted deposition of silver from a silver enhancer solution to obtain the sensitive electrochemical-stripping signal of the AgNPs. This multiplexed immunoassay method shows acceptable precision and wide linear ranges over four orders of magnitude with detection limits down to 0.093 and 0.061 pg mL^{-1} for CEA and AFP, respectively [91].

QDs exhibit sharp and well-resolved stripping voltammetry signals due to the well-defined oxidation potentials of the metal components. A CdTe QDs functionalized poly(styrene-co-acrylic acid) microbead as novel nanoparticle label has been used to amplify the electrochemical signal of DNA hybridization.

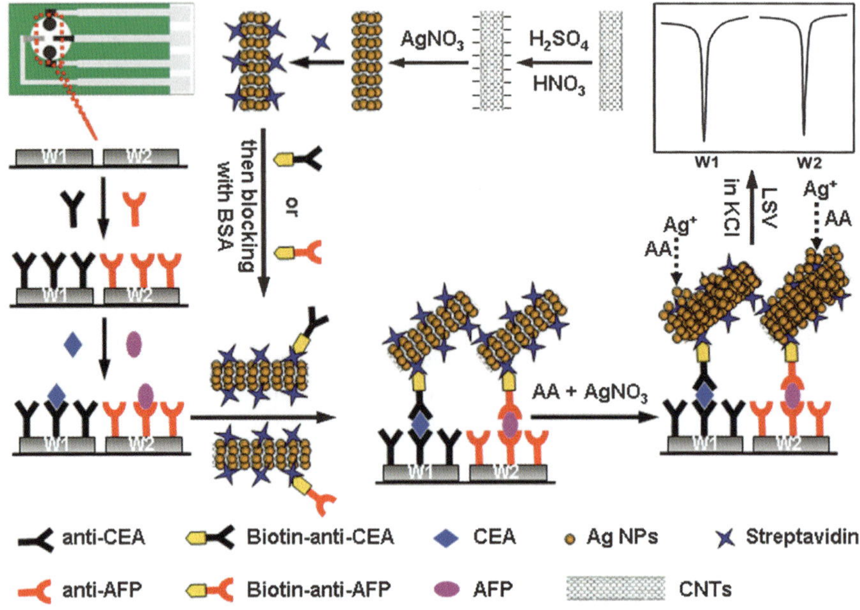

Fig. 7 Schematic representation of preparation of immunosensor array and trace tag, and detection strategy by linear-sweep stripping voltammetric analysis of AgNPs on the immunosensor surface. Reprinted with permission from Lai et al. [91]. © 2011, Wiley

The mean quantum-dot coverage is $(9.54 \pm 1.2) \times 10^3$ per polybead. By square-wave voltammetry of Cd^{2+} after the dissolution of the CdTe tags with HNO_3. The detection of the DNA hybridization process is achieved with a detection limit of 0.52 fmol L^{-1} and a dynamic range spanning five orders of magnitude [92]. Since the signals from multiple metal sulfide nanoparticles can be resolved by anodic stripping voltammetry, a method can possibly be extended to detect a multitude of hybridization events with 100 aM sensitivity by detecting the amplified electrochemical signal [93].

Further, combining the rolling circle amplification technique with oligonucleotide-functionalized QDs, a cascade signal amplification strategy has been proposed for detection of protein target at ultralow concentration. The designed strategy can quantitatively detect protein down to 16 molecules in a 100 μL sample with a linear calibration range from 1 aM to 1 pM and is amenable to quantification of protein target in complex biological matrixes [94].

A sensitive electrochemical aptasensor for the detection of thrombin has been prepared by the amplification of nanoparticles and the usage of differential pulse voltammetry for the detection of dissolved Cd^{2+} in the solution. This assay can directly detect thrombin with a low detection limit of 0.55 fM [95]. Coupling of aptamers with the coding and amplification features of multiple metal sulfide nanoparticles, a highly sensitive and selective panel has been designed for simultaneous detection of several protein targets [96]. By measurement of Au^{3+} ions, a simple electrochemical DNA probe based on submicrometer-size latex spheres with AuNPs is developed for detection of DNA hybridization with a detection limit of 0.5 fM [97].

The ECL emission of QDs is another way to sensitively detect target DNA concentration and sequence by using QDs as the tags. A sensitive DNA ECL sensor based on both the quenching and enhancement of ECL from CdS:Mn NCs by AuNPs in one assay has been constructed. The favorable response of 50 aM target DNA indicates a remarkably low detection limit. Such energy transfer in ECL systems opens a new way for transduction of biological recognition events [98].

Recently, photoelectrochemical assays based on the functional NPs have been quickly developed. These techniques hold the advantages of both optical and electrochemical detections. A photoelectrochemical biosensing platform has been constructed for the detection of biomolecules at relatively low applied potentials using porphyrin-functionalized TiO_2 nanoparticles. The proposed photoelectrochemical method can detect glutathione ranging from 0.05 to 2.4 mmol L^{-1} with a detection limit of 0.03 mmol L^{-1} at a signal-to-noise ratio of 3 [99]. Similarly, a photoelectrochemical platform based on free-base-porphyrin-functionalized Zinc oxide nanoparticles is developed for photoelectrochemical detection of cysteine with a linear range of 0.6–157 mmol L^{-1} in physiological media [100].

5.3 *Electrocatalysis of Nanoparticles*

It is the alternative way to realize the ultrasensitive detection based on electrocatalysis of nanoparticle. Typically, an ultrasensitive and simple electrochemical method for signal amplification is achieved by catalytic reduction of *p*-nitrophenol to *p*-aminophenol using gold-nanocatalyst labels. The detection limit of this assay is 1 fg mL^{-1} for mouse IgG corresponding to ca. 7 aM. This assay also achieves 1 fg mL^{-1} detection limit for PSA, which is comparable to that of the bio-barcode assay [101].

A sandwich-type DNA sensor has been proposed by employing PdNPs as electrocatalytic label. To achieve low level of nonspecific binding of DNA-conjugated PdNPs, indium–tin oxide (ITO) electrode is firstly modified with a silane copolymer containing poly(ethylene glycol) and carboxylic acid. Then, amine-terminated capture probe is covalently attached to a silane copolymer-modified ITO electrode. After target DNA is hybridized with capture probe on the electrode, the fast catalytic hydrolysis of $NaBH_4$ on PdNPs generates many atomic hydrogens, which are rapidly absorbed into Pd NPs, leading to the rapid enhancement of electrocatalytic activity of Pd NPs. PdNP-based ultrasensitive DNA sensor shows an ultralow detection limit (10 aM) and a wide detection range (ten orders of magnitude) [102].

Fe_3O_4 magnetic NPs are highly effective as a catalyst, with a higher binding affinity for the substrate 3,3,5,5-tetramethylbenzidine than HRP and a 40-fold higher level of activity at the same molar concentration of catalyst [103]. A simple and sensitive biosensor has been developed for the detection of DNA hybridization based on flow injection-CL and signal amplification by bio-bar-code functionalized magnetic nanoparticle labels, in which a large amount of metal ions are released from the magnetic NPs. Thus, an ultrasensitive detection of DNA hybridization is achieved by the luminol–H_2O_2–Fe^{3+} CL system with the detection limit as low as 0.32 fM without any preconcentration process [104].

The amplification strategy based on platinum nanoparticles (PtNPs) catalyzing a hydrogen evolution reaction has been developed for the ultrasensitive electrochemical immunosensing. After a typical immuno-sandwich protocol, the signal readout is obtained electrochemically via a PtNPs-catalyzed hydrogen evolution reaction in an acidic aqueous medium containing 10 mM of HCl and 1 M of KCl [105]. A multiple amplification immunoassay has also been proposed to detect AFP, which is based on ferrocenemonocarboxylic–HRP conjugated on PtNPs as labels for rolling circle amplification. The enzymatic amplification signal can be produced by the catalysis of HRP and PtNPs with the addition of H_2O_2, resulting in high sensitivity of the immunoassay with the detection limit of 1.7 pg mL^{-1} [106].

6 Conclusions and Perspectives

Based on the unique properties of nanomaterials, a wide variety of nanoscaled materials with different sizes, shapes, and compositions have been introduced into biosensing for signal amplification. The nanoparticles can be functionalized with biomolecules via noncovalent interaction and covalent route for specific recognition. The biofunctional nanoparticles can produce a synergic effect among catalytic activity, conductivity, and biocompatibility. Therefore, the biofunctional NPs have been used as carriers or tracers for design of a new generation of electronic, optical, and photoelectrochemical biosensing devices. Many considerations such as the good biocompatibility, the sufficient binding sites for functionalization, capacity in the multiple analysis, and so on should be emphasized in the development of ultrasensitive bioassay based on the biofunctional nanomaterials systems. In addition, the photoelectrochemical assays, which hold the advantages of both optical and electrochemical detections, should be a promising direction for constructing a ultrasensitive tool. Signal amplification strategies based on nanomaterials not only provide an ultrasensitive assay in detection of trace analytes but also a concept for basic research in nanobiosensing.

References

1. Goesmann H, Feldmann C (2010) Nanoparticulate functional materials. Angew Chem Int Ed 49:1362–1395
2. Scrimin P, Prins LJ (2011) Sensing through signal amplification. Chem Soc Rev 40:4488–4505

3. Patolsky F, Weizmann Y, Willner I (2004) Long-range electrical contacting of redox enzymes by SWCNT connectors. Angew Chem Int Ed 43:2113–2117
4. Krishnan S, Mani V, Wasalathanthri D et al (2011) Attomolar detection of a cancer biomarker protein in serum by surface plasmon resonance using superparamagnetic particle labels. Angew Chem Int Ed 50:1175–1178
5. Gill R, Zayats M, Willner I (2008) Semiconductor quantum dots for bioanalysis. Angew Chem Int Ed 47:7602–7625
6. Kim SN, Rusling JF, Papadimitrakopoulos F (2007) Carbon nanotubes for electronic and electrochemical detection of biomolecules. Adv Mater 19:3214–3228
7. Chen RJ, Zhang YG, Wang DW et al (2001) Noncovalent sidewall functionalization of single-walled carbon nanotubes for protein immobilization. J Am Chem Soc 123:3838–3839
8. Tu WW, Lei JP, Zhang SY et al (2010) Characterization, direct electrochemistry, and amperometric biosensing of graphene by noncovalent functionalization with picket-fence porphyrin. Chem Eur J 16:10771–10777
9. Munge B, Liu GD, Collins G et al (2005) Multiple enzyme layers on carbon nanotubes for electrochemical detection down to 80 DNA copies. Anal Chem 77:4662–4666
10. Goodwin AP, Tabakman SM, Welsher K et al (2009) Phospholipid-dextran with a single coupling point: a useful amphiphile for functionalization of nanomaterials. J Am Chem Soc 131:289–296
11. Martin AL, Li B, Gillies ER (2009) Surface functionalization of nanomaterials with dendritic groups: toward enhanced binding to biological targets. J Am Chem Soc 131:734–741
12. Wu BH, Hu D, Kuang YJ et al (2009) Functionalization of carbon nanotubes by an ionic-liquid polymer: dispersion of Pt and Pt–Ru nanoparticles on carbon nanotubes and their electrocatalytic oxidation of methanol. Angew Chem Int Ed 48:4751–4754
13. Zhang CY, Yeh HC, Kuroki MT et al (2005) Single-quantum-dot-based DNA nanosensor. Nat Mater 4:826–831
14. Ding L, Cheng W, Wang XJ et al (2008) Carbohydrate monolayer strategy for electrochemical assay of cell surface carbohydrate. J Am Chem Soc 130:7224–7225
15. Nikitin MP, Zdobnova TA, Lukash SV et al (2010) Protein-assisted self-assembly of multifunctional nanoparticles. Proc Natl Acad Sci USA 107:5827–5832
16. Yu X, Munge B, Patel V et al (2006) Carbon nanotube amplification strategies for highly sensitive immunodetection of cancer biomarkers. J Am Chem Soc 128:11199–11205
17. Cheng W, Ding L, Lei JP et al (2008) Effective cell capture with tetrapeptide-functionalized carbon nanotubes and dual signal amplification for cytosensing and evaluation of cell surface carbohydrate. Anal Chem 80:3867–3872
18. Cai WB, Shin DW, Chen K et al (2006) Peptide-labeled near-infrared quantum dots for imaging tumor vasculature in living subjects. Nano Lett 6:669–676
19. Li J, Song SP, Liu XF et al (2008) Enzyme-based multi-component optical nanoprobes for sequence-specific detection of DNA hybridization. Adv Mater 20:497–500
20. Schladt TD, Shukoor MI, Schneider K et al (2010) Au@MnO nanoflowers: hybrid nanocomposites for selective dual functionalization and imaging. Angew Chem Int Ed 49:3976–3980
21. Zhang J, Song SP, Zhang LY et al (2006) Sequence-specific detection of femtomolar DNA via a chronocoulometric DNA sensor (CDS): effects of nanoparticle-mediated amplification and nanoscale control of DNA assembly at electrodes. J Am Chem Soc 128:8575–8580
22. Haun JB, Devaraj NK, Hilderbrand SA et al (2010) Bioorthogonal chemistry amplifies nanoparticle binding and enhances the sensitivity of cell detection. Nat Nanotechnol 5:660–665
23. Kolb HC, Finn MG, Sharpless KB (2001) Click chemistry: diverse chemical function from a few good reactions. Angew Chem Int Ed 40:2004–2021
24. Krovi SA, Smith D, Nguyen ST (2010) "Clickable" polymer nanoparticles: a modular scaffold for surface functionalization. Chem Commun 5277–5279
25. Kamphuis MMJ, Johnston APR, Such GK et al (2010) Targeting of cancer cells using click-functionalized polymer capsules. J Am Chem Soc 132:15881–15883

26. Xu W, Xue XJ, Li TH et al (2009) Ultrasensitive and selective colorimetric DNA detection by nicking endonuclease assisted nanoparticle amplification. Angew Chem Int Ed 48:6849–6852
27. Kowalczyk B, Walker DA, Soh S et al (2010) Nanoparticle supracrystals and layered supracrystals as chemical amplifiers. Angew Chem Int Ed 48:5737–5741
28. Sendroiu IE, Gifford LK, Lupták A et al (2011) Ultrasensitive DNA microarray biosensing via in situ RNA transcription-based amplification and nanoparticle-enhanced SPR imaging. J Am Chem Soc 133:4271–4273
29. Wang J, Shan Y, Zhao WW et al (2011) Gold nanoparticle enhanced electrochemiluminescence of CdS thin films for ultrasensitive thrombin detection. Anal Chem 83:4004–4011
30. Deng CY, Chen JH, Nie LH et al (2009) Sensitive bifunctional aptamer-based electrochemical biosensor for small molecules and Protein. Anal Chem 81:9972–9978
31. Kong RM, Zhang XB, Zhang LL et al (2009) An ultrasensitive electrochemical "turn-on" label-free biosensor for Hg^{2+} with AuNP-functionalized reporter DNA as a signal amplifier. Chem Commun 5633–5635
32. Zhu ZQ, Su YY, Li J et al (2009) Highly sensitive electrochemical sensor for mercury(II) ions by using a mercury-specific oligonucleotide probe and gold nanoparticle-based amplification. Anal Chem 81:7660–7666
33. Zhang XZ, Jiao K, Liu SF et al (2009) Readily reusable electrochemical DNA hybridization biosensor based on the interaction of DNA with single-walled carbon nanotubes. Anal Chem 81:6006–6012
34. Wang J, Liu GD, Jan MR (2004) Ultrasensitive electrical biosensing of proteins and DNA: carbon-nanotube derived amplification of the recognition and transduction events. J Am Chem Soc 126:3010–3011
35. Gao WC, Dong HF, Lei JP et al (2011) Signal amplification of streptavidin–horseradish peroxidase functionalized carbon nanotubes for amperometric detection of attomolar DNA. Chem Commun 5220–5222
36. Zhang J, Lei JP, Xu CL et al (2010) Carbon nanohorn sensitized electrochemical immunosensor for rapid detection of microcystin-LR. Anal Chem 82:1117–1122
37. Ding L, Ji QJ, Qian RC et al (2010) Lectin-based nanoprobes functionalized with enzyme for highly sensitive electrochemical monitoring of dynamic carbohydrate expression on living cells. Anal Chem 82:1292–1298
38. Zhang J, Lei JP, Pan R et al (2011) In situ assembly of gold nanoparticles on nitrogen-doped carbon nanotubes for sensitive immunosensing of microcystin-LR. Chem Commun 668–670
39. Cui RJ, Liu C, Shen JM et al (2008) Gold nanoparticle–colloidal carbon nanosphere hybrid material: preparation, characterization, and application for an amplified electrochemical immunoassay. Adv Funct Mater 18:2197–2204
40. Du D, Zou ZX, Shin Y et al (2010) Sensitive immunosensor for cancer biomarker based on dual signal amplification strategy of graphene sheets and multienzyme functionalized carbon nanospheres. Anal Chem 82:2989–2995
41. Wan Y, Wang Y, Wu JJ et al (2011) Graphene oxide sheet-mediated silver enhancement for application to electrochemical biosensors. Anal Chem 83:648–653
42. Xu YX, Zhao L, Bai H et al (2009) Chemically converted graphene induced molecular flattening of 5,10,15,20-tetrakis(1-methyl-4-pyridinio)porphyrin and its application for optical detection of cadmium(ii) ions. J Am Chem Soc 131:13490–13497
43. Wanunu M, Dadosh T, Ray V et al (2010) Rapid electronic detection of probe-specific microRNAs using thin nanopore sensors. Nat Nanotechnol 5:807–814
44. Fu AH, Hu W, Xu L et al (2009) Protein-functionalized synthetic antiferromagnetic nanoparticles for biomolecule detection and magnetic manipulation. Angew Chem Int Ed 48:1620–1624
45. Mani V, Chikkaveeraiah BV, Patel V et al (2009) Ultrasensitive immunosensor for cancer biomarker proteins using gold nanoparticle film electrodes and multienzyme-particle amplification. ACS Nano 3:585–594

46. Munge BS, Coffey AL, Doucette JM et al (2011) Nanostructured immunosensor for attomolar detection of cancer biomarker interleukin-8 using massively labeled superparamagnetic particles. Angew Chem Int Ed 50:7915–7918
47. Zhuo Y, Yuan PX, Yuan R et al (2009) Bienzyme functionalized three-layer composite magnetic nanoparticles for electrochemical immunosensors. Biomaterials 30:2284–2290
48. Tang DP, Su BL, Tang J et al (2010) Nanoparticle-based sandwich electrochemical immunoassay for carbohydrate antigen 125 with signal enhancement using enzyme-coated nanometer-sized enzyme-doped silica beads. Anal Chem 82:1527–1534
49. Bi S, Yan YM, Yang XY et al (2009) Gold nanolabels for new enhanced chemiluminescence immunoassay of alpha-fetoprotein based on magnetic beads. Chem Eur J 15 (4704):4704–4709
50. Guo YS, Jia XP, Zhang SS (2011) DNA cycle amplification device on magnetic microbeads for determination of thrombin based on graphene oxide enhancing signal-on electrochemiluminescence. Chem Commun 725–727
51. Osterfeld SJ, Yu H, Gaster RS et al (2008) Multiplex protein assays based on real-time magnetic nanotag sensing. Proc Natl Acad Sci USA 105:20637–20640
52. Galanzha EI, Shashkov EV, Kelly T et al (2009) In vivo magnetic enrichment and multiplex photoacoustic detection of circulating tumour cells. Nat Nanotechnol 4:855–860
53. Dong HF, Gao WC, Yan F et al (2010) Fluorescence resonance energy transfer between quantum dots and graphene oxide for sensing biomolecules. Anal Chem 82:5511–5517
54. Freeman R, Liu XQ, Willner I (2011) Chemiluminescent and chemiluminescence resonance energy transfer (CRET) detection of DNA, metal ions, and aptamer-substrate complexes using hemin/G-quadruplexes and CdSe/ZnS quantum dots. J Am Chem Soc 133:11597–11604
55. Yang MH, Li H, Javadi A et al (2010) Multifunctional mesoporous silica nanoparticles as labels for the preparation of ultrasensitive electrochemical immunosensors. Biomaterials 31:3281–3286
56. Wang J, Liu GD, Engelhard MH et al (2006) Sensitive immunoassay of a biomarker tumor necrosis factor-r based on poly(guanine)-functionalized silica nanoparticle label. Anal Chem 78:6974–6979
57. Zanarini S, Rampazzo E, Ciana LD et al (2009) $Ru(bpy)_3$ Covalently doped silica nanoparticles as multicenter tunable structures for electrochemiluminescence amplification. J Am Chem Soc 131:2260–2267
58. Li HX, Rothberg L (2004) Colorimetric detection of DNA sequences based on electrostatic interactions with unmodified gold nanoparticles. Proc Natl Acad Sci USA 104:14036–14039
59. Nam JM, Jang KJ, Groves JT (2007) Detection of proteins using a colorimetric bio-barcode assay. Nat Protoc 2:1438–1444
60. Zhao WA, Chiuman W, Lam JCF et al (2008) DNA aptamer folding on gold nanoparticles: from colloid chemistry to biosensors. J Am Chem Soc 130:3610–3618
61. Qu WS, Liu YY, Liu DB et al (2011) Copper-mediated amplification allows readout of immunoassays by the naked eye. Angew Chem Int Ed 50:3442–3445
62. Ji HX, Dong HF, Yan F et al (2011) Visual scanometric detection of DNA through silver enhancement regulated by gold-nanoparticle aggregation with a molecular beacon as the trigger. Chem Eur J 17:11344–11349
63. Zheng GF, Daniel WL, Mirkin CA (2008) A new approach to amplified telomerase detection with polyvalent oligonucleotide nanoparticle conjugates. J Am Chem Soc 130:9644–9645
64. Ding L, Xiao XR, Chen YL et al (2011) Competition-based transfer of carbohydrate expression information from a cell-adhered surface to a secondary surface. Chem Commun 3742–3744
65. Liang CH, Wang CC, Lin YC et al (2009) Iron oxide/gold core/shell nanoparticles for ultrasensitive detection of carbohydrate-protein interactions. Anal Chem 81:7750–7756

66. Bonomi R, Cazzolaro A, Sansone A et al (2011) Detection of enzyme activity through catalytic signal amplification with functionalized gold nanoparticles. Angew Chem Int Ed 50:2307–2312
67. Ambrosi A, Airò F, Merkoçi A (2010) Enhanced gold nanoparticle based ELISA for a breast cancer biomarker. Anal Chem 82:1151–1156
68. Kim JH, Chung BH (2010) Proteolytic fluorescent signal amplification on gold nanoparticles for a highly sensitive and rapid protease assay. Small 6:126–131
69. Mu CJ, LaVan DA, Langer RS et al (2010) Self-assembled gold nanoparticle molecular probes for detecting proteolytic activity in vivo. ACS Nano 4:1511–1520
70. Li JS, Zhang TR, Ge JP et al (2009) Fluorescence signal amplification by cation exchange in ionic nanocrystals. Angew Chem Int Ed 48:1588–1591
71. Agrawal A, Deo R, Wang GD et al (2008) Nanometer-scale mapping and single-molecule detection with color-coded nanoparticle probes. Proc Natl Acad Sci USA 105:3298–3303
72. Bailey VJ, Easwaran H, Zhang Y et al (2009) MS-qFRET: a quantum dot-based method for analysis of DNA methylation. Genome Res 19:1455–1461
73. Pu KY, Li K, Liu B (2010) Cationic oligofluorene-substituted polyhedral oligomeric silsesquioxane as light-harvesting unimolecular nanoparticle for fluorescence amplification in cellular imaging. Adv Mater 22:643–646
74. Brouard D, Viger ML, Bracamonte AG et al (2011) Label-free biosensing based on multilayer fluorescent nanocomposites and a cationic polymeric transducer. ACS Nano 5:1888–1896
75. Wu MS, Shi HW, Xu JJ et al (2011) CdS quantum dots/$Ru(bpy)_3^{2+}$ electrochemiluminescence resonance energy transfer system for sensitive cytosensing. Chem Commun 7752–7754
76. Zhou H, Liu J, Xu JJ et al (2011) Ultrasensitive DNA detection based on Au nanoparticles and isothermal circular double-assisted electrochemiluminescence signal amplification. Chem Commun 8358–8360
77. Kim NH, Lee SJ, Moskovits M (2010) Aptamer-mediated surface-enhanced Raman spectroscopy intensity amplification. Nano Lett 10:4181–4185
78. Rodríguez-Lorenzo L, Álvarez-Puebla RA, Pastoriza-Santos I et al (2009) Zeptomol detection through controlled ultrasensitive surface-enhanced Raman scattering. J Am Chem Soc 131:4616–4618
79. Lim DK, Jeon KS, Kim HM et al (2010) Nanogap-engineerable Raman-active nanodumbbells for single-molecule detection. Nat Mater 9:60–67
80. Li JF, Huang YF, Ding Y et al (2010) Shell-isolated nanoparticle-enhanced Raman spectroscopy. Nature 464:392–395
81. Huang HW, Qu CT, Liu XY et al (2011) Amplification of localized surface plasmon resonance signals by a gold nanorod assembly and ultra-sensitive detection of mercury. Chem Commun 6897–6899
82. Law WC, Yong KT, Baev A et al (2011) Sensitivity improved surface plasmon resonance biosensor for cancer biomarker detection based on plasmonic enhancement. ACS Nano 5:4858–4864
83. Singh AK, Senapati D, Wang SG et al (2009) Gold nanorod based selective identification of *Escherichia coli* bacteria using two-photon rayleigh scattering spectroscopy. ACS Nano 3:1906–1912
84. Heller DA, Jin H, Martinez BM et al (2009) Multimodal optical sensing and analyte specificity using single-walled carbon nanotubes. Nat Nanotechnol 4:114–120
85. Wu Z, Zhen Z, Jiang JH et al (2009) Terminal protection of small-molecule-linked DNA for sensitive electrochemical detection of protein binding via selective carbon nanotube assembly. J Am Chem Soc 131:12325–12332
86. Nie HG, Liu SJ, Yu RQ et al (2009) Phospholipid-coated carbon nanotubes as sensitive electrochemical labels with controlled-assembly-mediated signal transduction for magnetic separation immunoassay. Angew Chem Int Ed 48:9862–9866

87. Lai GS, Yan F, Ju HX (2009) Dual signal amplification of glucose oxidase-functionalized nanocomposites as a trace label for ultrasensitive simultaneous multiplexed electrochemical detection of tumor markers. Anal Chem 81:9730–9736
88. Cheng W, Ding L, Ding SJ et al (2009) A simple electrochemical cytosensor array for dynamic analysis of carcinoma cell surface glycans. Angew Chem Int Ed 48:6465–6468
89. Xiao Y, Patolsky F, Katz E et al (2003) "Plugging into enzymes": nanowiring of redox enzymes by a gold nanoparticle. Science 299:1877–1881
90. Lai GS, Yan F, Wu J et al (2011) Ultrasensitive multiplexed immunoassay with electrochemical stripping analysis of silver nanoparticles catalytically deposited by gold nanoparticles and enzymatic reaction. Anal Chem 83:2726–2732
91. Lai GS, Wu J, Ju HX et al (2011) Streptavidin-functionalized silver-nanoparticle-enriched carbon nanotube tag for ultrasensitive multiplexed detection of tumor markers. Adv Funct Mater 21:2938–2943
92. Dong HF, Yan F, Ji HX et al (2010) Quantum-dot-functionalized poly(styrene-co-acrylicacid) microbeads: step-wise self-assembly, characterization, and applications for sub-femtomolar electrochemical detection of DNA hybridization. Adv Funct Mater 20:1173–1179
93. Hansen JA, Mukhopadhyay R, Hansen JØ et al (2006) Femtomolar electrochemical detection of DNA targets using metal sulfide nanoparticles. J Am Chem Soc 128:3860–3861
94. Cheng W, Yan F, Ding L et al (2010) Cascade signal amplification strategy for subattomolar protein detection by rolling circle amplification and quantum dots tagging. Anal Chem 82:3337–3342
95. Ding CF, Ge Y, Lin JM (2010) Aptamer based electrochemical assay for the determination of thrombin by using the amplification of the nanoparticles. Biosens Bioelectron 25:1290–1294
96. Hansen JA, Wang J, Kawde AN et al (2006) Quantum-dot/aptamer-based ultrasensitive multi-analyte electrochemical biosensor. J Am Chem Soc 128:2228–2229
97. Pinijsuwan S, Rijiravanich P, Somasundrum M et al (2008) Sub-femtomolar electrochemical detection of DNA hybridization based on latex/gold nanoparticle-assisted signal amplification. Anal Chem 80:6779–6784
98. Shan Y, Xu JJ, Chen HY (2009) Distance-dependent quenching and enhancing of electrochemiluminescence from a CdS:Mn nanocrystal film by Au nanoparticles for highly sensitive detection of DNA. Chem Commun 905–907
99. Tu WW, Dong YT, Lei JP et al (2010) Low-potential photoelectrochemical biosensing using porphyrin-functionalized TiO_2 nanoparticles. Anal Chem 82:8711–8716
100. Tu WW, Lei JP, Wang P et al (2011) Photoelectrochemistry of free-base-porphyrin-functionalized zinc oxide nanoparticles and their applications in biosensing. Chem Eur J 17:9440–9447
101. Das J, Aziz MA, Yang H (2006) A nanocatalyst-based assay for proteins: DNA-free ultrasensitive electrochemical detection using catalytic reduction of *p*-nitrophenol by gold-nanoparticle labels. J Am Chem Soc 128:16022–16023
102. Das J, Kim H, Jo K et al (2009) Fast catalytic and electrocatalytic oxidation of sodium borohydride on palladium nanoparticles and its application to ultrasensitive DNA detection. Chem Commun 6394–6396
103. Gao LZ, Zhuang J, Nie L et al (2007) Intrinsic peroxidase-like activity of ferromagnetic nanoparticles. Nat Nanotechnol 5:577–583
104. Bi S, Zhou H, Zhang SS et al (2009) Bio-bar-code functionalized magnetic nanoparticle label for ultrasensitive flow injection chemiluminescence detection of DNA hybridization. Chem Commun 5567–5569
105. Zhang J, Ting BP, Khan M et al (2010) Pt nanoparticle label-mediated deposition of Pt catalyst for ultrasensitive electrochemical immunosensors. Biosens Bioelectron 26:418–423
106. Su HL, Yuan R, Chai YQ et al (2010) Ferrocenemonocarboxylic–HRP@Pt nanoparticles labeled RCA for multiple amplification of electro-immunosensing. Biosens Bioelectron 26:4601–4604

Nanomaterial-Based Electroanalytical Biosensors for Cancer and Bone Disease

Yeoheung Yun, Boyce Collins, Zhongyun Dong, Christen Renken, Mark Schulz, Amit Bhattacharya, Nelson Watts, Yongseok Jang, Devdas Pai, and Jag Sankar

Abstract With recent advances in novel nanomaterial development, electroanalytical biosensors are undergoing a paradigm shift. New nanomaterial-based electrochemical biosensors can detect specific biomolecules at previously unattainable ultra-low concentrations. This chapter lists the existing biosensor technologies,

Y. Yun (✉)
Bioengineering, Engineering Research Center, North Carolina A & T State University, 1601 E. Market Street, Fort IRC, Room 119, Greensboro, NC 27411, USA
e-mail: yyun@ncat.edu

B. Collins, Y. Jang, D. Pai and J. Sankar
Engineering Research Center, North Carolina A & T State University, Greensboro, NC 27411, USA
e-mail: becollin@ncat.edu; pooh3180@hotmail.com; pai@ncat.edu; sankar@ncat.edu

Z. Dong
Internal Medicine, College of Medicine, University of Cincinnati, Cincinnati, OH 45221, USA
e-mail: dongzu@ucmail.uc.edu

C. Renken
Life Sciences, Applied BioPhysics, Inc., 185 Jordan Road, Troy, NY 12180, USA
e-mail: renken@biophysics.com

M. Schulz
Nanoworld and Smart Materials and Devices Laboratory, College of Engineering, University of Cincinnati, Cincinnati, OH 45221, USA
e-mail: Mark.J.Schulz@email.uc.edu

A. Bhattacharya
Environmental Health, College of Medicine, University of Cincinnati, Cincinnati, OH 45267, USA
e-mail: bhattaat@uc.edu

N. Watts
Bone Health and Osteoporosis Center, College of Medicine, University of Cincinnati, Cincinnati, OH 45221, USA
e-mail: nelson.watts@uc.edu

A. Tuantranont (ed.), *Applications of Nanomaterials in Sensors and Diagnostics*, Springer Series on Chemical Sensors and Biosensors (2013) 14: 43–58
DOI 10.1007/5346_2012_43,
Published online: 8 December 2012

describes the mechanisms, and applications of two types of electroanalytical biosensors, and then identifies the barriers in developing these biosensors and concludes by illustrating how nanomaterials can help overcome these limitations. A key feature of the electrochemical impedance sensor is that biomolecules detection can occur in real time without any pre-labeling. Specifically, this chapter summarizes the state of knowledge of the impedance sensor as applied in cancer and bone disease studies, which are clinically relevant.

Keywords Biosensors, Bone disease, Electroanalysis, Electrochemical impedance, Nanomaterial

Contents

1 Introduction

With recent advances in novel nanomaterials and developments in instrument capability based on these nanomaterials, biosensing is entering a new era. Next-generation sensors are being developed that can detect ultra-low concentrations of analytes. From an electroanalytical perspective, new nanomaterials that have high electrical conductivity, nanoscale size, and high mechanical strength are ideal tools for the development of electrochemical-based biosensors. This chapter describes the mechanisms and advances in electroanalytical methods in relation to recent advances in the development of nanomaterial-driven biosensors.

Significant advances in biosensor technology have occurred in recent years [1–10]. These advances have occurred as a direct consequence of interdisciplinary research, in which specialists from disparate traditional disciplines contribute to advances in specialized systems designed to solve complex problems. Figure 1 is a schematic diagram of nanomaterial-based biosensors, which consist of three parts: (1) biomaterials, (2) nano-interface, and (3) transducer. Current approaches to biosensor technology owe much to the foundation laid by researchers focused upon these three categories to such an extent that many devices thought futuristic 20 years ago are being realized and even becoming commonplace today. The three components of modern biosensors are briefly discussed next.

The optimal function of biomaterials is to provide an interface between the biological activity of interest and an artificial system designed to communicate with a human-operated device. Ideally, the biomaterial will interact with specificity with a target molecule that is part of a biological system or process. Specificity in

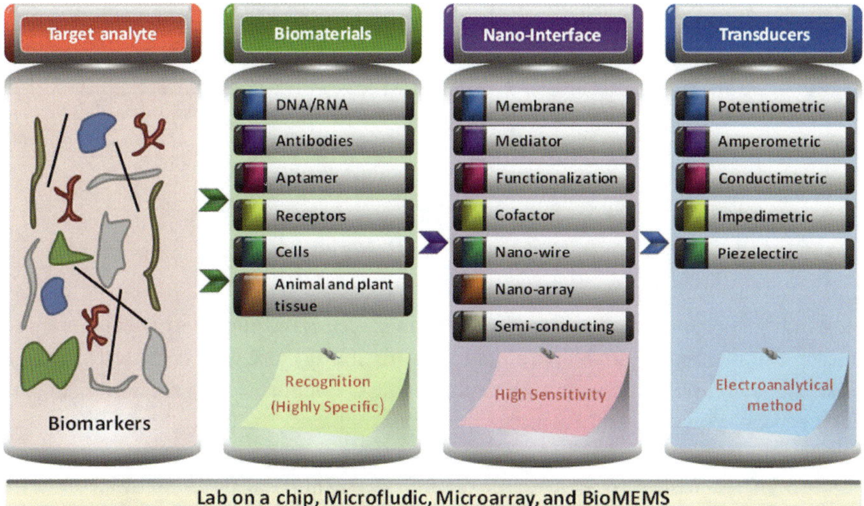

Fig. 1 Schematic diagram of nanomaterial-based biosensors

solution is possible due to the unique selectivity of our evolved biology. DNA, antibodies, aptamers, and cells can be used as the recognition layer, since they only react with a specific DNA sequence or protein and thus can produce a unique trace of binding event due to a change in conformation, mass, or electronic characteristics. Synthetic biology and molecular biology have advanced in recent years [6] such that the recognition layer can include artificial elements such as a single-chain variable fragment, highly specific to a targeted protein and provide a link to the second part of the biosensor, the nano-interface.

The nano-interface is a chemical and/or materials interface to measure and amplify the binding event using a distinguishing property such as a change in the electrical conductivity or resistance or mass. One great advance in recent years is the development of novel carbon nanomaterials. Carbon nanotubes, wires, and other similar forms synthesized on conducting substrates for electroanalytical purposes during the last decade have dramatically increased the sensitivity and lowered the detection limits of biosensors. For example, carbon nanomaterials, including nanotubes and graphene, have high electrical conductivity and high mechanical strength, providing an ideal interface between biomaterial sensor materials and circuit-based electronics. Physical and chemical modifications of these nanomaterials to form membranes, array patterns, porous structures, and catalytic reaction have been explored for sensor devices [6–12].

The last part of sensor development is to transduce the physical or chemical signal typically to an electrical signal for quantitative measurement. Again, recent advances in carbon nanomaterials [10–13] are providing materials that can efficiently perform as electrical wires and interface between biomaterial and human interaction device such as a computer chip. Efficient instrument design and signal conditioning, resulting from perturbations in the nanophase interface with the biomaterial, are

Table 1 Types of electroanalytical biosensors with measurement types and corresponding analytes

Measurement type	Transducer	Analytes
Potentiometric method	Ion selective electrode	Metal ions
	Gas electrode	K^+, Cl^-, Ca^{2+}, F^-, Na^+
	Metal electrode	Redox species
Voltammetric/amperometric method	Enzyme electrode	Sugars, alcohols, O_2
Conductometric method	Interdigitated electrode	Charged species
Impedimetric method	Metal electrode	Antigens, proteins

required to interpret the signaling events. Robust electronic engineering is required to interpret signals and to develop a signal transmission system, with regard to sampling time, signal amplification, and electromagnetic induction (EMI) shielding. In this chapter, we consider only electroanalytical methods, including potentiometric, amperometric, impedimetric, and piezoelectric methods.

Modern fabrication practices allow that biosensors can be integrated into one chip for cost and efficacy considerations. Fabrication techniques using advanced micro-electro-mechanical-system (MEMS) technologies including microelectronics, microfabrication, injection molding, screen-printed electrodes, and micromachining are now commonly applied to the biomedical research devices. Integration of microfluidic systems has several advantages including automatic small-volume sampling, pre-separation, pre-concentration, multi-analyte detection, and waste treatment (waste minimization). The microscale design provides the opportunity for multiple-protein or multiple-DNA detection as immobilizing different recognition proteins or DNA strands on the microarray surface and multiplexing over the sensor elements are compatible with modern technologies. Using multiple sensing elements provides more quantitative and accurate results.

For all its promise, challenges do remain. Integration of a sensor device into a lab-on-a-chip often makes it difficult to obtain repeatable and reliable results. Thus, calibration of sensors should be carefully investigated. Design of a biosensor for a clinical purpose requires careful consideration of several factors: sensitivity across a range of detection limits, linearity range, selectivity to minimize interference from possible confounding chemicals, the stability of the biomaterials and nano-interface materials, reproducibility, and the response time of the sensor. Thus, the sensor data should be clinically meaningful and practical. Based on this biosensor concept, the following sections will provide a very brief review of electrochemical analytical methods using unique nanomaterials. Table 1 shows a summary of the electroanalytical methods currently used for biosensor development.

2 Potentiometric Biosensor

The method of measurement of the potential difference between two electrodes of a galvanic cell under the condition of zero current is called potentiometry [14–28]. The potential difference between the indicator electrode and reference electrode is

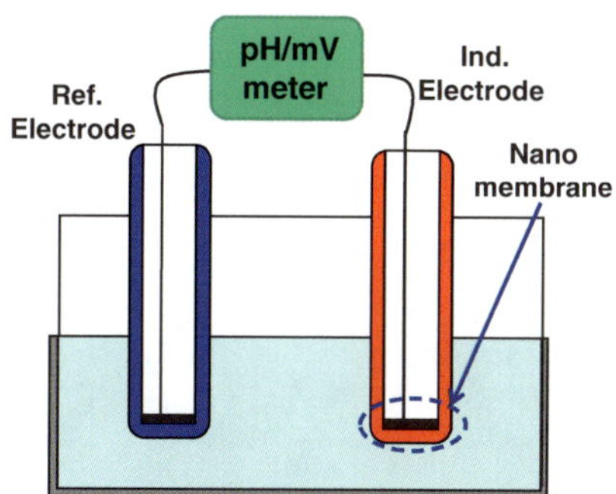

Fig. 2 Potentiometry system consisting of (a) reference electrode, (b) indicator electrode, and (c) analytic-specific membrane

measured by a pH or voltmeter (Fig. 2). The potentiometric sensor can provide accurate information regarding (a) the concentration of a specific chemical and its activity, and (b) the associated free energy change. Normally, three different types of detection mechanism are well established [14].

Further, the development of an ion-selective electrode (ISE), by adapting membranes which are specific for the analyte, has opened a new way in potentiometric sensor during the last decade. The membrane can be polymer, solid-state, and biocatalyticin nature. Ions including Pb^{2+}, Cu^{2+}, Cd^{2+}, Ti^{+}, K^{+}, Na^{+}, NH_4^{+}, Fe^{3+}, Cr^{4+}, Cl^{-}, Ca^{2+}, Mg^{2+}, Ba^{2+}, Ag^{-}, and NO_3^{-} are reliably detected in water-type solutions [15]. Kulapina et al. [16] reviewed the use of ISE for drug analysis. Their study shows that potentiometry can determine organic and inorganic ion levels. Cation drugs are generally detected by the associations between the cations and large ions such as tetraphenyl borate (TPB), tri(octylhydroxy)benzenesulfonate (TOBS), and molybdophosphate (MPA). Anion drugs are determined using electroactive substances with counter ions such as tertiary ammonium bases [17]. In particular, ISEs can be used for the pharmaceutical analytical investigation of bulk-drug materials, the intermediates in their synthesis, products of drugs, and biological samples containing the drugs and their metabolites [17], which eventually helps to maximize drug therapy efficiency and process optimization.

In vivo monitoring of various analytes using ISEs is important for many bioanalytical and biomedical applications [20]. The interaction of blood with ISE surface is undesirable, not only because of the possible interference on the sensor performance (i.e., biofouling of the sensor surface), but also because of the side effects on the patients due to the release of chemicals from the sensor. Gavalas et al. [20] suggested surface modification strategies to make them more blood-compatible. Anti-coagulant coating such as heparin, biomimetic coatings such as phosphorylcholine, and bioengineering coatings with endothelial cells which deposit an extracellular matrix of collagen and glycoproteins are suggested (Table 2).

With the advance of nanotechnology, ionophores which provide selectivity in ISE membranes have been developed, which provide ultra-sensitivity and ultra-low detection limits [21–24]. Patrycja Ciosek et al. [27] present the application of a potentiometric detection of urea and creatinine in the post-dialysate fluids. And also, their group monitored cell-culture media change, detected of the growth of

Table 2 Strategies employed to improve the blood compatibility of polymeric materials [20]

Concepts	Strategies
Minimization of interactions with blood proteins	Hydrophilic surfaces Heterogenic surfaces Negatively charged surfaces
Grafting of active agents	Heparin immobilization Albumin immobilization
Biomembranes	Phospholipids Molecular cilia
Cell seeding	Fibronectin Collagen

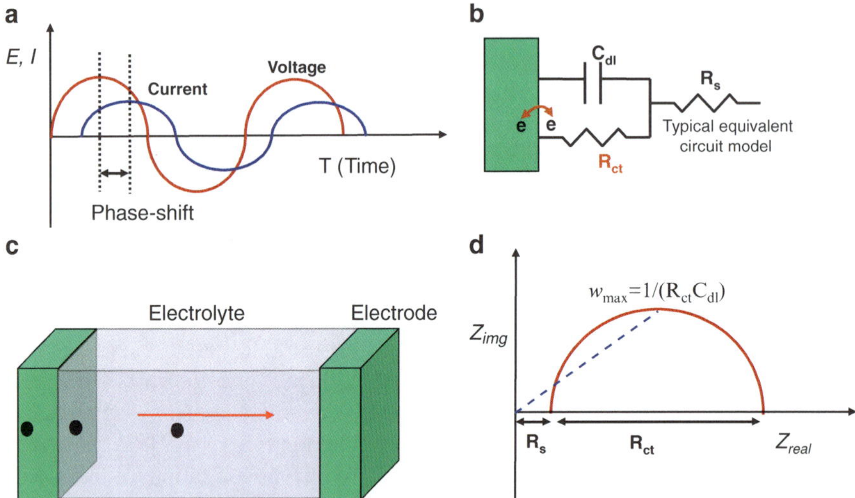

Fig. 3 Electrochemical impedance spectroscopy: (**a**) applied alternating-voltage and measured alternating-current with time, (**b**) equivalent circuit model with double-layer capacitance and electron transfer resistance, (**c**) current flow between two electrodes, and (**d**) Nyquist plot with an equivalent circuit model

various species, and performed toxicological studies with the use of cells using the potentiometric method. Different ions and urea in biological solutions such as serum, milk, and blood can be detected [24–28].

3 Impedimetric Biosensor

Electrochemical impedance is a method that determines the complex electrical resistance and capacitance behavior due to the application of small amplitude of alternating voltage by measuring current amplitude and phase response (Fig. 3). Because impedance (Z) is derived from scanning different frequencies and hence a

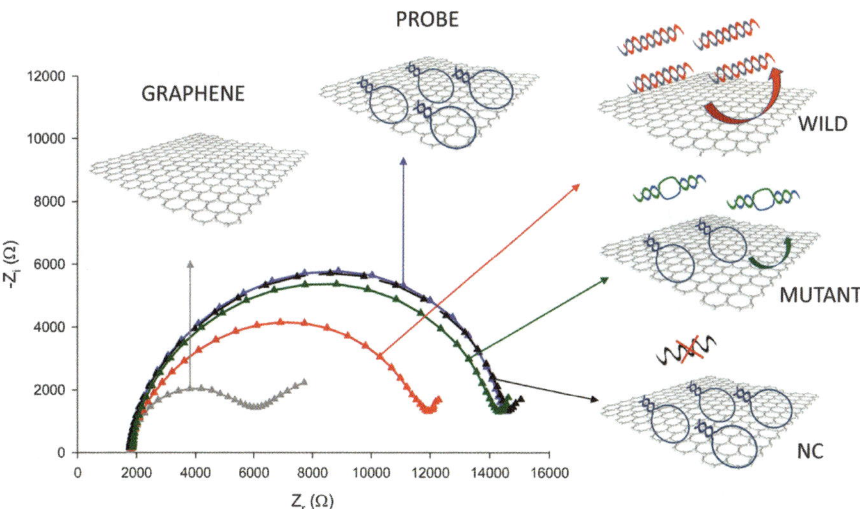

Fig. 4 Schematic of the protocol and Nyquist plots of the graphene surface (*gray*), hpDNA (*blue*), complementary target (*red*), 1-mismatch target (*green*), and negative control with a noncomplementary sequence (*black*) (concentration of the DNA probes, 1×10^{-5} M; concentration of the DNA target, 3×10^{-8} M). All measurements were performed in 0.1 M PBS buffer solution containing 10 mM $K_3[Fe(CN)_6]/K_4[Fe(CN)_6]$. Reprinted with permission from [34]

range of timescales, impedance spectroscopy can provide characterization of surface behavior of the diffusion process, and charge transfer in membrane layer reactions. Typically, the measured impedance output is modeled as an equivalent circuit which consists of virtual resistors and capacitors that model the electronic response, but ultimately represent different electrokinetic events, such as diffusion, in the biological system. The change of the components of resistance and capacitance can be a function of solution properties such as concentration of electroactive species or the electrode's reaction behavior. Nyquist ($Z_{\text{imaginary}}$ vs Z_{real}) and Bode plots (Parameter vs Frequency) are well-established methods to present and interpret impedance behavior. Several examples of impedance sensing follow [29–37].

Carbon nanomaterials such as carbon nanotubes and graphenes have been explored for electroanalytical purpose because of facile electrical conductance, high mechanical stiffness, high aspect ratio, chemical inertness, and possibility for functionalizing (chemical modification) to tune their intrinsic properties. Either as a single nanoelectrode or array electrode, nanoelectrodes with biomaterial conjugation through functionalization offer a viable pathway to lower detection limits and further increase sensitivity. Semiconducting processes can be used to fabricate graphene electrodes for biosensor development.

As an illustration of the utility of impedimetry, the detection of the single-nucleotide polymorphism of a DNA strand was done on a graphene platform. As shown in Fig. 4, a hairpin-DNA probe was immobilized on the graphene surface and achieved high selectivity for the rapid detection of single-nucleotide polymorphism correlated to the development of Alzheimer's disease. Figure 4 shows the

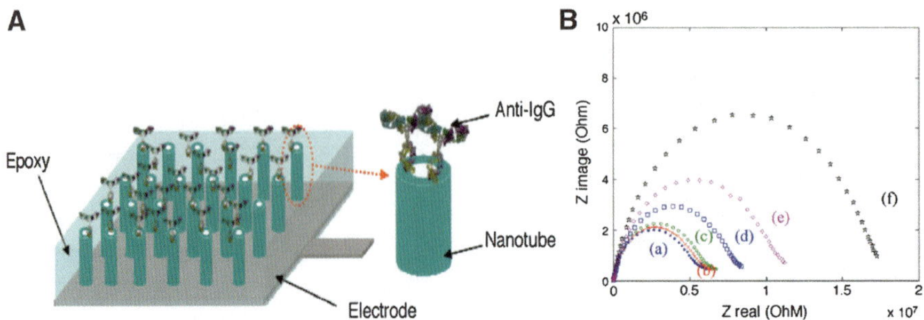

Fig. 5 (**A**) Schematic representation of antibody-immobilized carbon nanotube array electrode and (**B**) electrochemical impedance spectra of immunosensor response (Nyquist plot) after the addition of different concentrations of antigen: (*a*) without antigen; with (*b*) 500 ng/mL; (*c*) 1 μg/mL; (*d*) 5 μg/mL; (*e*) 10 μg/mL; and (*f*) 100 μg/mL of antigen. Reprinted with permission from [37]

schematic diagram of impedance change with different hair-pin DNA conjugations. The impedance method is being actively explored for many areas of biosensor development, such as protein binding kinetics [34].

Figure 5 shows a schematic representation of an antibody-immobilized carbon nanotube array electrode and the electrochemical impedance spectra of the immunosensor response after the addition of different concentrations of antigen. Carbon nanomaterial-based impedance sensing methods are dramatically increasing over the past years, and analytes such as proteins, antibodies, and nucleic acids are being successfully measured using a label-free electronic method.

Two applications of impedance sensors are introduced in their use as osteoporosis markers and cancer markers. Bone turnover markers are useful to predict fracture risk and indicate the need for preventive treatments. Bone turnover markers can measure the body's metabolic activity during the modeling and remodeling phase. Presently, most methods used to detect bone turnover markers are based on the enzyme-linked immunosorbent assay (ELISA). A point-of-care detection method, which allows dynamic measurement, has huge potential, since the level of these biochemical markers are changed with the patient's condition and environments such as day and night.

Biochemical markers which reflect bone remodeling include bone formation markers, for example, alkaline phosphatase and osteocalcin [4, 8]. Also, resorption markers are associated with collagen cross-links such as cross-linked telopeptides. Among several bone turnover markers, reported as listed in Table 3, type I collagen accounts for more than 90% of the organic matrix of bone [4, 38]. During the remodeling process of bone matrix, type I collagen is degraded into small peptide fragments that are eventually secreted into the blood stream. In addition to pyridinium cross-links, cross-linked N-terminal telopeptides of type I collagen and C-terminal telopeptides of type I collagen are released into urine.

Figure 5 is a schematic representation of a label-free immunosensor for bone turnover marker detection. The sensor is based on a gold electrode. Electrochemical

Table 3 List of bone markers in either serum or urine that might be used in biosensors

Bone formation	Bone resorption
Total alkaline phosphatase	Cross-linked telopeptides (NTx, CTx)
Bone alkaline phosphatase	Pyridinoline (pyridinoline, deoxypyridinoline)
Osteocalcin	Hydroxyproline/hydoxylysine
Procollagen type I propeptides	Deoxypyridinoline
	Cathepsin K
	Bone sialoprotein BSP

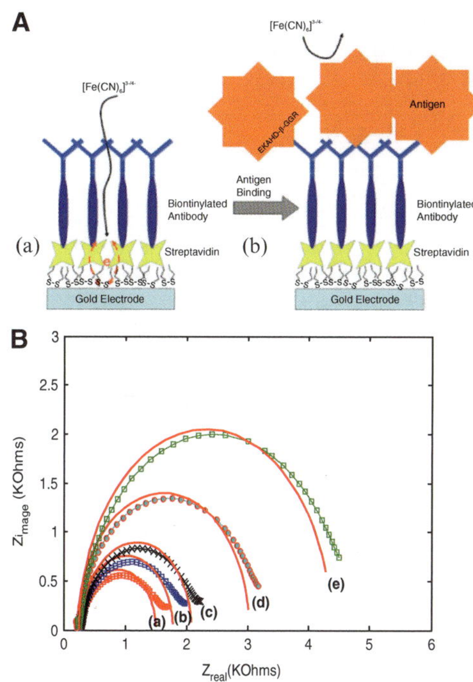

Fig. 6 (**A**) Schematic representation of a label-free immunosensor for bone turnover maker detection. Part **A** (*a*) shows a self-assembled monolayer of dithiodipropionic acid deposited on a gold surface with streptavidin immobilized next as a self-assembled monolayer. Then the biotinylated antibody was bound to the streptavidin. Part **A** (*b*) illustrates the antigen-antibody binding event and how it hinders the interfacial electron transfer reaction of $[Fe(CN)_6]^{3-/4-}$. (**B**) Electrochemical impedance spectra (*EIS*) response recorded at the biotinylated antihuman C-terminal telopeptide antibody-modified electrode in the presence of increasing concentration of human C-terminal telopeptide: 0 μg/mL (*a*), 0.2 μg/mL (*b*), 0.5 μg/mL (*c*), 1 μg/mL (*d*), and 10 μg/mL (*e*) concentration of antigen. EIS was done at a DC potential of 0.2 V at frequencies between 0.1 Hz and 300 kHz. The sinusoidal potential magnitude was ±20 mV in 5.0 mM $K_3Fe(CN)_6$ and 5.0 mM of $K_4Fe(CN)_6$ in PBS (pH 7.0). Reprinted with permission from [4, 8]

impedence spectra (EIS) response recorded at the biotinylated antihuman C-terminal telopeptide antibody modified electrode increased with increasing concentration of human C-terminal telopeptide antigen concentration (Fig. 6).

Table 4 Known biomarkers associated with cancer diagnosis and prognosis

Cancer disease	Markers
Prostate cancer	PSA, PAP
Breast cancer	CA125, BRCA2, CEA, MUC, ING-1
Ovarian	CA125
Liver	AFP, CEA
Metastasis	Circulating tumor cells
Lung	CEA, CA125, CYFRA

The early diagnosis of cancer is the most critical factor for its successful treatment [4] and patient survival. Biological substances called cancer markers are synthesized and released or produced by the host in response to the tumor's presence. The detection of cancer markers with an ultra-low detection threshold in a solid tumor, in peripheral blood, in a lymph node, in urine, and in bone marrow is critical for early diagnosis of cancer. Cancer markers can be detected using a number of techniques including standard immunoassays using samples from blood, urine, and tissue biopsy. A label-free detection method using electroanalytical methods will help to establish diagnosis, monitor treatment, surveillance for recurrence, and target therapy. Correlation of the sensor-reported data along with results from other invasive and noninvasive techniques such as MRI (which tells the tumor status such as the location, grade, and stage of the tumor) will allow the tumor to be properly identified and treated. Array-type carbon nanomaterial sensors can be employed for detection of multiple markers and can quantitatively describe the status of the cancer. Lab-on-chip sensors can thus provide point-of-care service as a noninvasive technique.

Specific protein markers for prostate, liver, lung, breast, and colon cancers are established and listed in Table 4. Detection of protein markers or circulating tumor cells in serum, urine, and saliva are of primary importance for such point-of-care devices. DNA markers of genetic abnormalities such as germ line RB, p53, BRCA I & II, APC, and MMR genes are also suggested as markers for early diagnosis of cancers as well. Detection limits, specificity, cancer heterogeneity, and false negatives are among the factors that should be carefully considered in the sensor development process.

4 Electrochemical-Cell Impedimetric Sensor

Another application for the electrochemical impedance method is to measure the change of cell morphology in real time. An impedance-based sensing system offers a label-free, noninvasive means to electrochemically monitor cell behavior in real time, which eventually allows for the time-lapse study of cell morphology, migration, proliferation, adhesion strength to substrate, cell spreading, cell barrier function, and cytotoxicity. An impedance sensing system can quantitatively monitor in real time the cell morphology change in relation to the metallic substrate as well as

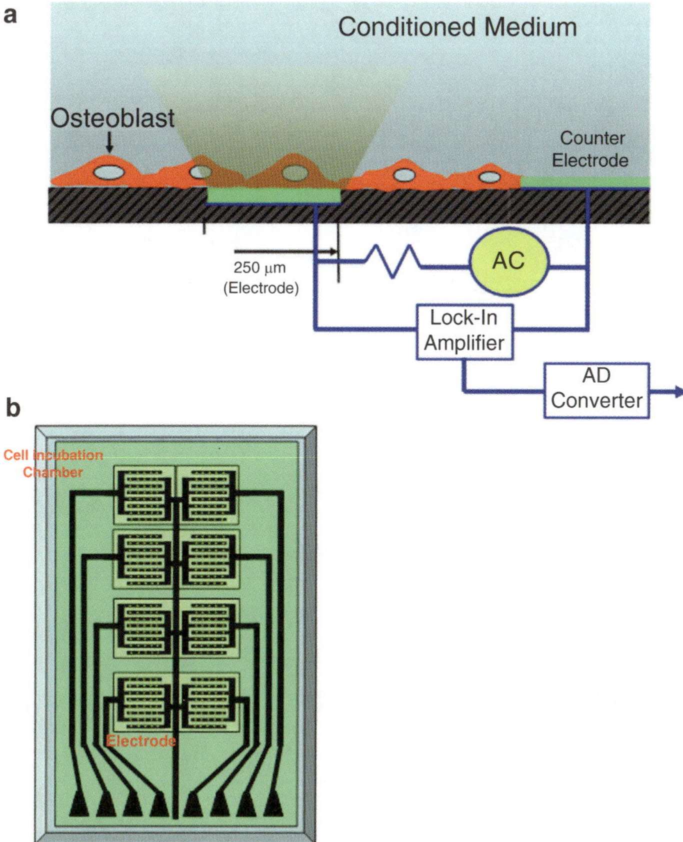

Fig. 7 (**a**) Schematic representation of the cell-impedance sensing method called electrochemical impedance spectroscopy to measure cell adhesion on electrode surface in a conditioned medium. (**b**) 8-well interdigitated array electrode sensors for high-throughput experimentation

neighboring cell interaction. Figure 7 schematically shows how electrochemical impedance is changing with osteoblast cell adhesion. This method was designed in an 8-well cell-culture plate for a high-throughput experiment (Fig. 7b). Interdigitated array electrodes are used to enhance and average signal.

As an example, Fig. 8a shows the electrochemical impedance results with different magnesium ion (Mg^{2+}) concentrations for cytotoxicity studies in cell-culture medium. The impedance gradually decreased with increase in ion concentration. These results were also confirmed with optical imaging of cells on the electrode surface. In this case, we only measured the impedance change at a single frequency.

A critical issue in cancer biology is metastasis, in which the cancer cell detaches and invades other organs or bones. The study of this invasion process requires knowledge of the cell's micro-environmental parameters, such as pH, enzyme, extracellular matrix, and growth factors. The three typical steps of the invasion

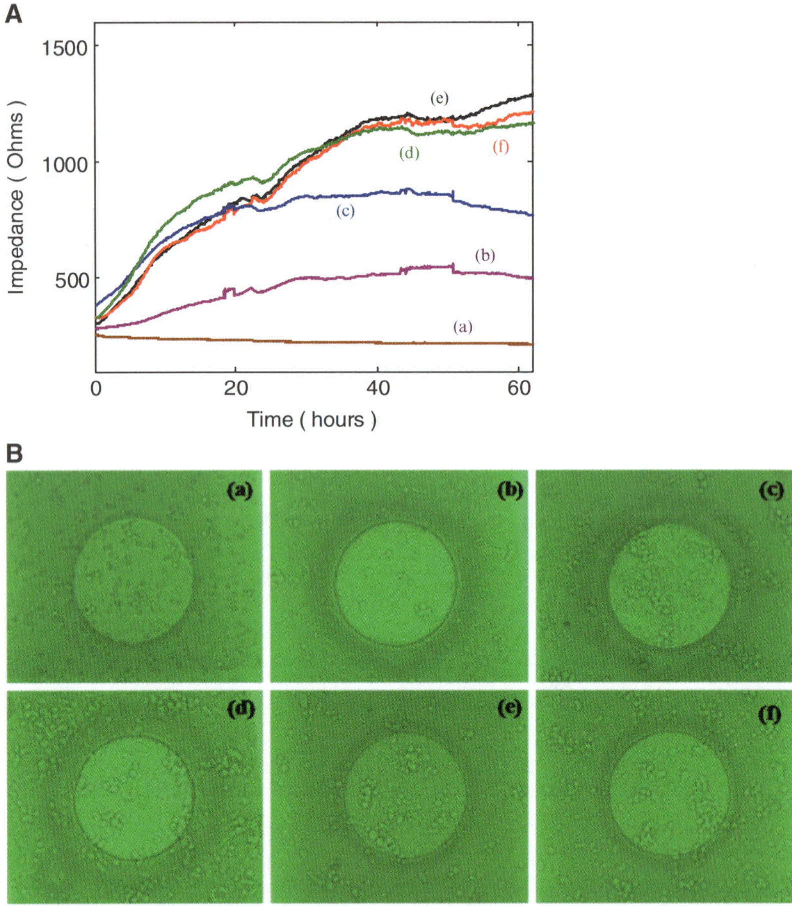

Fig. 8 Electrochemical impedance sensing systems for toxicity study; (**A**) impedance results under different magnesium ion concentrations of (*a*) 200 mM, (*b*)100 mM, (*c*) 50 mM, (*d*) 10 mM, (*e*) 1 mM, (*f*) control of $MgSO_4$. (**B**) optical images of cells on the electrode surface

process, including matrix attachment, matrix degradation, and locomotion, can be measured using an electrochemical impedance spectroscopy (EIS) in vitro. Figure 9a shows the schematic representation of an invasion assay and Fig. 9b shows the invasion process of different prostate cancer cells: Prostate endothelial cells were first incubated for 24 h and cancer cells were added next. Impedances are dramatically decreased with different slopes, which reflect the invasive ability of each cancer cell. PC-3 is a human prostate cancer cell line which is androgen receptor negative. C2 (TRAMP-C2) is a cell line established from a TRAMP tumor. RE3 (full name TRAMP-C2RE3) is a cell line derived from C2 by recycling thrice in murine prostate glands. L5 is derived from RE3, recycled five times by injection into the prostate and the collection of lymph node metastasis.

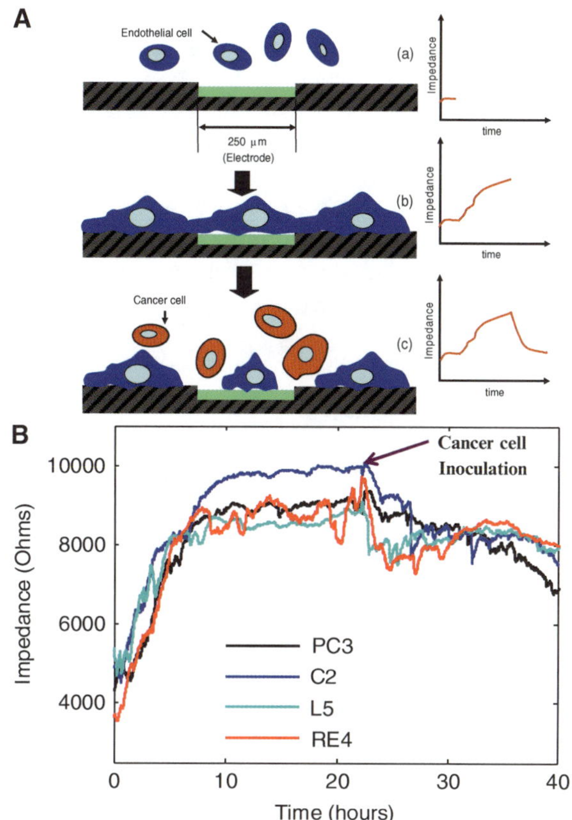

Fig. 9 Electrochemical impedance study for cancer invasion study; (**A**) schematic representation of an invasion assay and (**B**) invasion of different prostate cancer cells. Reprinted with permission from [4]

5 Future In-Body Electrochemical Nanosensors

Conventional approaches diagnose and treat illness and diseases by making measurements external to the body or by taking blood or a tissue sample. Nanotechnology can provide new type of sensors that can be implanted into the human body for real-time human health monitoring [4, 8, 39–41]. For example, these include carbon nanotube-based electrochemical orthopedic implant sensors, sensors for monitoring the degradation of biodegradable implants [42], monitoring bone healing [4, 8, 39–41], smart orthopedic implant sensors, anti-infection sensors, contact lens sensors for diabetics, diabetic sensors, inflammation and infection sensors, brain-derived self-adapting sensors, implanted neural sensors, nanoporous membranes, tissue-healing sensor, brain hemorrhaging sensor, acid reflux sensors, chemical sensors, protein sensors, and others. The future of electrochemical-based nanosensors is bright, because monitoring multiple variables simultaneously in the body can help uncover disease early and provide understanding of biological processes [43–45].

The real success of this personalized medicine depends on whether diagnostic techniques such as biosensors can be used to obtain reliable and repeatable results in a timely manner. Point-of-care devices should be modified and improved for this concept [43]. For example, biosensors can be wearable, implantable, provide real-time monitoring, communicate with drug injection instrumentation, and measure a drug-release rate [4, 40, 41]. Also, adaptable biosensors can scale their sensitivity range based on the first measurement. Another challenging topic for biosensing will be the non-specific binding problem. Sensors should either simultaneously measure non-specific binding and target proteins or actively remove the non-specific binding proteins by applying external energy.

6 Conclusions

This chapter has presented some fundamental considerations concerning modern biosensor fabrication, with specific emphasis given to electrochemical methods, potentiometry, and impedimetry. Several examples were presented that highlighted the design considerations for these biosensors, drawing on the knowledge and tools from many disciplines such as biochemistry, electronic engineering, and clinical care. The advances in modern micro-scale fabrication, coupled with advances in electronics and molecular biology, provide a solid foundation upon which current researchers are expanding the boundaries of what is possible in modern preventive medicine. It is an exciting time to conduct research.

Acknowledgments This research was partially supported by BAA11-001 Long Range Board Agency for Navy and Marine Corps Science and Technology Program, National Science Foundation, and the Korean Small and Medium Business Administration (Project no.00042172-1).

References

1. Theavenoti RD, Toth K, Durst AR, Wilson SG (1999) Electrochemical biosensors: recommended definitions and classification. Pure Appl Chem 71(12):2333–2348
2. Wang J (2005) Nanomaterial-based electrochemical biosensors. Analyst 130:421–426
3. Li H, Liu S, Dai Z et al (2009) Applications of nanomaterials in electrochemical enzyme biosensors. Sensors 9(11):8547–8561
4. Yun Y, Eteshola E, Bhattacharya A et al (2009) Tiny medicine: nanomaterial-based biosensors. Sensors 9(11):9275–9299
5. Yun Y, Dong Z, Shanov V et al (2007) Nanotube electrodes and biosensors. Nanotoday 2(6):30–37
6. Yun Y, Conforti L, Muganda P, Sankar J (2011) Nanomedicine-based synthetic biology, editorial. Nanomed Biotherapeutic Discovery 1:1
7. Guo X, Yun Y, Shanov NV et al (2011) Determination of trace metals by anodic stripping voltammetry using a carbon nanotube tower electrode. Electroanalysis 23:1252–1259

8. Yun Y, Bhattacharya A, Watts BN, Schulz JM (2009) A label-free electronic biosensor for detection of a bone turnover marker. Sensors 9:7957–7969
9. Yun Y, Bange A, Dong Z et al (2008) Electrodeposition of gold nanoparticles on carbon nanotube array: electrochemical characterization and evaluation. Sens Actuators B133 (1):208–212
10. Yun Y, Dong Z, Shanov NV, Schulz JM (2007) Carbon nanotube array for electrochemical impedance measurement of prostate cancer cells under microfludic channel. Nanotechnology 18:465505
11. Shao Y, Wang J, Wu H et al (2010) Graphene based electrochemical sensors and biosensors: review. Electroanalysis 22(10):1027–1036
12. Yang W, Ratinac RK, Ringer PS et al (2010) Carbon nanomaterials in biosensors: should you use nanotubes or graphene? Angew Chem Int Ed 49:2114–2138
13. Tang Z, Wu H, Cort RJ et al (2010) Constraint of DNA on functionalized graphene improves its biostability and specificity. Small 6(11):1205–1209
14. Heineman RW, Strovel AH (1989) Potentiometric methods. In: Chemical instrumentation: a systematic approach, 3rd edn. (Chap. 28):1000–1053
15. Bratov A, Abramova N, Ipatov A (2010) Recent trends in potentiometric sensor arrays-a review. Anal Chim Acta 678:149–159
16. Kulapina EG, Barinova OV (1997) Structure of chemical compounds, methods of analysis and process control, Ion selective electrodes in drug analysis. Pharm Chem J 31(12):667–672
17. Gupta KV, Nayak A, Agarwal S, Singhal B (2011) Recent advances on potentiometric membrane sensors for pharmaceutical analysis. Comb Chem High Throughput Screen 14:284–302
18. Dimeski G, Badrick T, John SA (2010) Ion selective electrodes (ISEs) and interferences - a review. Clin Chim Acta 411:309–317
19. Bakker E, Pretsch E (2007) Modern potentiometry. Angew Chem Int Ed 46:5660–5668
20. Gavalas GV, Berrocal JM, Bachas GL (2006) Enhancing the blood bompatibility of ion-selective electrodes. Anal Bioanal Chem 384:65–72
21. Liu B, Rieck D, Van Wie JB et al (2009) Bilayer lipid membrane (BLM) based ion selective electrodes at the meso-, micro-, and nano-scales. Biosens Bioelectron 24:1843–1849
22. Lai C, Fierke AM, Costa RC et al (2010) Highly selective detection of silver in the low ppt range with ion-selective electrodes based on ionophore-doped fluorous membranes. Anal Chem 82:7634–7640
23. Bakkera E, Chumbimuni-Torres K (2008) Modern directions for potentiometric sensors. J Braz Chem Soc 19(4):621–629
24. Amemiya S, Kim Y, Ishimatsu R, Kabagambe B (2011) Electrochemical heparin sensing at liquid/liquid interfaces and polymeric membranes. Anal Bioanal Chem 399:571–579
25. Mottram T, Rudnitskaya A, Legin A et al (2007) Evaluation of a novel chemical sensor system to detect clinical mastitis in bovine milk. Biosens Bioelectron 22:2689–2693
26. Correia D, Magalhães J, Machado A (2008) Array of potentiometric sensors for multicomponent analysis of blood serum. Microchim Acta 163:131–137
27. Ciosek P, Grabowska I, Brzózka Z, Wróblewski W (2008) Analysis of dialysate fluids with the use of a potentiometric electronic tongue. Microchim Acta 163:139–145
28. Ciosek P, Zawadzki K, Skolimowski J et al (2009) Monitoring of cell cultures with LTCC microelectrode array. Anal Bioanal Chem 393:2029–2038
29. Yun Y, Bange A, Shanov NV et al (2007) Highly sensitive carbon nanotube needle biosensors. J Nanosci Nanotechnol 7:2293–2300
30. Yun Y, Shanov V, Tu Y et al (2006) Synthesis of long aligned multi-wall carbon nanotube arrays by water-assisted chemical vapor deposition. J Phys Chem B 110(47):23920–23925
31. Yun Y, Shanov V, Schulz JM et al (2006) High sensitivity carbon nanotube tower electrodes. Sens Actuators B120:298–304
32. Yun Y, Shanov V, Tu Y et al (2005) A multi-wall carbon nanotube tower electrochemical actuator. Nano Lett 6(4):689–693

33. Yun Y, Gollapudi R, Shanov V et al (2007) Carbon nanotubes grown on stainless steel to form plate and probe electrodes for chemical/biological sensing. J Nanosci Nanotechnol 7:891–897
34. Bonanni A, Pumera M (2011) Graphene platform for hairpin-DNA-based impedimetric genosensing. ACS Nano 5:2356–2361
35. Lisdat F, Schäfer D (2008) The use of electrochemical impedance spectroscopy for biosensing. Anal Bioanal Chem 391:1555–1567
36. Yun Y, Bange A, Shanov NV et al (2006) A nanotube composite microelectrode for monitoring dopamine levels using cyclic voltammetry and differential pulse voltammetry. Proc Inst Mech Eng Part N J Nanoeng Nanosyst 220:53–60
37. Yun Y, Bange A, Heineman RW et al (2007) A nanotube array immunosensor for direct electrochemical detection of antigen-antibody binding. Sens Actuators B123:177–182
38. Carey JJ, Licata AA, Delaney MF (2006) Biochemical markers of bone turnover. Clin Rev Bone Miner Metab 4(3):197–212
39. Yun Y, Dong Z, Yang D, Sfeir C, Kumta P et al (2009) Biodegradable magnesium metal for possible bone implant: electrochemical and in vitro evaluation for biocompatibility study. J Mater Sci Eng C 29:1814–1821
40. Yun Y, Conforti L, Muganda P, Sankar J (2011) Nanomedicine-based synthetic biology. J Nanomedic Biotherapeu Discover 1:1
41. Yun Y, Pixley S, Cui XT et al (2012) Carbon nanomaterials: from therapeutics to regenerative medicine. J Nanomedic Biotherapeu Discover 2:1000104
42. NSF ERC for Revolutionizing Metallic Biomaterials. http://erc.ncat.edu/
43. Tothill IE (2009) Biosensors for cancer markers diagnosis. Semin Cell Dev Biol 20:55–62
44. Yogeswaran U, Chen SM (2008) A review on the electrochemical sensors and biosensors composed of nanowires as sensing material. Sensors 8:290–313
45. Zhao Y, Wei J, Vajtai R et al (2011) Iodine doped carbon nanotube cables exceeding specific electrical conductivity of metals. Sci Rep 83(1):1–5

Integration of CNT-Based Chemical Sensors and Biosensors in Microfluidic Systems

Pornpimol Sritongkham, Anurat Wisitsoraat, Adisorn Tuantranont, and Mithran Somasundrum

Abstract We describe and discuss the different components necessary for the construction of a microfluidic system including micropump, microvalve, micromixer and detection system. For the microfluidic detector, we focus on carbon nanotube (CNTs) based electrochemical sensors. The properties, structure and nomenclature of CNTs are briefly reviewed. CNT modification and the use of CNTs in conjunction with electrochemical microfluidic detection are then extensively discussed.

Keywords Biosensors, Carbon nanotubes, Chemical sensors, Electrochemical sensor, Microfluidic electrochemical system

Contents

P. Sritongkham (✉)
Department of Biomedical Engineering, Mahidol University, Nakornpathom, Thailand
e-mail: pornpimol.srt@mahidol.ac.th

A. Wisitsoraat and A. Tuantranont
National Electronics and Computer Technology Center, National Science and Technology Development Agency, Pathumthani, Thailand

M. Somasundrum
National Center for Genetic Engineering and Biotechnology, National Science and Technology Development Agency, Pathumthani, Thailand

A. Tuantranont (ed.), *Applications of Nanomaterials in Sensors and Diagnostics*, Springer Series on Chemical Sensors and Biosensors (2013) 14: 59–102
DOI 10.1007/5346_2012_42,
Published online: 8 December 2012

1 Introduction

In recent years there has been great interest in miniaturized analysis systems for chemical and biological sensing applications. These systems and devices, also well known as lab-on-a-chip (LOC), offer several advantages over conventional scale analytical methods, including low sample/reagent consumption, fast analysis, high throughput and automation capability [1–16]. In this chapter, microfluidic systems and the various microfluidic components are concisely described and discussed. Finally, we discuss the properties which make carbon nanotubes (CNTs) ideal modifiers for improving the sensitivity and/or selectivity of electrochemical detectors, and describe how such detectors can be integrated into microfluidic systems.

2 Microfluidic System

2.1 Classes of Microfluidic Systems

Microfluidic systems can generally be divided into two main classes, continuous-flow and droplet-based [17–19]. Continuous-flow is based on the manipulation of a continuous flow of fluid in micron-sized channels, which may be driven by mechanical fluid pressure [20, 21] or by capillary forces and electrokinetic actuation [22, 23]. In a typical continuous flow system, fluids including carrier fluid, buffer and analytes from reservoirs are driven into nanoport inlets by either external pumps or integrated on-chip micropumps through microchannels at controlled flow rates [24]. Fluid manipulations such as mixing and focusing are then conducted using microvalves, micromixers and other microfluid manipulators depending on the application [25]. In some specific applications such as the polymerase chain reaction, fluid will be confined in reaction microchambers that include special-purpose components such as a microheater and microsensors [26]. Processed fluids will be then analyzed by either on-chip or off-chip detectors [27]. In advanced biochemical applications, fluids may be separated along a microcolumn by methods including capillary electrophoresis and liquid chromatography using a specially treated microcolumn and microelectrodes [28, 29]. Finally, analyzed fluids will exit outlets and be collected at waste sinks.

Continuous microfluidic operation is a common technique that is easy to implement, suitable for definite and uncomplicated biochemical applications and is necessary for particular flow-based functions such as separation. However, it suffers from several limitations including relatively high sample/analyte consumption, limited flexibility/reconfigurability, poor fault tolerance capability and poor scalability because of the permanent closed-channel configuration and flow-component-system interdependency. Droplet-based microfluidics has recently been developed to overcome these shortcomings. In these systems, droplets of samples/reagent are formed instead of a continuous stream, which results in lower sample/reagent consumption.

There are two main types of droplet-based systems, namely continuous and discrete [30]. In the continuous type, droplets are generated and transported along microfluidic channels by flowing sample/reagent into an immiscible carrier fluid that separates and encapsulates the continuous solution stream into microdroplets. This system can be considered an extension of the continuous flow scheme and hence it is essentially the continuous flow systems that include a special mixer specially designed to generate a stream of micro-droplets. The main benefit of this method is very high throughput and lower sample/reagent consumption. However, it also requires additional carrier fluids and it still suffers from similar shortcoming to a continuous flow system, particularly complicated fluid manipulation, limited flexibility/reconfigurability, poor fault tolerance capability, and poor scalability.

In the discrete droplet-based approach, samples/reagent are formed, manipulated and analyzed as isolated microdroplets with no need of carrier and microchannel [19, 30, 31]. The use of discrete droplets can greatly reduce sample/reagent consumption when compared to continuous systems. Due to the architectural similarities with digital microelectronic systems, droplet-based systems are often referred to as "digital microfluidics". In typical digital systems, individual droplets are generated, moved, merged, split and mixed by a system of structures operated based on various droplet actuation mechanisms including electrowetting-on-dielectric (EWOD), surface acoustic wave (SAW), dielectrophoresis, thermocapillary forces and magnetic forces. These techniques allow parallel processing ability, high architectural flexibility, scalability and dynamic reconfigurability because the force driving the droplets may be controlled by software-driven electronics and the droplets can be freely manipulated in an open platform. In addition, the fabrication of the digital systems is relatively simple and lower priced because complex microchannels and mechanical components are eliminated. Thus, digital microfluidic systems have increasingly gained interest in various applications. Nevertheless, droplet-based systems also have some limitations. For instance, they cannot provide some important functions such as chemical separation, which is very important for biochemical sensing applications. In addition, integration of detection components in the system requires further technological development. Therefore, we will focus here on continuous flow microfluidic systems as they are still more widely used in biochemical sensing applications.

2.2 *Microfluidic Components*

The key microfluidic components for a general integrated microfluidic system are concisely discussed below.

2.2.1 Micropump

A micropump is a miniaturized pumping device that can be integrated into a microfluidic system for driving fluid at a desired flow rate. Micropumps can be divided into two main classes, mechanical or non-mechanical [32].

Mechanical Micropump

Mechanical micropump operation is based on mechanical moving parts such as membranes or diaphragms and check valves, which are periodically actuated by one or more physical mechanisms to generate fluid flow. Major types of mechanical micropumps include reciprocating diaphragm, peristaltic, syringe and rotary structures [20]. A reciprocating diaphragm micropump typically contains a pumping chamber, flexible actuating diaphragm, mechanical actuator, inlet valve and outlet valve as illustrated in Fig. 1a. Fluid flow is generated by the periodic movement of the diaphragm, which draws in fluid through the inlet valve when the diaphragm retracts (under pressure state) and drives it out when diaphragm extends (over pressure state) [33]. The actuators move the diaphragm back and forth between the two states, causing a volume change called the stroke volume, which is proportional to the generated pumping pressure and pumping rate. The actuation must go against the remaining dead volume of fluid in the chamber and thus the pumping efficiency is proportional to the ratio of stroke to dead volume, called the compression ratio (ε).

Passive inlet and outlet check valves [34], allowing flow only in one direction, are normally used to dictate the flow direction as shown in Fig. 1a. These check valves may be replaced by nozzle/diffuser elements to yield valveless micropumps as illustrated in Fig. 1b. These elements are taper pipes that allow more fluid to enter via the inlet than to leave at the outlet during the under pressure state and vice versa during over pressure state [35]. The main advantage of a valveless structure is simpler fabrication, while the key disadvantage is relatively lower efficiency due to some backward flow.

A peristaltic micropump [36, 37] is a valveless bi-directional micropump containing three actuating diaphragms arranged in series as illustrated in Fig. 1c. Sequential peristaltic movement of diaphragms can be programmed to transfer the fluid from one port to the other with no valve or diffuser/nozzle element and the flow direction can be inverted by reversing the order of diaphragm movement [33, 38]. In contrast, syringe micropumps utilize actuators located in a reservoir

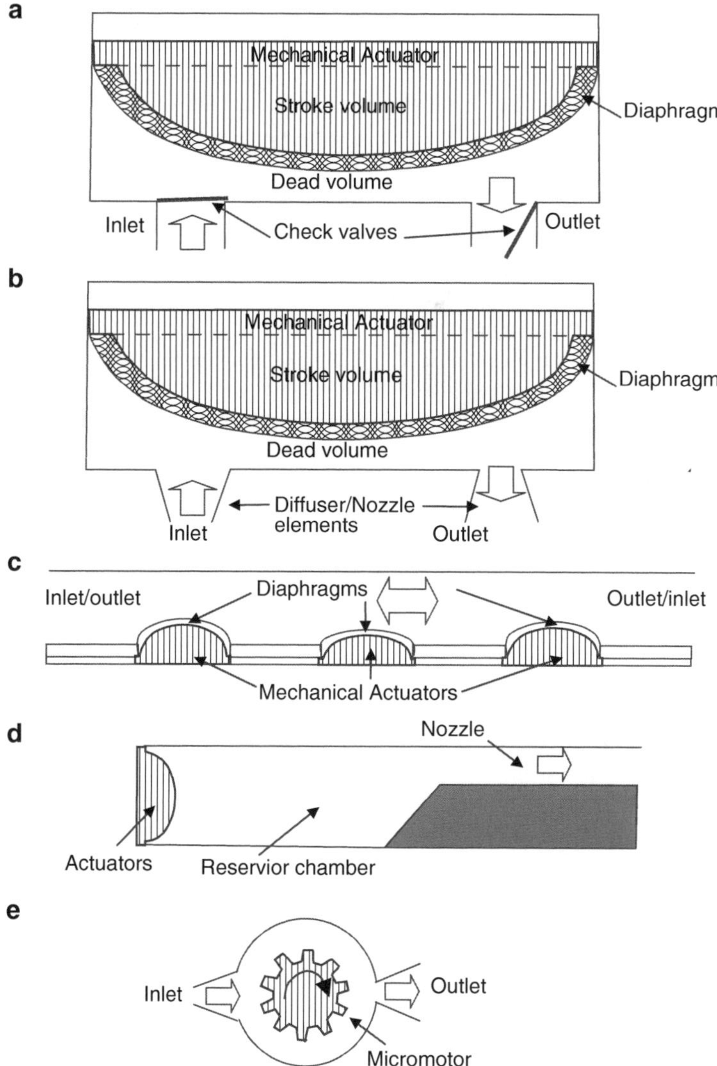

Fig. 1 Structures of (**a**) basic reciprocating, (**b**) valveless reciprocating, (**c**) peristaltic, (**d**) syringe and (**e**) rotary micropumps

chamber to drive the fluid into a nozzle as shown in Fig. 1d. The key feature of syringe micropump is ability to provide accurate, steady and non-turbulent flow control [39, 40]. Rotary micropumps employ a rotating microgear motor to transfer fluid from inlet to outlet, which may employ nozzle/diffuser elements, as illustrated in Fig. 1e. Micromotor is typically driven by either electromagnetic or electrostatic actuation [41]. A rotary micropump can provide a large pumping rate, however its fabrication involves a complicated micromachining process.

The mechanical actuator is the most critical part of a mechanical micropump because it primarily dictates the pumping performance. Important actuation mechanisms employed in mechanical micropumps include piezoelectric, shape memory alloy (SMA), bimorph, ionic conductive polymer film (ICPF), electrostatic, electromagnetic, pneumatic, thermopneumatic and phase change [26, 32]. Table 1 summarizes the main structures, advantages and disadvantages of mechanical micropumps. These mechanical actuators can be further divided into two groups according to application method.

In the first group, comprising piezoelectric, SMA, bimorph and ICPF, the actuator is employed in the micropump by direct attachment of the whole actuator body to the diaphragm. A piezoelectric actuator comprises a piezoelectric layer sandwiched between two metallic films, which are deformed due to strain induced in the piezoelectric crystals under high applied voltage [42–44]. An SMA actuator consists of a microheater and a layer of special metal alloy that has shape memory effect, such as Au/Cu, In/Ti, and Ni/Ti [45, 46]. With this effect, the SMA structure is deformed and reformed between two states as a result of phase transformation between the austenite phase at high temperature and martensitic phase at low temperature. An SMA structure that is mechanically constrained by a diaphragm will exert a large actuation force and displacement by heating and cooling. A bimorph actuator contains a microheater and a bilayer of two dissimilar materials. The bilayer structure will expand or contract upon heating or cooling due to the difference in the thermal expansion coefficients of the two materials [47]. An ICPF actuator is a polyelectrolyte film with both sides chemically plated with platinum and can be actuated by a stress gradient due to ionic movement under an applied electric field [48]. During ionic movement, cations attached to polymer molecules move and simultaneously take solvent shell water molecules to the negative electrode (cathode), causing cathode expansion and anode shrinkage.

In the second group of micropumps, the diaphragm of the micropump is a part of the actuator system. Electrostatic actuators are used in micropumps by making the diaphragm the movable electrode of an electrostatic capacitor. The movable electrode is separated from a fixed plate by a micron-sized air gap and is deflected by Coulomb attraction forces under high applied voltage between the two electrodes [49, 50]. Similarly, electromagnetic actuation can be used for micropumping by bonding a permanent magnet to the diaphragm. The magnetic diaphragm is then actuated by an electromagnetic field induced by applying current to coils around soft magnetic cores [51, 52]. Pneumatic micropump utilizes a compliant diaphragm that is a wall of actuation chamber and is expanded or contracted by air/vacuum pressure, which is manipulated by external solenoid-valve-controlled oil-less diaphragm vacuum pump/compressors [53, 54]. Typically, PDMS membrane, an elastomer with low Young's modulus and high reversible strain, is employed as diaphragm for pneumatic actuation. PDMS provides large stroke, low leakage performance in close position and excellent adherence to glass/silicon substrate. The use of an external air/vacuum system makes the overall system become large. Thermopneumatic principle that uses thermally-induced volume change of air in the sealed chamber has been utilized to replace the external air/vacuum part with

Table 1 Summary of structure, advantages and disadvantages of mechanical micropumps

Type	Structure	Advantages	Disadvantages
Piezoelectric	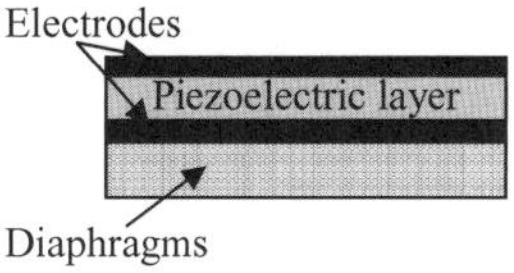 Electrodes Piezoelectric layer Diaphragms	Large actuation force, fast response	Difficult fabrication, high actuation voltage and small actuation stroke
SMA	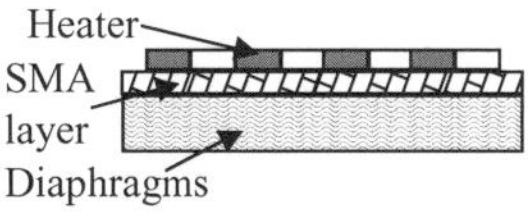 Heater SMA layer Diaphragms	Large actuation force/stroke, simple and small	High power consumption and slow response
Bimorph	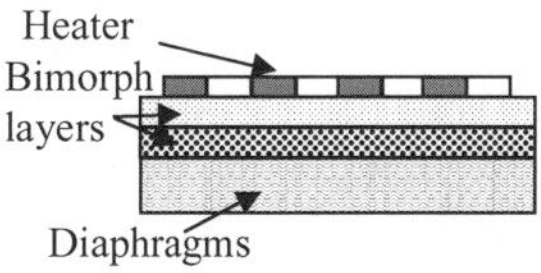 Heater Bimorph layers Diaphragms	Large actuation force, simple and small	High power consumption, slow response, small actuation stroke
ICPF	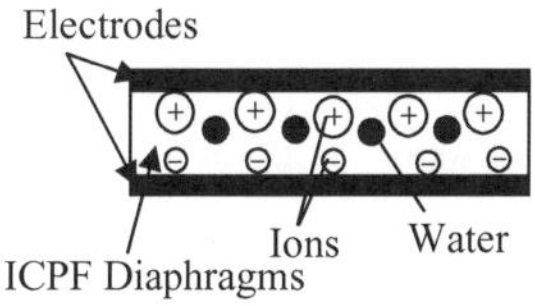 Electrodes Ions Water ICPF Diaphragms	Large deflection and low voltage/power consumption	Low actuation force, complex fabrication process
Electrostatic	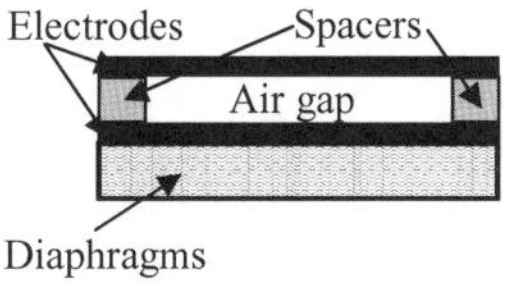 Electrodes Spacers Air gap Diaphragms	Low power consumption and fast response	High actuation voltage and small actuation stroke
Electromagnetic	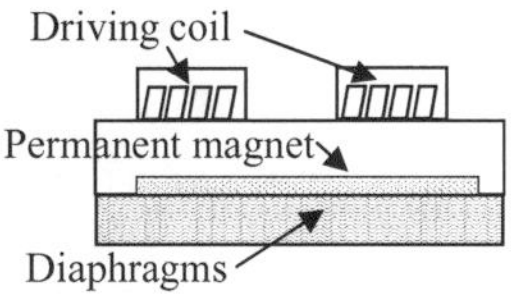 Driving coil Permanent magnet Diaphragms	Large actuation force/stroke, fast response	High power consumption and difficult and expensive fabrication
Pneumatic	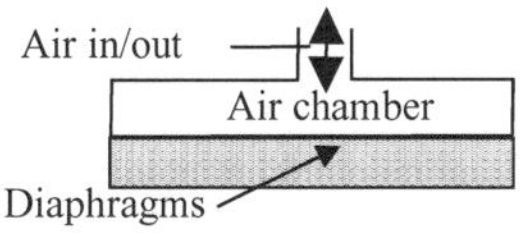 Air in/out Air chamber Diaphragms	Large stroke, fast response	External pump/air flow system required

(continued)

Table 1 (continued)

Type	Structure	Advantages	Disadvantages
Thermo-pneumatic	Heater Air chamber Diaphragms	Large induced pressure and large stroke	High power consumption and slow response
Phase change	Heater Liquid chamber Diaphragms	Large actuation force	High power consumption, slow response, small stroke

integrated microheater/refrigerator [55, 56]. Thermopneumatic system is thus better in term of system integration. Phase-change actuation is similar to thermopneumatic actuation except that it uses a liquid-phase fluid and an external heater (outside actuation chamber). The diaphragm is expanded and contracted by vaporization and condensation of the working fluid [57, 58].

From Table 1, it can be seen that actuation mechanisms which require heating, including SMA, bimorph, thermopneumatic and phase change will require high power and give a slow response but can produce a large actuation stroke. Electromagnetic actuation is a special category in that it has high power consumption due to high current but offers a fast response. In contrary, actuation mechanisms that are based on electric fields, including electrostatic, piezoelectric and ICPF have low power consumption and provide a fast response but have a small actuation stroke. The choice of mechanical micropump depends on flow rate, pumping fluid force, power consumption, operating frequency and fabrication/cost requirements.

Non-Mechanical Micropump

Non-mechanical micropumps are based on micro-scale phenomena that convert non-mechanical energy to kinetic fluidic energy. Geometry design and fabrication techniques for these kinds of pump are simpler due to the absence of moving parts. Non-mechanical phenomena used in micropumps include magnetohydrodynamic (MHD), ferrofluidic, electrohydrodynamic (EHD), electroosmotic, electrowetting, thermo bubble, flexural planar wave (FPW), electrochemical and evaporation [26, 32]. The structure, advantages and disadvantages of non-mechanical micropumps are summarized in Table 2.

MHD actuation is based on the Lorentz force acting on electrically conductive fluids (conductivity >1 s/m) under magnetic and electrical fields [59, 60]. A typical MHD micropump comprises of a rectangular microchannel with two opposite walls being electrodes for applying an electric field and the other two walls being

Table 2 Summary of diagram, advantages and disadvantages of non-mechanical micropumps

Type	Diagram	Advantages	Disadvantages
MHD	Fluid flow, Electrodes, Insulator, Magnets	Large actuation force, fast response	Bubble generation, limited to conductive fluid and Required permanent magnets
Ferrofluidic magnetic	Ferrofluid, Moving magnet, Sample plug	Precise fluid control, fast response	External moving magnet required, low fluid actuation pressure
EHD	Permeable electrodes, Travelling wave, Induction electrodes	High fluid pressure produced, fast response	Limited to low conductivity fluid, High fluid and electric field interaction

(continued)

Table 2 (continued)

Type	Diagram	Advantages	Disadvantages
Electro-osmosis	Electrodes; Guoy-Chapman layer; Stern layer on channel wall	Uniform fluid flow, fast response and ease of fabrication	Limited to conductive fluid and required high applied voltage
Electro-wetting	Mercury droplet; Sample plug; Electrodes	Fast response and low power consumption	Low actuation force and complex fabrication process
Thermo bubble	Microheater; Inlet; Outlet; Nozzle/ diffuser; Bubbles; Nozzle/ diffuser	High fluid pressure and no fluid limitation	High power consumption and slow response

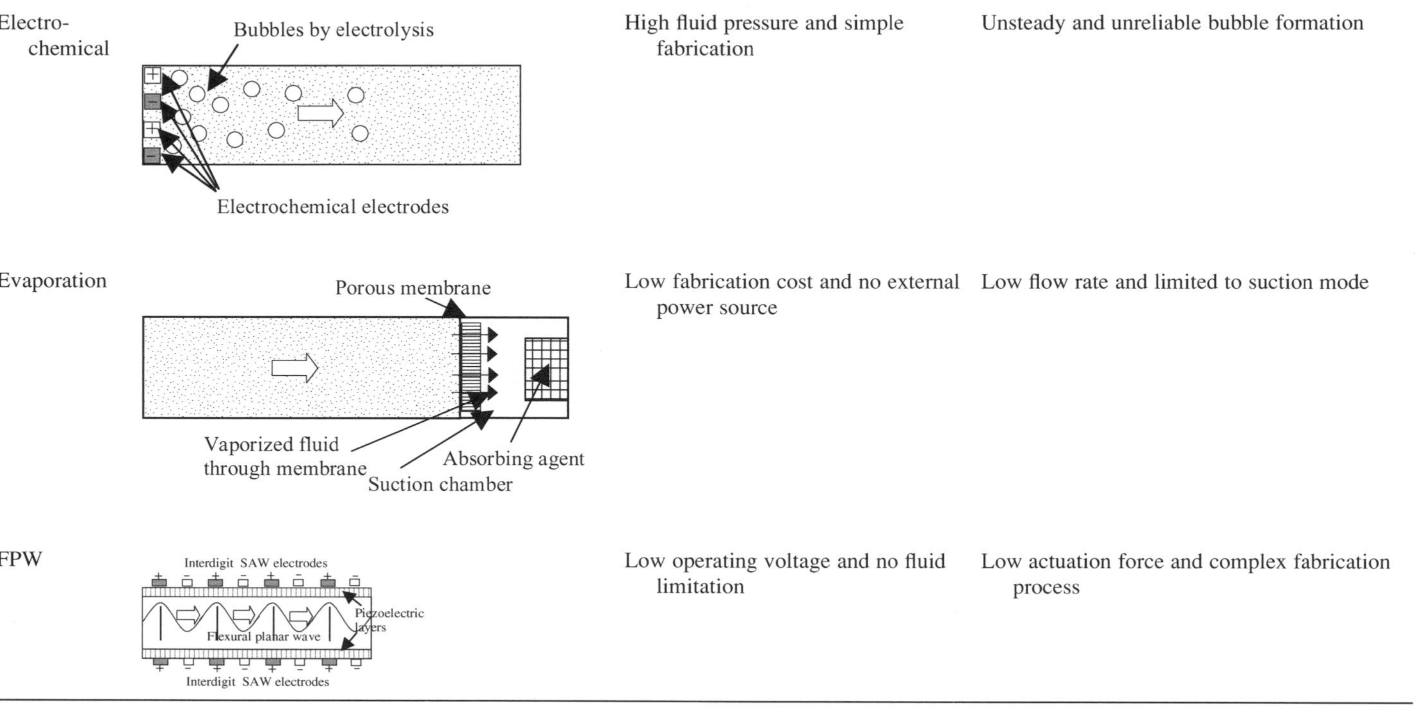

Electro-chemical		High fluid pressure and simple fabrication	Unsteady and unreliable bubble formation
Evaporation		Low fabrication cost and no external power source	Low flow rate and limited to suction mode
FPW		Low operating voltage and no fluid limitation	Low actuation force and complex fabrication process

permanent magnets of opposite polarity for providing the magnetic field. Ferrofluid magnetic actuation utilizes the movement of ferrofluid plug, which is a carrier fluid such as water, hydrocarbons or fluorocarbons with suspended ferromagnetic nanoparticles. The ferrofluid movement is induced by moving magnetic field mechanically using motor controlled magnet [61].

EHD actuation, an electrokinetic effect, relies on Coulombic force acting on electric-field-induced charges in a low conductivity or dielectric fluid [62, 63]. Since charges must be induced, the pumping liquid is directly affected by EHD actuation. In an EHD micropump, ions in the fluid move between two permeable electrodes (emitters and collectors directly in contact with fluid), resulting in an EHD pressure gradient. Electroosmosis is another electrokinetic effect, in which ionic liquid is moved by viscous force as a result of boundary ion movement relative to the stationary charged surfaces of a microchannel under an applied electric field [64, 65]. Stationary negative charges are induced due to interaction between an ionic solution and a microchannel surface such as fused silica and PDMS. The induced charged surface attracts the positively charged ions to form electric double layers, which comprise a fixed-charge Stern layer and cation-rich Guoy–Chapman boundary layer, in which ions can move under an electric field. The movement of these ions leads to viscous motion of neutral fluid in the channel (Table 2). In contrast, electrowetting provides fluid movement through a change in surface tension under applied electric potential via the dielectric layer. The change in electrical potential results in a change in interfacial energy between the liquid and dielectric interface [66, 67]. A continuous electrowetting micropump utilizes the electrowetting effect to manipulate two immiscible liquids such as liquid-phase mercury droplet/electrolyte The surface tension difference between left and right droplet surfaces, due to potential drop across the electric double layer, pushes the droplet from left and right (Table 2). The droplet movement in electrolyte is then used to actuate a micropump diaphragm.

Thermo bubble is the generation and expansion of a bubble by heating under a fluid [68]. The bubble expansion creates a large pressure and actuation force that can be used to drive fluid in a micropump. It can be used to form a diaphragm less and valveless micropump when used along with nozzle/diffuser element (see Table 2). A net flow is generated from nozzle to diffuser by bubble expansion and extinction periodically controlled by heating voltage. Electrochemical reactions such as the electrolysis of water can be used instead to generate gas bubbles to provide the fluid driving force [40, 69]. In an electrochemical micropump, an electrode containing water in an actuation chamber is used to generate bubbles that drive the diaphragm to pump fluid into the main chamber (Table 2).

The FPW micropump utilizes ultrasonically driven acoustic streaming to initiate fluid flow. FPW produced by an array of piezoelectric or SAW transducers propagate along a thin wall of microchannel to transfer its momentum to fluid [70, 71]. Lastly, evaporation-based micropumps exploit controlled liquid evaporation through a membrane into a gas space containing a sorption agent and flow-induced capillary forces, which are generated as a result of evaporation [72]. The vapor pressure in the gas chamber is kept below saturation by gas absorber and evaporated liquid is continuously replaced by the flow of liquid through the microfluidic system.

All non-mechanical micropumps share the common advantages of no moving parts making design and fabrication relatively simple, while there are different pros/cons for each type of micropump. From Table 2, it can be seen that mechanisms involving the application of an electrical potential, including MHD, ferrofluidic, EHD, electroosmosis, electrowetting and electrochemical, provide fast responses, low power consumption and are simple to fabricate but are limited to conducting fluids. Thermo-bubble requires high power consumption and gives a slow response but can produce a large actuation stroke. FPW is a unique method that requires low operating power and can work with any fluid but suffers from a low actuation force and has a complex fabrication process. Evaporation is a simple and novel method that requires no external energy but is limited to suction operation. The choice of non-mechanical micropump depends on type of application and other operational parameters such as flow rate, pumping fluid, power consumption, operating frequency and cost. For example, an electroosmosis pump is essential for capillary electrophoresis.

2.2.2 Microvalves

A microvalve is a miniaturized switching device that can be integrated into microfluidic systems for valving fluid flow direction, and is an important element of mechanical micropumps as well as fully integrated microfluidic chips. Microvalves can be divided into two main classes, passive and active microvalves and each class may be further subdivided into mechanical or non-mechanical types. They are summarized in Table 3 and discussed below.

Passive Microvalves

Passive microvalves, which are often used in mechanical micropumps, do not employ any kind of external actuation. Most common passive microvalves are mechanical flap valves with various structures including cantilever, disc and spherical ball (Table 3) [58, 73]. Disc and ball structures are more preferred for low leakage during closing operation. In addition, valve seat having a mesa pit structure may be used in these structures to reduce any pressure-drop during opening operation by diminishing the structural stiction on the valve anchor. They use a pressure difference between the inlet and outlet of the valve to overcome spring force and control the valve opening. In contrast, passive plug (PP) mechanical microvalve, consisting of snug and airtight stub and stem components, utilizes stub weight and flow control elevation mechanism inside the access port to control fluid flow [74]. The fluid flow is controlled by relative input/output pressure, height of elevation and stub weight. In addition, tight seal (fully close) operation may be attained manually using the snug fit and additional weight above the PP stub.

Table 3 Summary of diagram, advantages and disadvantages of various types of microvalves

Type	Diagram	Advantages	Disadvantages
Passive flap	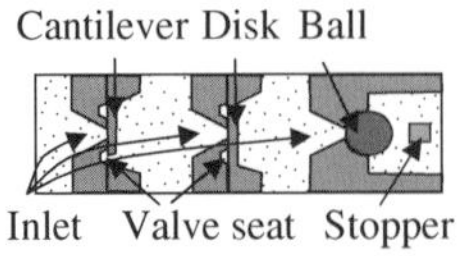 Cantilever Disk Ball Inlet Valve seat Stopper	Simple operation, efficient, low leakage	Complex fabrication process, slow response
Passive plug	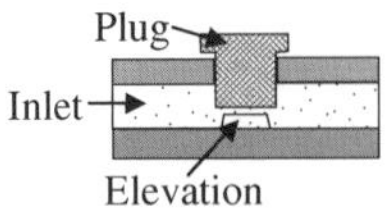 Plug Inlet Elevation	Simple operation, efficient, low leakage, mechanically robust	Low degree of automation, slow response, complex fabrication process
Passive in-line polymerized gel	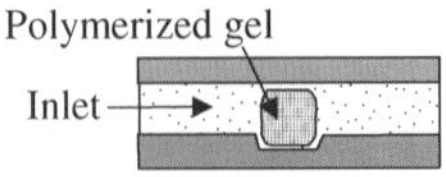 Polymerized gel Inlet	Simple structure, simple operation, low leakage	Non-reversible operation, limited to a group of suitable fluids
Passive hydrophobic	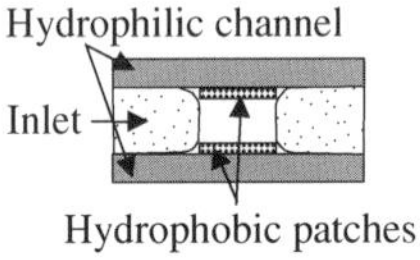 Hydrophilic channel Inlet Hydrophobic patches	Simple fabrication, low cost	High leakage, low fluid pressure, poor stability, limited to a group of suitable fluids
Active diaphragm based mechanical	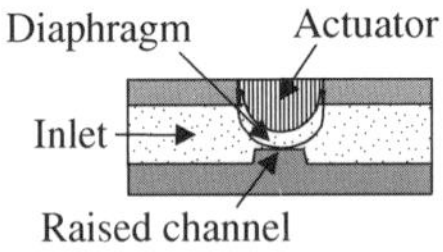 Diaphragm Actuator Inlet Raised channel	Simple and effective, low leakage	High fabrication cost
Active phase-change non-mechanical	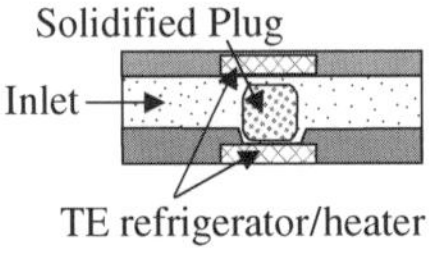 Solidified Plug Inlet TE refrigerator/heater	Simple and effective, very low leakage	High power consumption, low speed

In-line polymerized gel microvalve is a mechanical passive valve that utilizes gel photopolymerization by a thermally induced pressure difference to create local gel plugs to prevent bulk liquid flow [75, 76]. The microfluidic gel valve is typically fabricated from a mixture of Tris–HCl, acrylamide/bic and 1-hydroxyl-cyclohexylphenylketone (HCPK) photoinitiator. Polymerization will occur when passing fluid provides a suitable temperature and pH to the gel. The leakage characteristic of the gel plug mainly depends on the degree of gel cross-linking, which can be optimized to yield minimal leakage at high hydrostatic pressure. Hydrophobic microvalve is a non-mechanical passive valve that utilizes hydrophobic patches in otherwise hydrophilic microchannels to confine fluid within hydrophilic regions by surface tension force [77]. Hydrophobic patches can be formed by coating the regions selectively with a hydrophobic substance such as Teflon and EGC-1700. For instance, passive stop valves can be made by coating access ports with a hydrophobic film. The hydrophobic microvalve is easy to fabricate but it seriously suffers from poor stability due to possible desorption of the hydrophobic film, or reaction with the working fluid.

Active Microvalves

Active microvalves are operated by actuating force and offer higher performance than passive ones but they also require more complexity and fabrication cost. The majority of active microvalves are mechanical valves that employ a sealing diaphragm actuated by various mechanical mechanisms to interpose between the two ports of microchannel as illustrated in Table 3. When the diaphragm is not actuated, fluid flows normally under a driving pressure between inlet and outlet ports. Under actuation, the diaphragm is pressed onto the valve seat and stops fluid flow. Mechanical actuation mechanisms utilized in micropumps such as electrostatic, electromagnetic, piezoelectric, SMA and thermopneumatic can all be used in microvalves [78–82]. Diaphragm based mechanical microvalve offers simple and effective operation with low leakage in close mode and low resistance in open mode but they require relatively high fabrication cost. The speed and power consumption are dependent on actuation mechanisms, which can be selected and optimized according to application requirements.

Another important group of active microvalves is non-mechanical phase change type operated based on solid–liquid phase transition effects [83]. The phase-change microvalves typically consist of a microchannel, a cavity, a phase change plug material and a thermoelectric (TE) microheater/refrigerator as depicted in Table. 3. In operation, phase change material including hydrogel, sol–gel, paraffin and ice are thermally actuated to transform their phase from liquid to solid and vice versa to stop and pass fluid flow. The advantages of this type of microvalves include very low leakage, simple fabrication and ease of large-scale integration. However, they suffer from high power consumption and slow response.

A stimuli-responsive, or smart, hydrogel is an effective phase change material that can change its volume reversibly and reproducibly by more than one order of magnitude from a very small change of input parameters such as pH, glucose, temperature, electric field, light, carbohydrates or antigen [84]. At temperatures above the hydrogel critical temperature T_c (32°C), the hydrogel is unexpanded and the microchannel is open. When the temperature is decreased below T_c by a thermoelectric (TE) refrigerator, the hydrogel swells and blocks the flow. Hydrogel plugs offer perfect sealing, relatively high pressure tolerance and self-actuated open-loop control abilities. Pluronics sol–gel is another promising phase change material [85]. The pluronic polymer forms a cubic liquid crystalline solid gel plug that stop fluid flow at room temperature. When cooled below the pluronics gel transition point (5°C), the gel is liquefied, allowing fluid flow. The advantages of pluronics gel include good sealing and simple implementation by one-shot injection.

Paraffin is another attractive phase change material due to low cost. In paraffin microvalve, the paraffin plug is liquefied by thermal heating, then moved in the microchannel to a fluid port by upstream fluid pressure or external pneumatic air/vacuum actuation and conformally solidified on the port wall, forming a leak-proof solid seal [86, 87]. Heating can be done globally or locally by isothermal or spatial gradient actuation, respectively. Local heating allows much shorter actuation time and lower power consumption than global heating. For reversible close–open operation, external pneumatic air/vacuum actuation is required to move the molten paraffin between fluid ports, making the system fairly complicated. Lastly, water can be used as phase transformation material. Ice plugs can be formed by TE cooling below water freezing point (0°C) to non-invasively close aqueous solution flow, and ice melting by heating can restore the flow [88]. Ice valves are advantageous for bio/chemical microfluidic systems due to biocompatibility and contamination-less property, but require high power consumption and have long response times. These may be reduced by pre-cooling at the expense of power consumption.

2.2.3 Micromixer

A micromixer is a miniaturized merging device that can be integrated in microfluidic systems for mixing two or more fluid flow. Fluid mixing in microchannels is generally difficult due to the low Reynolds number laminar flow effects of fluids in micron-sized conduits [23, 89]. Thus, various strategies have been developed to achieve effective micromixing in fully integrated microfluidic chips. Micromixers can be divided into two main classes, passive and active [26].

Passive Micromixers

The mixing process of passive micromixers mainly relies on chaotic diffusion and convection, which may be achieved by a properly designed microchannel's geometry and surface topography [90]. The Y/T-type flow configurations are the simplest mixing structures that combine two or more Y and T mixing junctions to induce chaotic diffusion [91–93]. For instance, four T-type and one Y-type micromixer combination have been shown to provide high mixing efficiency. However, large combinations of Y–T junctions are not desirable due to size, complexity and cost.

To avoid using large Y/T-type micromixers, a single Y or T mixing junction may be used in combination with other passive micromixing strategies including droplet formation and movement with recirculation flow [94–96]. Droplet formation from two or more fluids to be mixed increases the fluid interface between them, while droplet movement generates an internal flow field inside the turning and elongated droplet, resulting in internal mixing and enhanced mixing efficiency. For successful recirculation, multi-phase droplets must move as one with non-slip boundary conditions. Transient adsorption of droplets to the microchannel walls should be prevented by the use of a suitable surfactant. Droplets can be generated by pressure-driven mixing of multiple immiscible phases with large differences in surface tension, such as an aqueous solution and oil in a microchannel or by fluid mixing driven by capillary effects including thermocapillary or electrocapillary (electrowetting) that employ temperature difference or electric field to bring two or more fluids to mix at a junction. After droplet formation, thermocapillaries or electrocapillaries will then be used again to move droplets around the microchannel to induce recirculation mixing. Another effective droplet mixing strategy is to use the shear force between an additional immiscible carrier liquid and the sample droplet to accelerate the mixing process.

Active Micromixers

Active micromixers employ some external agitating energy to create turbulent mixing and are considerably more effective than passive mixers, but high cost and complexity make them less attractive for practical microfluidic systems. Common actuation mechanisms employed in an active micromixer include electrokinetic, acoustic, MHD and mechanical actuation by micropumps/microvalves. Electrokinetically-driven micromixing provides significant diffusion enhancement from local circulation zones and stretching/folding buck flow induced by a fluctuating applied electric potential that results in a variation of zeta potential on the microchannel walls [97, 98]. The electrokinetic method offers very high mixing speed but it requires a high operating voltage. Acoustic agitation or ultrasonication is a well-known and effective fluid mixing technique that utilizes an acoustic field to induce frictional forces at air/liquid interface of an air bubble in a liquid medium, resulting in bulk fluid flow called cavitation or acoustic microstreaming [99]. The bubble-induced streaming will be most effective if bubbles are excited near their resonance frequency, which depends mainly on bubble size and distribution. Acoustic fields can be induced by an integrated bulk wave or SAW actuator. Acoustic microstreaming offers many advantages including fast operation, simple apparatus, ease of implementation, low power consumption (~2 mW) and low cost.

MHD micromixing use Lorentz force to roll and fold electrolyte solutions in a mixing chamber [100]. In contrast to a MHD micropump, MHD electrode arrays are deposited on the microchannel's surface in the transverse direction, instead of parallel to the microchannel walls. In addition, complex flow fields such as cellular motion, which provide enhanced fluid mixing interface, can be induced under applied AC potential differences across pairs of electrodes with appropriately designed patterns. Mechanical actuation by integrated microvalves/micropumps can also provide highly effective micromixing [54]. Various type of actuation mechanisms have been reported for micromixing. For example, a pneumatic peristaltic rotary micropump has been used for fixed volume as well as continuous-flow on-chip mixing. In fixed volume mixing, fluid injected into the chamber are circulated and quickly mixed with the rotary pump. In a circular loop, different fluids are moving at different velocities with the parabolic profile of the Poiseuille flow and their interfaces will stretch into long and thin streams and finally wrap around into a complete mix. In continuous-flow mixing, the mixing efficiency increases as the overall fluid flow rate decreases and the rotary mixer loop increases.

2.2.4 Detection System

Detection system is an analysis apparatus that may be applied to and/or integrated in microfluidic systems for chemical and biosensing. It is a very important part of micro total analysis system. Sensors or detectors in microfluidic systems must be much more sensitive than conventional methods because of small sample volume, short sampling time and low analyte concentration. The most widely used detection strategies are electrochemical, optical and mass spectroscopic [101, 102].

Electrochemical detections including potentiometric [103, 104], voltametric [105–107], amperometric [108–110], coulometric [111] and AC impedance [112–114] techniques utilize potential, current or charge signals from reducing and/or oxidizing reaction at electrodes, which can be either two-electrode system that comprises working and counter electrodes or three-electrode system that includes an additional reference electrode. Electrochemical detectors are the easiest to integrate into microfluidic devices, due to the fact that the technology for the fabrication of microelectrodes and the placement of microelectrodes within or at the outlets of microchannels is now well established [115–118]. Additionally, electrochemical detection can provide relatively low limits of detection without requiring expensive transduction methods [119]. A number of microfluidic systems have been integrated with on-chip micro-electrochemical sensors using different working electrode materials, including gold, platinum, polymer, CNTs and other nanocomposites. Conventional metal thin film electrodes tend to suffer from insufficient sensitivity and poor limit of detection. The limit of detection for electrochemical measurement can often be improved by modifying the electrode surface with a catalyst of some kind. The improvement can be due to either increasing the rate of the electrode reaction, or by lowering the working potential and thus lowering the background current. The electrode surface can also be modified by a biological recognition element, which can be used to improve the selectivity of detection and/or extend detection to electro-inactive analytes when used in conjunction with an electrochemical label (for DNA or antibody-based detection). As outlined in the following sections, CNTs are ideally suited both to use as electrocatalysts and as nano-dimensional supports for the construction of electrochemical labels. As will be detailed below, both the CNT modification of electrodes and the covalent and non-covalent modification of CNTs by chemical or biological catalysts, have been successfully reported. Another type of electrochemical sensor, which is relatively less commonly used, is ion-sensitive field effect transistor

(ISFET). It is relatively difficult to be integrated in microfluidic system due to relatively complicated fabrication process. Nevertheless, its integration in microfluidic systems has been successfully demonstrated [120–122].

Optical detections including fluorescence [123–127], chemiluminescence [128–131], electrochemiluminescence [132, 133], bioluminescence [134–136], surface plasmon resonance (SPR) [137–145], surface enhanced Raman scattering (SERS) [146–150], absorbance [151], transmittance [152] and other microscopic imaging methods [153, 154] have widely been applied for biochemical sensing in microfluidic systems [155]. Typical optical detection system consists of laser, lens/mirror system and photodetectors. In general, optical components are normally located off-chip but may be assembled or attached to a chip. However, they have relatively large size and may not be fabricated directly on microfluidic chips. The integration still has not been widely attainable and optical-based microfluidic systems remain relatively expensive and bulky. In addition to miniaturization issue, optical signal amplification for sensitivity and limit of detection improvement are also challenging for optical microfluidic based systems [123, 156, 157].

Mass spectroscopic (MS) detection is a promising alternative due to its high detection capability. Typically, MS system comprises ionization chamber, quadruple mass selector and ion detectors. Similar to optical methods, its integration with microfluidic systems is difficult because it is challenging to miniaturize and integrate MS components with microfluidic system. Nevertheless, there have been several microfluidic systems with MS detection [158–163]. In addition to these common methods, many other sensors and detection methods have been demonstrated in microfluidic systems. For instance, Microfluidic chip calorimeter has been developed based on catalytic enzyme reaction detected by calorimetry [164, 165]. In addition, nanomechanical resonant sensors has also recently been integrated within microfluidic channels [166, 167]. Capacitively-coupled contactless conductivity detection is another new scheme recently developed in microfluidic systems [168, 169]. Nuclear magnetic resonance [170] as well as acoustic [171] based sensors have also been recently employed and integrated microfluidic platforms. Furthermore, multiple detection strategies, which combine more than one detection methods such as optical and electrochemical methods for selective and multiplexed detections, have also been widely studied [172–175].

2.3 Fabrication of Microfluidic System

Fabrication of microfluidic systems may be mainly divided into traditional processes based on microelectromechanical system (MEMS) or other alternative processes. Traditional MEMS processes include thick/thin deposition by physical vapor deposition (PVD) such as sputtering and evaporation or chemical vapor deposition (CVD) such as low-pressure and plasma enhanced CVD, photolithographic patterning, wet chemical etching, dry etching such as reactive ion etching (RIE) and bonding [176]. These processes are mainly used for fabrication of silicon- and glass-based continuous or digital microfluidic chip. Although silicon-based microfluidic devices offer full functional capability, their high fabrication/material cost and poor biocompatibility makes them less attractive for chemical and biosensing applications. Glass, polymer and paper are much more attractive materials for microfluidics due to low price, biocompatility and low cost fabrication. Fabrication of microfluidic chip based on these material involve the combination of some standard technology such as photolithography and other alternative processes including solution processing, replication by casting, embossing or injection molding, oxygen-plasma-assisted bonding, adhesive bonding, tape grafting, and inkjet or offset printing. The fabrication process for glass/polymer-based [37, 56, 177, 178] and paper-based devices [179–181] has been rapidly developed to realize cheap, portable and high throughput microfluidic chips. However, most of these microfluidic systems still cannot be implemented with full functionality due to low-temperature, low-structural-complexity and other fabrication limitations. Thus, new fabrication approaches and hybrid technologies are still being developed to achieve low-cost and fully functional integrated microfluidic devices.

3 Carbon Nanotubes

3.1 Introduction

Since the first report of their synthesis by Iijima [182, 183], CNTs have generated much interest as electrode surface modifiers in the construction of bio and chemical sensors. This interest comes from the fact that they have a high tensile strength, a high value of Young's modulus (a measure of stiffness), often good electrical conductivity, exhibit electrocatalytic properties to a number of analytes, can undergo chemical fictionalization, and can be used as carriers for redox mediators, metal nanomaterials or biological recognition elements. CNTs can be made by arc discharge [182], laser ablation [184] or CVD [185]. The later technique provides lower yields than the two former, but has the advantage of both producing high quality CNTs and allowing control of CNT position and alignment.

Structurally, the CNTs is a graphene sheet rolled into a cylinder. Hence, the carbon atoms in the nanotube are all sp^2 hybridized. Iijima initially reported the synthesis of structures consisting of tens of concentric graphitic cylinders [182], referred to as multi-walled carbon nanotubes (MWNTs). These typically have an internal cylinder spacing of 0.3–0.4 nm and external diameters of 2–500 nm. Iijima and Ichihashi [183] and Bethune et al. [186] later synthesized (single-walled carbon nanotubes (SWNTs), consisting of individual graphitic cylinders. More recently, structures consisting of two cylinders, double-walled carbon nanotubes (DWNTs), have been reported [187]. Recent reviews can be found for both the general properties of CNTs [188–192] and their use in sensors [193–196].

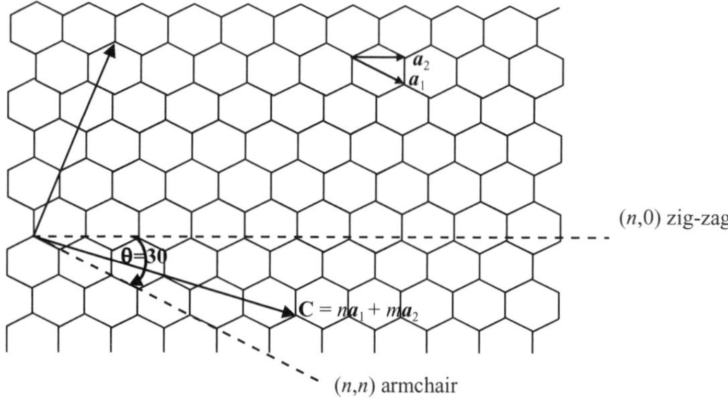

Fig. 2 Use of vector notation to describe CNT structure. The graphene sheet is rolled such that the two end points of the vector C are superimposed. All angles of C lying between zig-zag and armchair give rise to chiral CNTs

3.2 *Structure of CNTs*

As noted above, if a single graphene sheet is rolled into a cylinder the product is a CNTs. Although this is not physically how CNTs are made, as shown in Fig. 2, the process can be imagined as a means to understanding the different possible CNT structures. These are:

1. *Zig-zag carbon nanotubes*: These are produced by rolling up the graphene sheet parallel to the horizontal vector $\boldsymbol{a}_1$. The resulting nanotube has a "zig-zag"-shaped edge.
2. *Armchair carbon nanotubes*: In this case the graphene sheet is rolled on an axis 30° to $\boldsymbol{a}_1$. The result is that the edge of the nanotube consists of the sides of one row of six-membered rings.
3. *Chiral carbon nanotubes*: These are produced by rolling the graphene sheet at an angle θ to $\boldsymbol{a}_1$ such that $0° < \theta < 30°$. A line drawn parallel to $\boldsymbol{a}_1$ would then spiral upwards around the tube. Hence there are two enantiomeric forms of any given chiral CNTs.

The easiest and most common way to describe the structure of a given nanotube is by means of vector notation. As shown in Fig. 2, we can specify a vector $\boldsymbol{C}$ which will join two equivalent points on the graphene plane, i.e. when the graphene plane is rolled into the cylinder the two end-points of $\boldsymbol{C}$ will be superimposed. The structure of the resulting nanotube is then described by the direction and length of $\boldsymbol{C}$, expressed as:

$$C = na_1 + ma_2 \tag{1}$$

where $\boldsymbol{a}_1$ and $\boldsymbol{a}_2$ are the unit vectors shown in Fig. 2. As illustrated in Fig. 3, the integer pairs (n, m) correspond to points on the graphene lattice, following the notation of Dresselhaus and co-workers [197]. Hence, in Fig. 2 a vector $\boldsymbol{C}$ drawn to

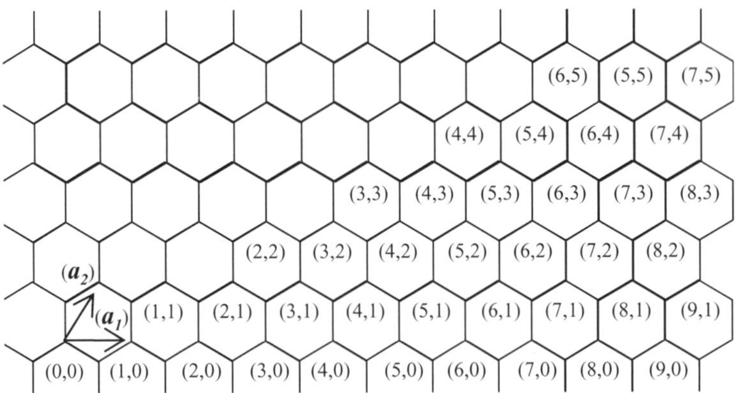

Fig. 3 Graphene sheet with atoms labeled according to the (n, m) notation of Dresselhaus and co-workers [197]

any (n, m) coordinate represents a particular nanotube structure. All nanotubes of the form $(n,0)$ must be zig-zag, while all nanotubes where $n = m$ must be armchair. All other (n, m) coordinates will be chiral. Based on the C–C bond lengths in graphene $\boldsymbol{a}_1 = \boldsymbol{a}_2 = 0.246$ nm. From which the dimensions of the nanotube can be calculated using:

$$\text{nanotube circumference (nm)} = 0.246\ \sqrt{n^2 + nm + m^2} \tag{2}$$

$$\text{nanotube diameter (nm)} = (0.246/\pi)\ \sqrt{n^2 + nm + m^2} \tag{3}$$

$$\text{chirality angle } \theta\ (^\circ) = \sin^{-1}\left[\frac{\sqrt{3}\,m}{2\sqrt{n^2 + nm + m^2}}\right] \tag{4}$$

All armchair SWNTs are metallic conductors, while for zig-zag and chiral SWNTs, approximately one-third of the possible structures are metallic and the other two thirds are semi-conducting.

3.3 *Properties of Carbon Nanotubes*

3.3.1 Inherent Electrode Properties

Due to the fact that CNTs conduct, a CNT film on an electrode will increase the effective area available to the analyte, and hence the sensitivity of the response can be expected to increase. For example, comparative cyclic voltammograms with

$Fe(CN_6)^{3-}$ showed that modification of a glassy carbon electrode with SWNTs increased the electrode area by a factor of approximately 1.8 [198]. In addition to this, CNTs have been shown to have an electrocatalytic effect to the oxidation of many analytically-important substances, including hydrogen peroxide [199], NADH [200], dopamine [201], ascorbic acid [201, 202], uric acid [202] and norepinephrine [203]. Crompton and co-workers have concluded this catalytic effect occurs via reaction at the ends of the CNTs. This is based on the fact that edge-plane pyrolytic graphite electrodes (pyrolytic graphite cut so that the graphite layers are perpendicular to the electrode surface) show similar electrochemistry and catalytic effects to CNTs, where as basal plane graphite (pyrolytic graphite cut so that the graphite layers are parallel to the surface) shows quite different properties, but can be made to behave like edge-plane electrodes if coated with a CNT film [204].

Another useful inherent property of CNTs is that they can promote direct electron transfer between a redox protein and an electrode surface. This has been reported for peroxidases [205], cytochrome c [206], myoglobin [207], catalase [208] and glucose oxidase [209]. The last case is exceptional as the FAD/$FADH_2$ redox couple is buried deep within the enzyme shell, and therefore direct communication with an electrode is usually difficult to achieve. The realization of electron transfer was probably assisted by the fact that the nanotube arrays were vertically aligned on the electrode surface, as Gooding et al. have shown in the case of peroxidases that direct electrochemistry came predominantly from the enzyme immobilized at the ends of the nanotubes, and that the electron transfer rates were independent of nanotube length [205].

3.3.2 Properties Through Modification

As described below, CNTs can be modified to improve dispersion in a solvent, to impart redox properties, and to allow binding to a particular biological recognition element such as an enzyme, antibody or DNA sequence.

Noncovalent Modification

Various ionic and non-ionic polymers can be "wrapped" around CNTs due to a combination of Van der Waals forces and hydrophobic interactions. Typically the polymer has a recurring functional group which interacts with the wall of the CNT, causing the wrapping to occur. Examples of such polymers include poly(*p*-phenylenevinylene) [210], poly{(*m*-phenylenevinylene)-*co*-[2,5-dioctyloxy-(−*p*-phenylenevinylene]} [211] and biopolymers such as amylose [191]. Essentially, the forces of attraction which occur between the CNTs, causing bundle formation, are replaced by forces of attraction to the polymer. Hence, the wrapping process solubilised the nanotubes. This is extremely useful if the nanotubes are to be deposited on an electrode in an uniform and reproducible manner. If the solubilisation is performed by surfactant molecules such as sodium dodecyl sulphate (SDS) [212] or sodium dodecyl benzene sulphate (SDBS) [213] then the nanotubes can be dispersed in aqueous solutions. In the case of Nafion-solublisation of CNTs, it was shown that the presence of the Nafion did not prevent the CNTs catalytic effect on the redox of hydrogen peroxide [199].

Redox behavior can be imparted to CNTs by the adsorption of aromatic compounds. Due to consisting of sp^2 hybridized carbon atoms, a CNT will possess an extended system of delocalized π electrons and this can interact with the π-system of an aromatic molecule. However, the molecule needs to be larger than a single benzene ring for adsorption to occur. If the molecule possesses redox properties then these will hence be imparted to the nanotube conjugate. For example, methylene blue (MB) has been strongly adsorbed onto CNTs purely by sonication [214], and the resulting structures showed a fast rate of electron transfer when adsorbed on glassy carbon electrodes [214–216]. Acridine orange (the MB molecule with the sulphur atom replaced by carbon) has also been adsorbed onto CNTs by the same method [216].

Covalent Modification

The covalent modification of CNTs is a widely researched field and a detailed review of all the methods of covalent nanotube attachment is beyond the scope of this chapter. Some comprehensive reviews are given in refs [217–219]. When nanotubes are treated with strong acids [220], or strong acids mixed with an oxidizing agent [221], for purification and opening, the nanotube ends and defects become functionalized by carboxyl groups. These groups can then undergo further reaction [222–225].

In contrast to tube-end modification, reaction with the side walls of CNTs proceeds by fictionalization of “pristine”, i.e. unoxidised, tubes. Highly reactive reagents are usually necessary. Nanotubes with smaller diameters are likely to be more amenable to reaction, due to their greater curvature causing a greater distortion of the π-electron cloud, thus presenting a richer π-electron surface. Methods of fictionalization include fluorination [226], addition reactions with carbenes, such as dipyridyl imidazolium carbene [227], or with nitrenes [228], reaction with ylides [229], solution phase ozonolysis [230] or silylation [231]. Electrochemistry can also be used as a means of fictionalization, either by performing the electrode reaction of a compound (e.g. a diazonium salt [232]) which then reacts with the nanotube, or by electrooxidation of the nanotubes themselves. All of these methods can in principle be used to attach an appropriately modified redox molecule to the nanotube, and hence there is huge scope for modifying nanotube redox behavior.

Biological Modification

The biological modification of CNTs covers both covalent and noncovalent methods. The first such modifications were performed in the mid to late 1990s

[233, 234] and since then a large amount of research has been performed in this field. Some recent reviews are given in ref [235, 236]. In the case of modification by enzymes, both covalent and non-covalent attachment are possible, although non-covalent attachment is thought to cause less change to the enzyme confirmation [237], and is thus more likely to retain enzyme activity. The simplest form of non-covalent attachment is direct adsorption to the nanotube. The forces responsible for the attachment can be a combination of hydrophobic interactions and π–π stacking between the delocalized π-system of the nanotube and residues on the enzyme containing aromatic groups (tryptophan, phenylalanine etc.) [237]. Electrostatic attraction can also promote direct adsorption [238] in cases where a nanotube has been carboxylated, provided the protons of the carboxylic acids are dissociated and the pH is below the enzyme isoelectric point, causing it to have a positive charge. Carboxylation of CNTS can also promote direct adsorption via hydrogen bonding [239]. Note that although structural changes upon adsorption can be expected to be less than with covalent attachment, they still occur. For example, horseradish peroxidase has been shown to lose about 35% of the helical content of its secondary structure when adsorbed on SWNTs [240].

Alternative to direct adsorption, the CNTs can first be modified by polymers or surfactants as described in Sect. 3.3.2.1. This has the advantage of imparting good aqueous dispersibility to the nanotubes and thus allowing them to be well mixed with the enzymes to be immobilized. The enzymes are taken up mainly through electrostatic attractions [241–243]. In some cases, such as the adsorption of glucose oxidase onto polyaniline-coated MWNTs, this method has enabled direct electrochemistry [242]. Surfactants which have been used for this purpose include SDS and Eastman AQ (for HRP) [242] and Triton X-100 (for biliverdin IXβ reductase) [244].

A popular method of modifying surfaces with polymers is the layer-by-layer (l-b-l) technique first described by Decher and co-workers [245]. This involves the deposition of alternating layers of oppositely charged polyelectrolytes. The adsorption of each layer causes a charge over-compensation to occur, making it possible to then adsorb a layer of the opposite charge [246]. The method can be adapted to the deposition of enzymes, provided the pH used is not equal to the enzymes' isoelectric point, i.e. provided the enzyme has an overall charge. Based on this method sensitive glucose sensors were developed using the immobilization of glucose oxidase (GOx) on SWNTs [247] as well as sensitive immunoassays using HRP with chemiluminescence detection [248], and alkaline phosphatase (ALP) with electrochemical detection [249].

In terms of covalent attachment of enzymes to nanotubes, this can be performed either by directly attaching the enzyme, usually by forming an amide bond between enzyme amine groups and carboxyl groups on the CNT sidewalls [250], or by using a linking molecule. The linking molecule is adsorbed to the CNT through hydrophobic and π–π interactions, and can then be bound to the enzyme through amino groups [251] or reaction with succinimidy ester groups [252], for example.

Since antibodies are also proteins, much of the above discussion for the enzyme modification of CNTs can also be applied to modification by antibodies. One of the simplest methods of antibody attachment is direct adsorption [253–255]. This is thought to proceed by hydrophobic interactions [256]. Probably due to the similarity in hydrophobicities, CNTs coated by aromatic redox compounds, such as methylene blue, can also take up antibodies by direct adsorption. This enables the modified CNTs to act as immunoassay redox labels. Not all of the CNTs communicate with the electrode in such a configuration, but when a solution phase mediator is added, to carry charge from CNT to electrode, fg mL^{-1} concentrations can be detected [257]. Alternatively, antibodies can be coated onto CNTs by the l-b-l method [258] or by using a succinimidyl link [259, 260].

A simple demonstration of DNA immobilization on CNTs can be performed by sonicating bundles of CNTs in water in the presence of single strand DNA (ssDNA). The result is that the CNTs are dispersed, due to the helical wrapping of the ssDNA around the nanotubes. The bases are attracted to the nanotube sidewalls by π–π interactions, and so the ionized sugar-phosphate backbone points outwards, imparting solubility in aqueous media [261]. This type of immobilization can be used to form sensors, based on the fact that metal ions induce a structural change in the DNA, which can be detected as a change in IR spectra [262]. A spectral change is also associated with hybridization, and hence the complementary DNA strand can be detected [263]. CNTs can also be covalently attached to DNA. Acid treatment can be used to produce carboxylic acid groups on CNT sidewalls and these can be reacted with amine-tagged DNA to form amide bonds [264]. If the CNTs are modified by a redox couple then the DNA hybridization can be detected, as has been demonstrated using ferrocene carboxaldehyde [265]. Alternatively, streptavidin can be adsorbed onto the CNTs and then attached to DNA bearing a biotin tag [266]. In this case hybridization detection came from the fact that the CNT was also functionalized by ALP. Thus, enzyme immobilization on a nanotube can be combined with either antibody or DNA immobilization, to produce a highly sensitive electrochemical label.

3.4 Carbon Nanotube Modification of Electrodes

If CNTs are dispersed in solution by one of the coating methods described above, then they can be cast directly onto the electrode. The advantage of this procedure is its simplicity, the disadvantage is the lack of uniformity that will occur over the electrode surface and the degree of variation from one electrode to the next. Casting can also be used to deposit composites of nanotubes mixed with other electroactive materials. A thorough review of this is given by Agui et al. [267]. The most commonly used materials for composite production are conducting polymers and metal nanoparticles. In the case of conducting polymers, the main purpose if to provide the CNTs with good electrode adhesion within a conducting matrix. If the conducting polymer is electrochemically deposited onto the CNT layer, then the polymer quantity can be fairly accurately regulated by controlling the charge passed during the deposition. Analyte reaction can possibly occur at the polymer as well as at the CNTs, depending on the potential and polymer used. Examples of composites of this kind include polyaniline/MWNT films on Au for nitrite oxidation [268], polypyrrole/SWNT films on glassy carbon for ascorbic acid, dopamine and uric acid [269], and poly3-methylthiophene/MWNT composites for NADH oxidation [270].

The construction of nanoparticle (NP)/CNTs composites is mainly performed to increase electrocatalytic activity to a particular analyte. After casting a nanotube film, metal nanoparticles can be electrochemically deposited onto the film, as has been demonstrated for Pt-NP/MWNT coatings for estrogen detection [271]. Alternatively, the metal nanoparticles can be synthesised in the presence of CNTs, which results in adsorption onto the nanotube surface. Examples of this include FeCo-NP/MWNT composites for cathodic hydrogen peroxide detection [272] and Ag/MWNT composites for thiocyanate detection [273]. Nanotube electrode coatings bearing multiple species of nanoparticle are also possible [274]. Finally, after a nanoparticle–CNT composite has been synthesised and deposited on an electrode, a conducting polymer can be electropolymerised over the film. This has been demonstrated for the electropolymerisation of polythionine onto a Au-NP/MWNT coating for an electrode detecting DNA by guanine and adenine oxidation [275].

Another relatively simple electrode modification is to mix the nanotubes with mineral oil and pack then into a cavity above an electrical contact [276], analogous to the way carbon paste electrodes are made. This has been done using unmodified nanotubes for the detection of drugs such as pentoxifylline [277], urapidil [278], sulfamethoxazole [279], and clinically important analytes such as theophylline [280] and homocysteine [281]. Nanotubes have also been mixed with ionic liquids [282] and with a number of different mediators, which act as redox catalysts for an analyte. These have included hydroquinone [283] and a number of different metal complexes [284–289]. The combination in the paste of redox mediator, nanotube and enzyme has also been used to demonstrate biosensors for glucose [290] and lactate [291]. Nanotube pastes have also been mixed with a metal chelator for stripping analysis [292] and with Ag nanoparticles for electrocatalytic oxidation [293]. It has been found that if the metallic impurities on the nanotubes are removed prior to forming the paste, then the electrodes have better between batch precision [294]. This is due to the active role of the impurities in many electrochemical reactions, and the fact that the quantity of impurities can change from batch to batch. However, variations can still occur regarding the electrode surface, and similarly to carbon paste electrodes, we can expect a large amount of time will be required smoothing the electrode over a flat surface before use, to ensure reproducibility. An easier way to achieve reproducibility from roughly the same nanotube matrix is to use the CNTs to form an ink which can be screen printed [295].

A greater level of reproducibility can be achieved by modifying the electrodes with vertically aligned nanotubes. This has the advantage of presenting the highly catalytic end groups for reaction, and as noted in Sect. 3.3.1, can allow direct electrochemistry from enzymes. Alignment can be achieved by first generating carboxylic acid groups at the ends of the CNTs [296] and then reacting these to form functionalities that can be attached to an electrode surface. Examples of this include thiol-tagging for alignment on Au [297], metal ions as bridging agents between end groups and a modified surface [296], alignment to a modified surface by electrostatic attraction [298], and alignment through complexation [299]. Alternatively, aligned nanotubes can be grown onto a surface by photolithography [300].

3.5 Carbon Nanotube-Modified Electrodes in Microfluidics

CNTs have been used in microfluidic systems for the construction of both chemical and biosensors. The field has been reviewed relatively recently by Chen [301]. The simplest method of introducing CNTs to a microfluidic system is to deposit a CNT dispersion onto a working electrode that can be fitted into a microchip-based detector. CNT dispersions in Nafion have been deposited onto screen printed electrodes to form a chemical sensor for the detection, after separation, of hydrazine, phenol, purine and some amino acids [302], and deposited onto a glassy carbon electrode for use as an immunosensor. In the latter case the HRP label on an antibody specific to prostate specific antigen (PSA) was detected via the oxidation and then re-reduction at the electrode of 4-*tert*-butylcatechol [303]. Other electrodes modified by CNT deposition include indium tin oxide (ITO)-coated glass for dopamine measurement (CNTs dispersed in SDS and then spin coated onto the electrode) [304], Au electrodes for the oxidation of antibiotics [305], phenols [306], and catecholamines and their metabolites [307].

The use of CNT-composite electrodes has also been popular with microfluidic detectors. These have included CNTs mixed with Teflon [200], epoxy (thiols) [308], copper (carbohydrates) [309], dihydropyran (insulin) [310, 311], polystyrene (rutin) [311], and the polymer EpoTek H77A (free chlorine) [312]. The CNT composite can be cast onto a glassy carbon electrode surface [310, 311], but is more usually fitted into a cavity above an electrical contact. The same packing in a cavity procedure is used when CNT/carbon paste electrodes, made by mixing CNTs with mineral oil, are used in microfluidics [313]. Alternatively, covalent immobilization of an enzyme onto CNTs which are then formed into a specific microscale shape can be used to form a bioreactor which can then be fitted into a microfluidic device, as shown in Fig. 4 for co-immobilized glucose oxidase and HRP [314].

As noted in Sect. 3.1, CVD production of CNTs can allow control of position and alignment, and this has been used to grow CNTs directly on the sensing portion of the microchip [315–317]. Interest in this method stems from the fact that it is thought to provide a higher CNT density, better alignment and stronger substrate adhesion than casting CNT dispersions or forming CNT pastes. The technique has been applied to biosensor construction by immobilizing cholesterol oxidase (ChOx) onto the aligned nanotubes and measuring the oxidation of the redox mediator $Fe(CN)_6^{4-}$ [315]. The CNTs were grown on a Au film that had been sputter-coated in the form of a strip across the microchannel of the device. ChOx was immobilized on the nanotubes by using the microchannel to deliver a ChOx/$Fe(CN)_6^{4-}$/poly(vinyl alcohol) solution to the working electrode. A similar CNT deposition method, onto a Au strip sputter-coated across a microchannel, was used to construct a microfluidic device to measure salbutamol by its irreversible oxidation at the nanotubes [316]. As well as direct growth onto sputtered Au electrodes, CVD can also be used to form CNTs within the nanopores of a preformed layer of Al oxide. When used to detect iodide this method was found to have better signal to noise characteristics that that of aligned CNTs alone [317].

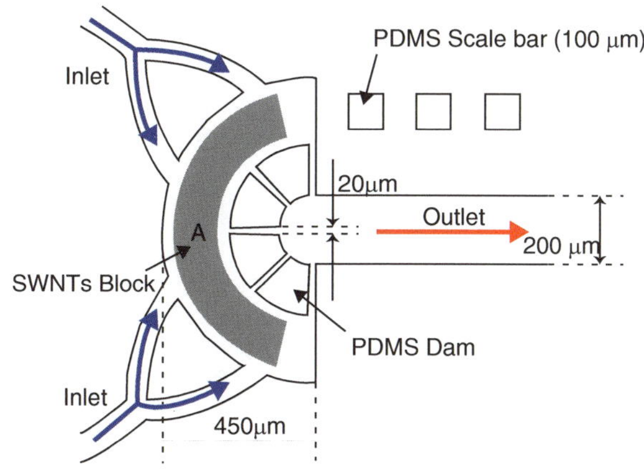

Fig. 4 Microreactor fabricated from polydimethylsiloxane (PDMS), incorporating microreactor of SWNTs functionalised with glucose oxidase and horseradish peroxidase. Taken with permission from [314] © Elsevier Ltd

4 Concluding Remarks

The use of microfluidic-based analytical devices has a number of advantages over analysis performed on a conventional scale. These include: only a low volume of samples and reagents are needed, reducing the cost of the analysis and the amount of waste generated; the large surface-to-volume ratio of the fluids used enhances mass and heat transfer, thus shortening analysis time; the portability of the device allows on-site analysis; multiple samples can be analyzed in parallel, allowing a high throughput. Of the detection systems used with microfluidics, electrochemistry is attractive due to providing relatively high sensitivities at relatively low cost. Also, where needed, microelectrode fabrication techniques are now well established. As described in this chapter, CNTs can be used to modify electrodes and can improve the response to particular analytes. This can be by: increasing the effective electrode area; catalyzing the electrode reaction of an analyte; enabling direct electron transfer from an enzyme that reacts with a particular analyte; acting as a conductive support for a redox mediator which can react with the analyte; forming the base of an electrochemical label to provide detection of an antibody-antigen or a DNA target-probe binding event. The well-characterized methods of chemical and adsorptive CNT modification mean that there is a wide scope for the further improvement of microfluidic capabilities by the use of CNTs in electrochemical detectors.

References

1. Bi H, Weng X, Qu H et al (2005) Strategy for allosteric analysis based on protein-patterned stationary phase in microfluidic chip. J Proteome Res 4(6):2154–2160
2. Blanco-Gomez G, Glidle A, Flendrig LM et al (2009) Integration of low-power microfluidic pumps with biosensors within a laboratory-on-a-chip device. Anal Chem 81(4):1365–1370
3. Chatrathi MP, Collins GE, Wang J (2007) Microchip-based electrochemical enzyme immunoassays. Methods Mol Biol (Clifton, NJ) 385:215–224

4. Dong H, Li CM, Zhang YF et al (2007) Screen-printed microfluidic device for electrochemical immunoassay. Lab Chip 7(12):1752–1758
5. Ferguson BS, Buchsbaum SF, Swensen JS et al (2009) Integrated microfluidic electrochemical DNA sensor. Anal Chem 81(15):6503–6508
6. Ferguson BS, Buchsbaum SF, Wu TT et al (2011) Genetic analysis of h1n1 influenza virus from throat swab samples in a microfluidic system for point-of-care diagnostics. J Am Chem Soc 133(23):9129–9135
7. He P, Greenway G, Haswell SJ (2010) Development of enzyme immobilized monolith micro-reactors integrated with microfluidic electrochemical cell for the evaluation of enzyme kinetics. Microfluid Nanofluidics 8(5):565–573
8. Huang CJ, Chien HC, Chou TC et al (2011) Integrated microfluidic system for electrochemical sensing of glycosylated hemoglobin. Microfluid Nanofluidics 10(1):37–45
9. Jiang JH, Bo ML, Jiang DC et al (2009) A novel fabrication route to integrating label-free detection of DNA hybridization in microfluidic channel. In: IFMBE proceedings, pp 140–143
10. Kwakye S, Baeumner A (2003) A microfluidic biosensor based on nucleic acid sequence recognition. Anal Bioanal Chem 376(7):1062–1068
11. Lin L, Cai Y, Lin R et al (2011) New integrated in vivo microdialysis-electrochemical device for determination of the neurotransmitter dopamine in rat striatum of freely moving rats. Microchim Acta 172(1):217–223
12. Liu RH, Yang J, Lenigk R et al (2004) Self-contained, fully integrated biochip for sample preparation, polymerase chain reaction amplification, and DNA microarray detection. Anal Chem 76(7):1824–1831
13. Njoroge SK, Chen HW, Witek MA et al (2011) Integrated microfluidic systems for DNA analysis. Top Curr Chem 304:203–260
14. Park S, Zhang Y, Lin S et al (2011) Advances in microfluidic PCR for point-of-care infectious disease diagnostics. Biotechnol Adv 29:830–839
15. Pavlovic E, Lai RY, Wu TT et al (2008) Microfluidic device architecture for electrochemical patterning and detection of multiple DNA sequences. Langmuir 24(3):1102–1107
16. Pereira SV, Bertolino FA, Messina GA et al (2011) Microfluidic immunosensor with gold nanoparticle platform for the determination of immunoglobulin g anti-*Echinococcus granulosus* antibodies. Anal Biochem 409(1):98–104
17. Abdelgawad M, Watson MWL, Wheeler AR (2009) Hybrid microfluidics: a digital-to-channel interface for in-line sample processing and chemical separations. Lab Chip 9(8):1046–1051
18. Srinivasan V, Pamula VK, Fair RB (2004) Droplet-based microfluidic lab-on-a-chip for glucose detection. Anal Chim Acta 507(1):145–150
19. Zhao Y, Chakrabarty K (2011) Fault diagnosis in lab-on-chip using digital microfluidic logic gates. J Electron Test 27(1):69–83
20. Laser DJ, Santiago JG (2004) A review of micropumps. J Micromech Microeng 14(6): R35–R64
21. Roman GT, Kennedy RT (2007) Fully integrated microfluidic separations systems for biochemical analysis. J Chromatogr A 1168(1–2):170–188
22. Pennathur S (2008) Flow control in microfluidics: are the workhorse flows adequate? Lab Chip 8(3):383–387
23. Squires TM, Quake SR (2005) Microfluidics: fluid physics at the nanoliter scale. Rev Mod Phys 77(3):977–1026
24. Atalay YT, Vermeir S, Witters D et al (2011) Microfluidic analytical systems for food analysis. Trends Food Sci Technol 22(7):386–404
25. Nguyen NT, Wu Z (2005) Micromixers – a review. J Micromech Microeng 15(2):R1–R16
26. Zhang C, Xing D, Li Y (2007) Micropumps, microvalves, and micromixers within PCR microfluidic chips: advances and trends. Biotechnol Adv 25(5):483–514
27. Crevillen AG, Avila M, Pumera M et al (2007) Food analysis on microfluidic devices using ultrasensitive carbon nanotubes detectors. Anal Chem 79(19):7408–7415

28. Jayarajah CN, Skelley AM, Fortner AD et al (2007) Analysis of neuroactive amines in fermented beverages using a portable microchip capillary electrophoresis system. Anal Chem 79(21):8162–8169
29. Revermann T, Götz S, Künnemeyer J et al (2008) Quantitative analysis by microchip capillary electrophoresis – current limitations and problem-solving strategies. Analyst 133 (2):167–174
30. Karuwan C, Sukthang K, Wisitsoraat A et al (2011) Electrochemical detection on electrowetting-on-dielectric digital microfluidic chip. Talanta 84(5):1384–1389
31. Paik P, Pamula VK, Fair RB (2003) Rapid droplet mixers for digital microfluidic systems. Lab Chip 3(4):253–259
32. Nisar A, Afzulpurkar N, Mahaisavariya B et al (2008) Mems-based micropumps in drug delivery and biomedical applications. Sens Actuators B Chem 130(2):917–942
33. Smits JG (1990) Piezoelectric micropump with three valves working peristaltically. Sens Actuators A Phys 21(1–3):203–206
34. Van de Pol FCM, Van Lintel HTG, Elwenspoek M et al (1990) A thermopneumatic micropump based on micro-engineering techniques. Sens Actuators A Phys 21(1–3):198–202
35. Stemme E, Stemme G (1993) A valveless diffuser/nozzle-based fluid pump. Sens Actuators A Phys 39(2):159–167
36. Hsu YC, Lin SJ, Hou CC (2008) Development of peristaltic antithrombogenic micropumps for in vitro and ex vivo blood transportation tests. Microsyst Technol 14(1):31–41
37. Jeong OC, Park SW, Yang SS et al (2005) Fabrication of a peristaltic PDMS micropump. Sens Actuators A Phys 123–124:453–458
38. Teymoori MM, Abbaspour-Sani E (2004) Design and simulation of a novel electrostatic peristaltic micromachined pump for drug delivery applications. Sens Actuators A Phys 117 (2):222–229
39. Lee SW, Sim WY, Yang SS (2000) Fabrication and in vitro test of a microsyringe. Sens Actuators A Phys 83(1):17–23
40. Suzuki H, Yoneyama R (2002) A reversible electrochemical nanosyringe pump and some considerations to realize low-power consumption. Sens Actuators B Chem 86(2–3):242–250
41. Dapper J, Clemens M, Ehrfeld W et al (1997) Micro gear pumps for dosing of viscous fluids. J Micromech Microeng 7(3):230–232
42. Feng GH, Kim ES (2005) Piezoelectrically actuated dome-shaped diaphragm micropump. J Microelectromech Syst 14(2):192–199
43. Kan J, Yang Z, Peng T et al (2005) Design and test of a high-performance piezoelectric micropump for drug delivery. Sens Actuators A Phys 121(1):156–161
44. Koch M, Harris N, Evans AGR et al (1998) A novel micromachined pump based on thick-film piezoelectric actuation. Sens Actuators A Phys 70(1–2):98–103
45. Benard WL, Kahn H, Heuer AH et al (1997) Titanium-nickel shape-memory alloy actuated micropump. In: Proceedings of the international conference on solid-state sensors and actuators, vol 1, pp 361–364
46. Benard WL, Kahn H, Heuer AH et al (1998) Thin-film shape-memory alloy actuated micropumps. J Microelectromech Syst 7(2):245–251
47. Zhan C, Lo T, Liu L et al (1996) A silicon membrane micropump with integrated bimetallic actuator. Chin J Electron 5(2):33–35
48. Guo S, Nakamura T, Fukuda T et al (1997) Development of the micro pump using ICPF actuator. In: Proceedings of the IEEE international conference on robotics and automation, vol 1, pp 266–271
49. Bourouina T, Bosseboeuf A, Grandchamp JP (1997) Design and simulation of an electrostatic micropump for drug-delivery applications. J Micromech Microeng 7(3):186–188
50. MacHauf A, Nemirovsky Y, Dinnar U (2005) A membrane micropump electrostatically actuated across the working fluid. J Micromech Microeng 15(12):2309–2316
51. Gong Q, Zhou Z, Yang Y et al (2000) Design, optimization and simulation on microelectromagnetic pump. Sens Actuators A Phys 83(1):200–207

52. Yamahata C, Lotto C, Al-Assaf E et al (2005) A PMMA valveless micropump using electromagnetic actuation. Microfluid Nanofluidics 1(3):197–207
53. Liao CS, Lee GB, Liu HS et al (2005) Miniature RT-PCR system for diagnosis of RNA-based viruses. Nucleic Acids Res 33(18)
54. Lien KY, Lee WC, Lei HY et al (2007) Integrated reverse transcription polymerase chain reaction systems for virus detection. Biosens Bioelectron 22(8):1739–1748
55. Jeong OC, Yang SS (2000) Fabrication and test of a thermopneumatic micropump with a corrugated p+ diaphragm. Sens Actuators A Phys 83(1):249–255
56. Kim JH, Na KH, Kang CJ et al (2005) A disposable thermopneumatic-actuated micropump stacked with PDMS layers and ito-coated glass. Sens Actuators A Phys 120(2):365–369
57. Boden R, Lehto M, Simu U et al (2005) A polymeric paraffin micropump with active valves for high-pressure microfluidics. In: Digest of technical papers – international conference on solid state sensors and actuators and microsystems, TRANSDUCERS '05, vol 1, pp 201–204
58. Sim WY, Yoon HJ, Jeong OC et al (2003) A phase-change type micropump with aluminum flap valves. J Micromech Microeng 13(2):286–294
59. Huang L, Wang W, Murphy MC et al (2000) Liga fabrication and test of a dc type magnetohydrodynamic (MHD) micropump. Microsyst Technol 6(6):235–240
60. Jang J, Lee SS (2000) Theoretical and experimental study of MHD (magnetohydrodynamic) micropump. Sens Actuators A Phys 80(1):84–89
61. Munchow G, Dadic D, Doffing F et al (2005) Automated chip-based device for simple and fast nucleic acid amplification. Expert Rev Mol Diagn 5(4):613–620
62. Darabi J, Rada M, Ohadi M et al (2002) Design, fabrication, and testing of an electrohydrodynamic ion-drag micropump. J Microelectromech Syst 11(6):684–690
63. Fuhr G, Hagedorn R, Muller T et al (1992) Pumping of water solutions in microfabricated electrohydrodynamic systems. 25–30
64. Chen L, Wang H, Ma J et al (2005) Fabrication and characterization of a multi-stage electroosmotic pump for liquid delivery. Sens Actuators B Chem 104(1):117–123
65. Zeng S, Chen CH, Mikkelsen JC Jr et al (2001) Fabrication and characterization of electroosmotic micropumps. Sens Actuators B Chem 79(2–3):107–114
66. Yun KS, Cho IJ, Bu JU et al (2002) A surface-tension driven micropump for low-voltage and low-power operations. J Microelectromech Syst 11(5):454–461
67. Zahn JD, Deshmukh A, Pisano AP et al (2004) Continuous on-chip micropumping for microneedle enhanced drug delivery. Biomed Microdevices 6(3):183–190
68. Tsai JH, Lin L (2001) A thermal bubble actuated micro nozzle-diffuser pump. In: Proceedings of the IEEE micro electro mechanical systems (MEMS), pp 409–412
69. Yoshimi Y, Shinoda K, Mishima M et al (2004) Development of an artificial synapse using an electrochemical micropump. J Artif Organs 7(4):210–215
70. Luginbuhl P, Collins SD, Racine GA et al (1997) Microfabricated lamb wave device based on PZT sol–gel thin film for mechanical transport of solid particles and liquids. J Microelectromech Syst 6(4):337–346
71. Nguyen NT, Meng AH, Black J et al (2000) Integrated flow sensor for in situ measurement and control of acoustic streaming in flexural plate wave micropumps. Sens Actuators A Phys 79(2):115–121
72. Effenhauser CS, Harttig H, Kramer P (2002) An evaporation-based disposable micropump concept for continuous monitoring applications. Biomed Microdevices 4(1):27–32
73. Carrozza MC, Croce N, Magnani B et al (1995) A piezoelectric-driven stereolithography-fabricated micropump. J Micromech Microeng 5(2):177–179
74. Prakash R, Kaler KVIS (2007) An integrated genetic analysis microfluidic platform with valves and a PCR chip reusability method to avoid contamination. Microfluid Nanofluidics 3(2):177–187
75. Fan ZH, Ricco AJ, Tan W et al (2003) Integrating multiplexed PCR with CE for detecting microorganisms. Micro Total Anal Syst 849

76. Koh CG, Tan W, Zhao MQ et al (2003) Integrating polymerase chain reaction, valving, and electrophoresis in a plastic device for bacterial detection. Anal Chem 75(17):4591–4598
77. Gong H, Ramalingam N, Chen L et al (2006) Microfluidic handling of PCR solution and DNA amplification on a reaction chamber array biochip. Biomed Microdevices 8(2):167–176
78. Kohl M, Dittmann D, Quandt E et al (2000) Thin film shape memory microvalves with adjustable operation temperature. Sens Actuators A Phys 83(1):214–219
79. Zdeblick MJ, Anderson R, Jankowski J et al (1994) Thermopneumatically actuated microvalves and integrated electro-fluidic circuits. In: Solid-state sensor and actuator workshop, pp 251–255
80. Felton MJ (2003) The new generation of microvalves. Anal Chem 75(19):429A–432A
81. Esashi M, Shoji S, Nakano A (1989) Normally closed microvalve and mircopump fabricated on a silicon wafer. Sens Actuators 20(1–2):163–169
82. Fahrenberg J, Bier W, Maas D et al (1995) A microvalve system fabricated by thermoplastic molding. J Micromech Microeng 5(2):169–171
83. Pal R, Yang M, Johnson BN et al (2004) Phase change microvalve for integrated devices. Anal Chem 76(13):3740–3748
84. Wang J, Chen Z, Mauk M et al (2005) Self-actuated, thermo-responsive hydrogel valves for lab on a chip. Biomed Microdevices 7(4):313–322
85. Liu Y, Rauch CB, Stevens RL et al (2002) DNA amplification and hybridization assays in integrated plastic monolithic devices. Anal Chem 74(13):3063–3070
86. Liu RH, Bonanno J, Yang J et al (2004) Single-use, thermally actuated paraffin valves for microfluidic applications. Sens Actuators B Chem 98(2–3):328–336
87. Pal R, Yang M, Lin R et al (2005) An integrated microfluidic device for influenza and other genetic analyses. Lab Chip 5(10):1024–1032
88. Gui L, Liu J (2004) Ice valve for a mini/micro flow channel. J Micromech Microeng 14(2):242–246
89. Stone HA, Stroock AD, Ajdari A (2004) Engineering flows in small devices: microfluidics toward a lab-on-a-chip. Annu Rev Fluid Mech 36:381–411
90. Stroock AD, Dertinger SKW, Ajdari A et al (2002) Chaotic mixer for microchannels. Science 295(5555):647–651
91. Hashimoto M, Barany F, Soper SA (2006) Polymerase chain reaction/ligase detection reaction/hybridization assays using flow-through microfluidic devices for the detection of low-abundant DNA point mutations. Biosens Bioelectron 21(10):1915–1923
92. Legendre LA, Bienvenue JM, Roper MG et al (2006) A simple, valveless microfluidic sample preparation device for extraction and amplification of DNA from nanoliter-volume samples. Anal Chem 78(5):1444–1451
93. Mastrangelo CH, Burns MA, Burke DT (1999) Integrated microfabricated devices for genetic assays. In: Microprocesses and nanotechnology conference, Yokohama, pp 1–2
94. Chang YH, Lee GB, Huang FC et al (2006) Integrated polymerase chain reaction chips utilizing digital microfluidics. Biomed Microdevices 8(3):215–225
95. Curcio M, Roeraade J (2003) Continuous segmented-flow polymerase chain reaction for high-throughput miniaturized DNA amplification. Anal Chem 75(1):1–7
96. Mohr S, Zhang YH, Macaskill A et al (2007) Numerical and experimental study of a droplet-based PCR chip. Microfluid Nanofluidics 3(5):611–621
97. Jacobson SC, McKnight TE, Ramsey JM (1999) Microfluidic devices for electrokinetically driven parallel and serial mixing. Anal Chem 71(20):4455–4459
98. Lee CY, Lee GB, Lin JL et al (2005) Integrated microfluidic systems for cell lysis, mixing/pumping and DNA amplification. J Micromech Microeng 15(6):1215–1223
99. Liu RH, Yang J, Pindera MZ et al (2002) Bubble-induced acoustic micromixing. Lab Chip 2(3):151–157
100. Bau HH, Zhong J, Yi M (2001) A minute magneto hydro dynamic (MHD) mixer. Sens Actuators B Chem 79(2–3):207–215

101. Gotz S, Karst U (2007) Recent developments in optical detection methods for microchip separations. Anal Bioanal Chem 387(1):183–192
102. Newman CID, Giordano BC, Cooper CL et al (2008) Microchip micellar electrokinetic chromatography separation of alkaloids with UV-absorbance spectral detection. Electrophoresis 29(4):803–810
103. Ibanez-Garcia N, Mercader MB, Mendes Da Rocha Z et al (2006) Continuous flow analytical microsystems based on low-temperature co-fired ceramic technology. Integrated potentiometric detection based on solvent polymeric ion-selective electrodes. Anal Chem 78(9):2985–2992
104. Rashid M, Dou YH, Auger V et al (2010) Recent developments in polymer microfluidic devices with capillary electrophoresis and electrochemical detection. Micro Nanosyst 2(2):108–136
105. Wang J, Polsky R, Tian B et al (2000) Voltammetry on microfluidic chip platforms. Anal Chem 72(21):5285–5289
106. Mishra NN, Retterer S, Zieziulewicz TJ et al (2005) On-chip micro-biosensor for the detection of human CD4+ cells based on ac impedance and optical analysis. Biosens Bioelectron 21(5):696–704
107. Triroj N, Lapierre-Devlin MA, Kelley SO et al (2006) Microfluidic three-electrode cell array for low-current electrochemical detection. IEEE Sens J 6(6):1395–1402
108. Moon BU, Koster S, Wientjes KJC et al (2010) An enzymatic microreactor based on chaotic micromixing for enhanced amperometric detection in a continuous glucose monitoring application. Anal Chem 82(16):6756–6763
109. Schwarz MA (2004) Enzyme-catalyzed amperometric oxidation of neurotransmitters in chip-capillary electrophoresis. Electrophoresis 25(12):1916–1922
110. Moreira NH, De Jesus De Almeida AL, De Oliveira Piazzeta MH et al (2009) Fabrication of a multichannel PDMS/glass analytical microsystem with integrated electrodes for amperometric detection. Lab Chip 9(1):115–121
111. Sassa F, Laghzali H, Fukuda J et al (2010) Coulometric detection of components in liquid plugs by microfabricated flow channel and electrode structures. Anal Chem 82(20):8725–8732
112. Wang R, Lin J, Lassiter K et al (2011) Evaluation study of a portable impedance biosensor for detection of avian influenza virus. J Virol Methods 178(1–2):52–58
113. Segerink LI, Sprenkels AJ, Ter Braak PM et al (2010) On-chip determination of spermatozoa concentration using electrical impedance measurements. Lab Chip 10(8):1018–1024
114. Sabounchi P, Morales AM, Ponce P et al (2008) Sample concentration and impedance detection on a microfluidic polymer chip. Biomed Microdevices 10(5):661–670
115. Crevillen AG, Pumera M, Gonzalez MC et al (2009) Towards lab-on-a-chip approaches in real analytical domains based on microfluidic chips/electrochemical multi-walled carbon nanotube platforms. Lab Chip 9(2):346–353
116. Hervas M, Lopez MA, Escarpa A (2011) Integrated electrokinetic magnetic bead-based electrochemical immunoassay on microfluidic chips for reliable control of permitted levels of zearalenone in infant foods. Analyst 136(10):2131–2138
117. Kovaehev N, Canals A, Escarpa A (2010) Fast and selective microfluidic chips for electrochemical antioxidant sensing in complex samples. Anal Chem 82(7):2925–2931
118. Shiddiky MJA, Park H, Shim YB (2006) Direct analysis of trace phenolics with a microchip: in-channel sample preconcentration, separation, and electrochemical detection. Anal Chem 78(19):6809–6817
119. Fritsch I, Aguilar ZP (2007) Advantages of downsizing electrochemical detection for DNA assays. Anal Bioanal Chem 387(1):159–163
120. Yang SY, DeFranco JA, Sylvester YA et al (2009) Integration of a surface-directed microfluidic system with an organic electrochemical transistor array for multi-analyte biosensors. Lab Chip 9(5):704–708
121. Zhang Q, Jagannathan L, Subramanian V (2010) Label-free low-cost disposable DNA hybridization detection systems using organic TFTS. Biosens Bioelectron 25(5):972–977

122. Sharma S, Buchholz K, Luber SM et al (2006) Silicon-on-insulator microfluidic device with monolithic sensor integration for μTAS applications. J Microelectromech Syst 15(2):308–313
123. Shin KS, Lee SW, Han KC et al (2007) Amplification of fluorescence with packed beads to enhance the sensitivity of miniaturized detection in microfluidic chip. Biosens Bioelectron 22 (9–10):2261–2267
124. Mecomber JS, Stalcup AM, Hurd D et al (2006) Analytical performance of polymer-based microfluidic devices fabricated by computer numerical controlled machining. Anal Chem 78 (3):936–941
125. Darain F, Gan KL, Tjin SC (2009) Antibody immobilization on to polystyrene substrate – on-chip immunoassay for horse IgG based on fluorescence. Biomed Microdevices 11(3):653–661
126. Matsubara Y, Murakami Y, Kobayashi M et al (2004) Application of on-chip cell cultures for the detection of allergic response. Biosens Bioelectron 19(7):741–747
127. Chandrasekaran A, Acharya A, You JL et al (2007) Hybrid integrated silicon microfluidic platform for fluorescence based biodetection. Sensors 7(9):1901–1915
128. Al Lawati HAJ, Suliman FEO, Al Kindy SMZ et al (2010) Enhancement of on chip chemiluminescence signal intensity of tris(1,10-phenanthroline)-ruthenium(ii) peroxydisulphate system for analysis of chlorpheniramine maleate in pharmaceutical formulations. Talanta 82(5):1999–2002
129. Lv Y, Zhang Z, Chen F (2002) Chemiluminescence biosensor chip based on a microreactor using carrier air flow for determination of uric acid in human serum. Analyst 127 (9):1176–1179
130. Thongchai W, Liawruangath B, Liawruangrath S et al (2010) A microflow chemiluminescence system for determination of chloramphenicol in honey with preconcentration using a molecularly imprinted polymer. Talanta 82(2):560–566
131. Heyries KA, Loughran MG, Hoffmann D et al (2008) Microfluidic biochip for chemiluminescent detection of allergen-specific antibodies. Biosens Bioelectron 23(12):1812–1818
132. Hosono H, Satoh W, Fukuda J et al (2007) On-chip handling of solutions and electrochemiluminescence detection of amino acids. Sens Actuators B Chem 122(2):542–548
133. Hosono H, Satoh W, Fukuda J et al (2007) Microanalysis system based on electrochemiluminescence detection. Sens Mater 19(4):191–201
134. Roda A, Guardigli M, Michelini E et al (2009) Bioluminescence in analytical chemistry and in vivo imaging. Trends Analyt Chem 28(3):307–322
135. Tran TH, Chang WJ, Kim YB et al (2007) The effect of fluidic conditions on the continuous-flow bioluminescent detection of ATP in a microfluidic device. Biotechnol Bioprocess Eng 12(5):470–474
136. Maehana K, Tani H, Kamidate T (2006) On-chip genotoxic bioassay based on bioluminescence reporter system using three-dimensional microfluidic network. Anal Chim Acta 560 (1–2):24–29
137. Liu J, Eddings MA, Miles AR et al (2009) In situ microarray fabrication and analysis using a microfluidic flow cell array integrated with surface plasmon resonance microscopy. Anal Chem 81(11):4296–4301
138. Escobedo C, Vincent S, Choudhury AIK et al (2011) Integrated nanohole array surface plasmon resonance sensing device using a dual-wavelength source. J Micromech Microeng 21(11)
139. Hsu WT, Hsieh WH, Cheng SF et al (2011) Integration of fiber optic-particle plasmon resonance biosensor with microfluidic chip. Anal Chim Acta 697(1–2):75–82
140. Huang C, Bonroy K, Reekmans G et al (2009) Localized surface plasmon resonance biosensor integrated with microfluidic chip. Biomed Microdevices 11(4):893–901
141. Okan M, Balci O, Kocabas C (2011) A microfluidic based differential plasmon resonance sensor. Sens Actuators B Chem 160(1):670–676
142. Hiep HM, Nakayama T, Saito M et al (2008) A microfluidic chip based on localized surface plasmon resonance for real-time monitoring of antigen-antibody reactions. Jpn J Appl Phys 47(2 Pt 2):1337–1341

143. Chien WY, Khalid MZ, Hoa XD et al (2009) Monolithically integrated surface plasmon resonance sensor based on focusing diffractive optic element for optofluidic platforms. Sens Actuators B Chem 138(2):441–445
144. Huang C, Bonroy K, Reekman G et al (2009) An on-chip localized surface plasmon resonance-based biosensor for label-free monitoring of antigen-antibody reaction. Microelectron Eng 86(12):2437–2441
145. Hemmi A, Usui T, Moto A et al (2011) A surface plasmon resonance sensor on a compact disk-type microfluidic device. J Sep Sci 34(20):2913–2919
146. Ashok PC, Singh GP, Tan KM et al (2010) Fiber probe based microfluidic raman spectroscopy. Opt Express 18(8):7642–7649
147. Lim C, Hong J, Chung BG et al (2010) Optofluidic platforms based on surface-enhanced Raman scattering. Analyst 135(5):837–844
148. Quang LX, Lim C, Seong GH et al (2008) A portable surface-enhanced Raman scattering sensor integrated with a lab-on-a-chip for field analysis. Lab Chip 8(12):2214–2219
149. Ackermann KR, Henkel T, Popp J (2007) Quantitative online detection of low-concentrated drugs via a sers microfluidic system. Chemphyschem 8(18):2665–2670
150. Chen L, Choo J (2008) Recent advances in surface-enhanced Raman scattering detection technology for microfluidic chips. Electrophoresis 29(9):1815–1828
151. Chandrasekaran A, Packirisamy M (2006) Absorption detection of enzymatic reaction using optical microfluidics based intermittent flow microreactor system. IEE Proc Nanobiotechnol 153(6):137–143
152. Gordon R, Sinton D, Kavanagh KL et al (2008) A new generation of sensors based on extraordinary optical transmission. Acc Chem Res 41(8):1049–1057
153. Zhao L, Cheng P, Li J et al (2009) Analysis of nonadherent apoptotic cells by a quantum dots probe in a microfluidic device for drug screening. Anal Chem 81(16):7075–7080
154. Stewart ME, Yao J, Maria J et al (2009) Multispectral thin film biosensing and quantitative imaging using 3D plasmonic crystals. Anal Chem 81(15):5980–5989
155. Luan L, Evans RD, Jokerst NM et al (2008) Integrated optical sensor in a digital microfluidic platform. IEEE Sens J 8(5):628–635
156. Fang X, Liu Y, Kong J et al (2010) Loop-mediated isothermal amplification integrated on microfluidic chips for point-of-care quantitative detection of pathogens. Anal Chem 82 (7):3002–3006
157. Lagally ET, Simpson PC, Mathies RA (2000) Monolithic integrated microfluidic DNA amplification and capillary electrophoresis analysis system. Sens Actuators B Chem 63 (3):138–146
158. Nguyen TH, Pei R, Qiu C et al (2009) An aptameric microfluidic system for specific purification, enrichment, and mass spectrometric detection of biomolecules. J Microelectromech Syst 18(6):1198–1207
159. Wei H, Li H, Mao S et al (2011) Cell signaling analysis by mass spectrometry under coculture conditions on an integrated microfluidic device. Anal Chem 83(24):9306–9313
160. Bai HY, Lin SL, Chan SA et al (2010) Characterization and evaluation of two-dimensional microfluidic chip-HPLC coupled to tandem mass spectrometry for quantitative analysis of 7-aminoflunitrazepam in human urine. Analyst 135(10):2737–2742
161. Thorslund S, Lindberg P, Andrén PE et al (2005) Electrokinetic-driven microfluidic system in poly(dimethylsiloxane) for mass spectrometry detection integrating sample injection, capillary electrophoresis, and electrospray emitter on-chip. Electrophoresis 26(24):4674–4683
162. Sikanen T, Tuomikoski S, Ketola RA et al (2007) Fully microfabricated and integrated SU-8-based capillary electrophoresis-electrospray ionization microchips for mass spectrometry. Anal Chem 79(23):9135–9144
163. Lazar IM, Grym J, Foret F (2006) Microfabricated devices: a new sample introduction approach to mass spectrometry. Mass Spectrom Rev 25(4):573–594

164. Leifheit M, Bergmann W, Greiser J (2008) Application of exchangeable biochemical reactors with oxidase-catalase-co-immobilizates and immobilized microorganisms in a microfluidic chip-calorimeter. Eng Life Sci 8(5):540–545
165. Zhang Y, Tadigadapa S (2004) Calorimetric biosensors with integrated microfluidic channels. Biosens Bioelectron 19(12):1733–1743
166. Waggoner PS, Tan CP, Craighead HG (2010) Microfluidic integration of nanomechanical resonators for protein analysis in serum. Sens Actuators B Chem 150(2):550–555
167. Ricciardi C, Canavese G, Castagna R et al (2010) Integration of microfluidic and cantilever technology for biosensing application in liquid environment. Biosens Bioelectron 26 (4):1565–1570
168. Mahabadi KA, Rodriguez I, Lim CY et al (2010) Capacitively coupled contactless conductivity detection with dual top-bottom cell configuration for microchip electrophoresis. Electrophoresis 31(6):1063–1070
169. Pumera M, Wang J, Opekar F et al (2002) Contactless conductivity detector for microchip capillary electrophoresis. Anal Chem 74(9):1968–1971
170. Lee H, Sun E, Ham D et al (2008) Chip-NMR biosensor for detection and molecular analysis of cells. Nat Med 14(8):869–874
171. Mitsakakis K, Gizeli E (2011) Detection of multiple cardiac markers with an integrated acoustic platform for cardiovascular risk assessment. Anal Chim Acta 699(1):1–5
172. Liu C, Yy M, Zg C et al (2008) Dual fluorescence/contactless conductivity detection for microfluidic chip. Anal Chim Acta 621(2):171–177
173. Yi C, Zhang Q, Li CW et al (2006) Optical and electrochemical detection techniques for cell-based microfluidic systems. Anal Bioanal Chem 384(6):1259–1268
174. Wellman AD, Sepaniak MJ (2007) Multiplexed, waveguide approach to magnetically assisted transport evanescent field fluoroassays. Anal Chem 79(17):6622–6628
175. James CD, McClain J, Pohl KR et al (2010) High-efficiency magnetic particle focusing using dielectrophoresis and magnetophoresis in a microfluidic device. J Micromech Microeng 20(4)
176. van Lintel HTG, van De Pol FCM, Bouwstra S (1988) A piezoelectric micropump based on micromachining of silicon. Sens Actuators 15(2):153–167
177. Pan T, McDonald SJ, Kai EM et al (2005) A magnetically driven PDMS micropump with ball check-valves. J Micromech Microeng 15(5):1021–1026
178. Gervais L, Delamarche E (2009) Toward one-step point-of-care immunodiagnostics using capillary-driven microfluidics and PDMS substrates. Lab Chip 9(23):3330–3337
179. Martinez AW, Phillips ST, Whitesides GM et al (2010) Diagnostics for the developing world: microfluidic paper-based analytical devices. Anal Chem 82(1):3–10
180. Nie Z, Nijhuis CA, Gong J et al (2010) Electrochemical sensing in paper-based microfluidic devices. Lab Chip 10(4):477–483
181. Carvalhal RF, Kfouri MS, De Piazetta MHO et al (2010) Electrochemical detection in a paper-based separation device. Anal Chem 82(3):1162–1165
182. Iijima S (1991) Helical microtubes of graphitic carbon. Nature 354:56–58
183. Iijima S, Ichihashi T (1993) Single-shell carbon nanotubes of 1-mm diameter. Nature 363:603–605
184. Thess A, Lee R, Nikolaev H et al (1998) Crystalline ropes of metallic carbon nanotubes. Science 273:483–487
185. Yacaman M, Yoshida M, Rendon L et al (1993) Catalytic growth of carbon microtubules with fullerene structure. Appl Phys Lett 62:202–204
186. Bethune D, Kiang C, de Vries M et al (1993) Cobalt-catalysed growth of carbon nanotubes with single-atomic-layer walls. Nature 363:605–607
187. Ci L, Rao Z, Zhou Z et al (2002) Double wall carbon nanotubes promoted by sulfur in a floating iron catalyst CVD system. Chem Phys Lett 359:63–67
188. Lehman J, Terrones M, Mansfield E et al (2011) Evaluating the characteristics of multiwall carbon nanotubes. Carbon 49:2581–2602
189. Popov V (2004) Carbon nanotubes: properties and application. Mat Sci Eng R Rep 43:61–102

190. Burghard M (2005) Electronic and vibrational properties of chemically modified single-wall carbon nanotubes. Surf Sci Rep 58:1–109
191. Krueger A (2010) Carbon materials and nanotechnology. Wiley, Weinheim, pp 123–281
192. Harris P (2009) Carbon nanotube science. Synthesis, properties and applications. Cambridge University Press, Cambridge
193. Ye J-S, Shen F-S (2007) Carbon nanotube-based sensor. In: Kumar C (ed) Nanomaterials for biosensors. Wiley, Weinheim
194. Fam D, Palaniappan A, Tok A (2011) A review on the technological aspects influencing commercialization of carbon nanotube sensors. Sens Actuators B Chem 157:1–7
195. Vashist S, Zheng D, Al-Rubeaan K et al (2011) Advances in carbon nanotube based electrochemical sensors for bioanalytical applications. Biotechnol Adv 29:169–188
196. Yanez-Sedeno P, Pingarron J, Riu J et al (2010) Electrochemical sensing based on carbon nanotubes. Trends Analyt Chem 29:939–953
197. Saito R, Fujita M, Dresselhaus G et al (1992) Electronic structure of chiral graphene tubules. Appl Phys Lett 60:2204–2206
198. Phanthong C, Somasundrum M (2009) Enhanced sensitivity of 4-chlorophenol detection by the use of nitrobenzene as a liquid membrane over a carbon nanotube-modified glassy carbon electrode. Electroanalysis 20:1024–1027
199. Wang J, Mussameh M, Lin Y (2003) Solubilization of carbon nanotubes by nafion toward the preparation of amperometric biosensors. J Am Chem Soc 125:2408–2409
200. Wang J, Musameh M (2003) Carbon nanotube/teflon composite electrochemical sensors and biosensors. Anal Chem 75:2075–2079
201. Wang Z, Liu J, Liang Q et al (2002) Carbon nanotube-modified electrodes for the simultaneous determination of dopamine and ascorbic acid. Analyst 127:653–658
202. Rubianes M, Rivas G (2003) Carbon nanotubes paste electrode. Electrochem Commun 5:689–694
203. Wang J, Li M, Shi Z et al (2002) Electrocatalytic oxidation of norepinephrine at a glassy carbon electrode modified with single wall carbon nanotubes. Electroanalysis 14:225–230
204. Banks C, Compton R (2006) New elements from old: from carbon nanotubes to edge plane pyrolytic graphite. Analyst 131:15–21
205. Gooding J, Wibowo R, Lin J et al (2003) Protein electrochemistry using aligned carbon nanotube arrays. J Am Chem Soc 125:9006–9007
206. Davies J, Coles R, Hill H (1997) Protein electrochemistry at carbon nanotube electrodes. J Electroanal Chem 440:279–282
207. Zhao G, Zhang L, Wei X et al (2003) Myoglobin on multi-walled carbon nanotubes modified electrode: direct electrochemistry and electrocatalysis. Electrochem Commun 5:825–829
208. Wang L, Wang J, Zhou F (2004) Direct electrochemistry of catalase at a gold electrode modified with single-wall carbon nanotubes. Electroanalysis 16:627–632
209. Patolsky F, Weizmann Y, Willner I et al (2004) Long-range electrical contacting of redox enzymes by SWCNT connectors. Angew Chem Int Ed 43:2113–2117
210. Star A, Stoddart J, Steurman D et al (2001) Preparation and properties of polymer-wrapped single-walled carbon nanotubes. Angew Chem Int Ed 40:1721–1775
211. Riggs J, Guo Z, Carroll D et al (2000) Strong luminescence of solubilized carbon nanotubes. J Am Chem Soc 122:5879–5880
212. Manne S, Cleveland J, Gaub H (1994) Direct visualization of surfactant hemimicelles by force microscopy of the electrical double layer. Langmuir 10:4409–4413
213. Islam M, Rojas E, Bergey D et al (2003) High weight fraction surfactant solubilization of single-wall nanotubes in water. Nano Lett 3:269–273
214. Yan Y, Zhang M, Gong K et al (2005) Adsorption of methylene blue dye onto carbon nanotubes:? A route to an electrochemically functional nanostructure and its layer-by-layer assembled nanocomposite. Chem Mater 17:3457–3463
215. Palangsuntikul R, Somasundrum M, Surareungchai W (2000) Kinetic and analytical comparison of horseradish peroxidase on bare and redox-modified single-walled carbon nanotubes. Electrochim Acta 56:470–475

216. Liu C-H, Li J-J, Zhang H-L et al (2008) Structure dependent interaction between organic dyes and carbon nanotubes. Colloids Surf A 313–314:9–12
217. Hirsch A, Vostrowsky O (2005) Functionalization of carbon nanotubes. Top Curr Chem 245:193–237
218. Tasis D, Tagmatarchis N, Bianco A (2006) Chemistry of carbon nanotubes. Chem Rev 106:1105–1136
219. Prato M, Kostarelos K, Bianco A (2008) Functionalized carbon nanotubes in drug design and discovery. Acc Chem Res 41:60–68
220. Tsang S, Chen Y, Harris P et al (1994) A simple chemical method of opening and filling carbon nanotubes. Nature 372:159–162
221. Hiura H, Ebbesen T, Tanigaki K (1995) Opening and purification of carbon nanotubes in high yields. Adv Mater 7:275–276
222. Chen J, Hamon M, Hu H et al (1998) Solution properties of single-walled carbon nanotubes. Science 282:95–98
223. Liu J, Rinzler A, Dai H et al (1998) Fullerene pipes. Science 280:1253–1256
224. Coleman K, Chakraborty A, Bailey S et al (2007) Iodination of single-walled carbon nanotubes. Chem Mater 19:1076–1081
225. Aizawa M, Shaffer M (2003) Silylation of multi-walled carbon nanotubes. Chem Phys Lett 368:121–124
226. Mickelson E, Huffman C, Rinnzler A (1998) Fluorination of single-wall carbon nanotubes. Chem Phys Lett 296:188–194
227. Holzinger M, Vostrowsky O, Hirsch A (2001) Sidewall functionalization of carbon nanotubes. Angew Chem Int Ed 40:4002–4005
228. Holzinger M, Abraham J, Whelan P (2003) Functionalization of single-walled carbon nanotubes with R-oxycarbonyl nitrenes. J Am Chem Soc 125:8566–8580
229. Georgakilas V, Kordatos K, Prato M (2002) Organic functionalization of carbon nanotubes. J Am Chem Soc 124:760–761
230. Banerjee S, Wong S (2002) Rational sidewall functionalization and purification of single-walled carbon nanotubes by solution-phase ozonolysis. J Phys Chem B 106:12144–12151
231. Hemraj-Benny T, Wong S (2006) Silylation of single-walled carbon nanotubes. Chem Mater 18:4827–4839
232. Bahr J, Yang J, Kosynkin D et al (2001) Functionalization of carbon nanotubes by electrochemical reduction of aryl diazonium salts: a bucky paper electrode. J Am Chem Soc 123:6536–6542
233. Tsang S, Davis J, Green M et al (1995) Immobilization of small proteins in carbon nanotubes: high resolution transmission electron microscopy study and catalytic activity. Chem Commun 1803–1804
234. Davis J, Green M, Hill H et al (1998) The immobilization of proteins in carbon nanotubes. Inorg Chim Acta 272:261–266
235. Feng W, Ji P (2011) Enzymes immobilized on carbon nanotubes. Biotechnol Adv. doi:101016/jbiotechadv201107007
236. Cui D (2007) Advances and prospects on biomolecules functionalized carbon nanotubes. J Nanosci Nanotechnol 7:1298–1314
237. Matsuura K, Saito T, Okazaki T et al (2006) Selectivity of water-soluble proteins in single-walled carbon nanotube dispersions. Chem Phys Lett 429:497–502
238. Nepal D, Geckeler K (2006) PH-sensitive dispersion and debundling of single-walled carbon nanotubes: lysozyme as a tool. Small 2:406–412
239. Yu C, Yen M, Chen L (2010) A bioanode based on MWCNT/protein assisted co-immobilization of glucose oxidase and 2,5-dihydroxybenzaldehyde for glucose fuel cells. Biosens Bioelectron 25:2515–2521
240. Das D, Das P (2009) Superior activity of strucurally deprived enzyme-carbon nanotube hybrids in cationic reverse micelles. Langmuir 25:2515–2521

241. Lee C, Tsai Y (2009) Preparation of multiwalled carbon nanotube-chitosan-alcohol dehydrogenase nanobiocomposite for amperometric detection of ethanol. Sens Actuators B 138:518–523
242. Lee K, Komathi S, Nam N et al (2010) Sulfonated polyaniline network grafted multiwall carbon nanotubes for enzyme immobilization, direct electrochemistry and biosensing of glucose. Microchem J 95:74–79
243. Wu X, Zhao B, Wu P et al (2009) Effects of ionic liquids on enzymatic catalysis of the glucose oxidase toward the oxidation of glucose. J Phys Chem B 113:13365–13373
244. Panhius M, Salvador-Morales C, Franklin E et al (2003) Characterization of an interaction between functionalized carbon nanotubes and an enzyme. J Nanosci Nanotechnol 3:209–213
245. Decher G (1997) Fuzzy nanoassemblies. Towards layered polymeric multicomposites. Science 277:1232–1237
246. Nagoka Y, Shiratori S, Einaga Y (2008) Photo-control of adhesion properties by detachment of the outermost layer in layer-by-layer assembled multilayer films of preyssler-type polyoxometalate and polyethyleneimine. Chem Mater 20:4004–4010
247. Wang Y, Joshi P, Hobbs K (2006) Nanostructured biosensors built by layer-by-layer electrostatic assembly of enzyme-coated single-walled carbon nanotubes and redox polymers. Langmuir 22:9776–9783
248. Bi S, Zhou H, Zhang S (2009) Multilayers enzyme-coated carbon nanotubes as biolabel for ultrasensitive chemiluminescence immunoassy of cancer biomarker. Biosens Bioelectron 24:2961–2966
249. Munge B, Liu G, Collins G et al (2005) Multiple enzyme layers on carbon nanotubes for electrochemical detection down to 80 DNA copies. Anal Chem 77:4662–4666
250. Huang W, Taylor S, Fu K et al (2002) Attaching proteins to carbon nanotubes via diimide-activated amide-activated amidation. Nano Lett 2:311–314
251. Pang H, Liu J, Hu D et al (2010) Immobilization of laccase onto l-aminopyrene functionalized carbon nanotubes and their electrocatalytic activity for oxygen reduction. Electrochim Acta 55:6611–6616
252. Kim B, Kang B, Bahk Y et al (2009) Immobilization of horseradish peroxidase on multiwalled carbon nanotubes and its enzymatic stability. Curr Appl Phys 9:e263–e265
253. Teker K, Sirdeshmukh R, Sivakumar K et al (2005) Applications of carbon nanotubes for cancer research. Nanobiotechnology 1:171–182
254. Liu G, Chen H, Peng H et al (2011) A carbon nanotube-based highly-sensitive electrochemical immunosensor for rapid and portable detection of clenbuterol. Biosens Bioelectron. doi:101016/jbios201107037
255. O'Conor M, Kim S, Killard A et al (2004) Mediated amperometric immunoassay using single walled carbon nanotube forests. Analyst 129:1176–1180
256. Villamizar R, Maroto A, Ruis F et al (2008) Fast detection of *Salmonella infantis* with carbon nanotube field effect transistors. Biosens Bioelectron 24:279–283
257. Chunglok W, Khownarumit P, Rijiravanich P et al (2011) Electrochemical immunoassay platform for high sensitivity protein detection based on redox-modified carbon nanotube labels. Analyst 136:2969–2974
258. Leng C, Wu J, Xu Q et al (2011) A highly sensitive disposable immunosensor through direct electro-oxidation of oxygen catalyzed by palladium nanoparticle decorated carbon nanotube label. Biosens Bioelectron 27:71–76
259. Liu S, Lin Q, Zhang X et al (2011) Electrochemical immunosensor for salbutamol detection based on CS-Fe3O4-PAMAM-GNPS nanocomposites and HRP-MWCNTS-Ab bioconjugates for signa amplification. Sens Actuators B Chem 156:71–78
260. Shao N, Lu S, Wickstrom E et al (2007) Integrated molecular targeting of IGF1R and HER2 surface receptors and destruction of breast cancer cells using single wall carbon nanotubes. Nanotechnology 18:315101
261. Zheng M, Jagota A, Strano M et al (2003) DNA-assisted dispersion and separation of carbon nanotubes. Nat Mater 2:338–342

262. Heller D, Jeng E, Yeung T et al (2006) Optical detection of DNA conformational polymorphism on single-walled carbon nanotubes. Science 311:508–511
263. Jeng E, Moll A, Roy A et al (2006) Detection of DNA hybridization using the near-infrared band-gap fluorescence of single-walled carbon nanotubes. Nano Lett 6:371–375
264. Dwyer C, Guthold M, Falvo M et al (2002) DNA-functionalized single-walled carbon nanotubes. Nanotechnology 13:601–604
265. He P, Li S, Dai L et al (2005) DNA-modified carbon nanotubes for self-assembling and biosensing applications. Synth Met 154:17–20
266. Wang J, Liu G, Jan M (2004) Ultrasensitive electrical biosensing of proteins and DNA: carbon nanotube derived amplification of the recognition and transduction events. J Am Chem Soc 126:3010–3011
267. Agui L, Yanez-Sedeno P, Pingarron JM (2008) Role of carbon nanotubes in electroanalytical chemistry. A review. Anal Chim Acta 622:11–47
268. Guo M, Chen J, Li J et al (2005) Fabrication of polyaniline/carbon nanotube composite modified electrode and its electrocatalytic property to the reduction of nitrite. Anal Chim Acta 532:71–77
269. Li Y, Wang P, Wang L et al (2007) Overoxidized polypyrrole film directed single-walled carbon nanotubes immobilization on glassy carbon electrode and its sensing applications. Biosens Bioelectron 22:3120–3125
270. Tang Q, Luo X, Wen R (2005) Construction of a heteropolyanion-containing polypyrrole/carbon nanotube modified electrode and its electrocatalytic property. Anal Lett 38:1445–1456
271. Lin X, Li Y (2006) A sensitive determination of estrogens with a Pt nano-clusters/multi-walled carbon nanotubes modified glassy carbon electrode. Biosens Bioelectron 22:253–259
272. Yuan PS, Wu HQ, Xu HY et al (2007) Synthesis characterization and electrocatalytic properties of FeCo alloy nanoparticles supported on carbon nanotubes. Mater Chem Phys 105:391–394
273. Yang P, Wei W, Tao C (2007) Determination of trace thiocyanate with nano-silver coated multi-walled carbon nanotubes modified glassy carbon electrode. Anal Chim Acta 585:331–336
274. Dai X, Wildgoose GG, Compton RG (2006) Designer electrode interfaces simultaneously comprising three different metal nanoparticle (Au, Ag, Pd)/carbon microsphere/carbon nanotube composites: progress towards combinatorial electrochemistry. Analyst 131:1241–1247
275. Liu H, Wang G, Chen D et al (2008) Fabrication of polythionine/NPAu/MWNTs modified electrode for simultaneous determination of adenine and guanine in DNA. Sens Actuators B Chem 128:414–421
276. Valentini F, Amine A, Orlanducci S et al (2003) Carbon nanotube purification: preparation and characterization of carbon nanotube paste electrodes. Anal Chem 75:5413–5421
277. Abbar JC, Malode SJ, Nandibewoor ST (2012) Electrochemical detection of a hemorheologic drug pentoxifylline at a multi-walled carbon nanotube paste electrode. Bioelectrochemistry 83:1–7
278. Zheng L, Song J (2007) Voltammetric behaviour of urapidil and its determination at multi-wall carbon nanotube paste electrode. Talanta 73:943–947
279. Arvand M, Ansari R, Heydari L (2011) Electrocatalyic oxidation and differential pulse voltammetric detection of sulfamethoxazole using carbon nanotube paste electrode. Mat Sci Eng C 31:1819–1825
280. Malode SJ, Shetti NP, Nandibewoor ST (2012) Voltammetric behaviour of theophylline and its determination at multi-wall carbon nanotube paste electrode. Colloids Surf B Biointerfaces 97:1–6
281. Lawrence NS, Deo RP, Wang J (2004) Determination of homocysteine at carbon nanotube paste electrodes. Talanta 63:443–449
282. Tavana T, Khalilzadeh MA, Karimi-Maleh H et al (2012) Sensitive voltammetric determination of epinephrine in the presence of acetaminophen at a novel ionic liquid modified carbon nanotubes paste electrode. J Mol Liq 168:69–74

283. Mazloum-Ardakani M, Ganjipour B, Beitollahi H et al (2011) Simultaneous determination of levodopa, carbidopa and tryptophan using nanostructured electrochemical sensor based on novel hydroquinone and carbon nanotubes: application to the analysis of some real samples. Electrochim Acta 56:9113–9120
284. Beitollahi H, Sheikhshoaie I (2011) Electrocatalytic and simultaneous determination of isoproterenol uric acid and folic acid at molybdenum (VI) complex-carbon nanotube paste electrode. Electrochim Acta 56:10259–10263
285. Patrascu D, David I, David V et al (2011) Selective voltammetric determination of electroactive neuromodulating species in biological samples using iron (III) phthalocyanine modified multi-wall carbon nanotubes paste electrode. Sens Actuators B Chem 156:731–736
286. Balan I, David IG, David V et al (2011) Electrocatalytic voltammetric determination of guanine at a cobalt phthalocyanine modified carbon nanotubes paste electrode. J Electroanal Chem 654:8–12
287. Zheng L, Song J-F (2009) Nickel (II)-baicalein complex modified multiwall carbon nanotube paste electrode and its electrcatalytic oxidation toward glycine. Anal Biochem 391:56–63
288. Shahrokhian S, Kamazadeh Z, Bezaatpour A et al (2008) Differential pulse voltammetric determination of *N*-acetylcysteine by the electrocatalytic oxidation at the surface of carbon nanotube-paste electrode modified with cobalt salophen complexes. Sens Actuators B Chem 133:599–606
289. Qu J, Zou X, Liu B et al (2007) Assembly of polyoxmetalates on carbon nanotubes paste electrode and its catalytic behaviours. Anal Chim Acta 599:51–57
290. Antiochia R, Gorton L (2007) Development of a carbon nanotube paste electrode osmium polymer-mediated biosensor for determination of glucose in alcoholic beverages. Biosens Bioelectron 22:2611–2617
291. Pereira AC, Aguiar M, Kisner A et al (2007) Amperometric biosensor for lactate based on lactate dehydrogenase and meldola blue coimmobilized on multi-wall carbon nanotube. Sens Actuators B Chem 124:269–276
292. Janegitz BC, Marcolino-Junior LH, Campana-Filho SP et al (2009) Anodic stripping voltammetric determination of copper (II) using a functionalized carbon nanotubes paste electrode modified with crosslinked chitosan. Sens Actuators B Chem 142:260–266
293. Tashkhourian J, Nezhad MRH, Khodavesi J et al (2009) Silver nanoparticles modified carbon nanotube paste electrode for simultaneous determination of dopamine and ascorbic acid. J Electroanal Chem 633:85–91
294. Merisalu M, Kruusma J, Banks CE (2010) Metallic impurity free carbon nanotube paste electrodes. Electrochem Commun 12:144–147
295. Wang J, Musameh M (2004) Screen-printed carbon nanotube paste electrodes. Analyst 129:512–515
296. Chattopadhyay D, Galeska I, Papadimitrakopoulos F (2001) Metal-assisted organization of shortened carbon nanotubes in monolayer and multilayer forest assemblies. J Am Chem Soc 26:9451–9452
297. Liu Z, Shen Z, Zhu T et al (2000) Organizing single-walled carbon nanotubes on gold using a wet chemical self-assembling technique. Langmuir 16:3569–3573
298. Kim B, Sigmund W (2003) Self-alignment of shortened multiwall carbon nanotubes on polyelectrolyte layers. Langmuir 19:4848–4851
299. Wu B, Zhang J, Zhong W (2001) Chemical alignment of oxidatively shortened single-walled carbon nanotubes on silver surface. J Phys Chem B 105:5075–5078
300. Yang Y, Huang S, He H et al (1999) Patterned growth of well-aligned carbon nanotubes: a photolithographic approach. J Am Chem Soc 121:10832–10833
301. Chen G (2007) Carbon nanotube and diamond as electrochemical detectors in microchip and conventional capillary electrophoresis. Talanta 74:326–332
302. Wang J, Chen G, Chatrathi M et al (2004) Capillary electrophoresis microchip with a carbon nanotube-modified electrochemical detector. Anal Chem 76:298–302

303. Panini N, Messina G, Salinas E et al (2008) Integrated microfluidic systems with an immunosensor modified with carbon nanotubes for detection of prostate specific antigen (PSA) in human serum samples. Biosens Bioelectron 23:1145–1151
304. Shi B-X, Wang Y, Zhang K et al (2011) Monitoring of dopamine release in single cell using ultrasensitive ito microsensors modified with carbon nanotubes. Biosens Bioelectron 26:2917–2921
305. Hong C-C, Wang C-Y, Peng K-T et al (2011) A microfluidic chip plat-form with electrochemical carbon nanotubes electrodes for pre-clinical evaluation of antibiotic nanocapsules. Biosens Bioelectron 26:3620–3626
306. Pumera M, Llopis X, Merkoci A et al (2006) Microchip capillary electrophoresis with a single-wall carbon nanotube/gold electrochemical detector for de-termination of aminophenols and neurotransmitters. Microchim Acta 152:261–265
307. Vlakova M, Schwarz M (2007) Determination of cationic neurotransmitters and metabolites in brain homogenates by microchip electrophoresis and carbon nanotube-modified amperometry. J Chromatogr A 1142:214–221
308. Chen G, Zhang L, Wang J (2004) Miniaturized capillary electrophoresis system with a carbon nanotube microelectrode for rapid separation and detection of thiols. Talanta 64:1018–1023
309. Wang J, Chen G, Wang M et al (2004) Carbon-nanotube/copper composite electrodes for capillary electrophoresis microchip detection of carbohydrates. Analyst 129:512–515
310. Snider R, Ciobanu M, Rue A et al (2008) A multiwall carbon nanotube/dihydropyran composite film electrode for insulin detection in a microphysiometer chamber. Anal Chim Acta 609:44–52
311. Xu J, Zhang H, Chen G (2007) Carbon nanotube/polystyrene composite electrode for microchip electrophoretic determination of rutin and quercetin in flos *Sophorae immaturus*. Talanta 73:932–937
312. Olive-Monllau R, Martinez-Cisneros C, Bartoli J et al (2011) Integration of a sensitive carbon nanotube composite electrode in a ceramic microanalyzer for the amperometric determination of free chlorine. Sens Actuators B Chem 151:416–422
313. Chicharro A, Sanchez E, Bermejo A et al (2005) Carbon nanotubes paste electrodes as new detectors for capillary electrophoresis. Anal Chim Acta 543:84–91
314. Kim J, Baek J, Kim H et al (2006) Integration of enzyme immobilised single-walled carbon nanotubes mass into the microfluidic platform and its application for the glucose detection. Sens Actuators A Phys 128:7–13
315. Wisitsoraat A, Sritongkham P, Karuwan C et al (2010) Fast cholesterol detection using flow injection microfluidic device with functionalized carbon nanotubes based electrochemical sensor. Biosens Bioelectron 26:1514–1520
316. Karuwan C, Wisitoraat A, Maturos T et al (2009) Flow injection based microfluidic device with carbon nanotube electrode for rapid salbutamol detection. Talanta 79:995–1000
317. Phokharatkul D, Karuwan C, Lomas T et al (2011) AAO-CNTs electrode on microfluidic flow injection system for rapid iodide sensing. Talanta 84:1390–1395

Graphene-Based Chemical and Biosensors

Anurat Wisitsoraat and Adisorn Tuantranont

Abstract Graphene is a novel and promising material for chemical and biosensing due to its extraordinary structural, electronic, and physiochemical properties. Recently, a large number of graphene-based chemical and biosensors with different structures and fabrication methods have been reported. In this chapter, graphene's synthesis methods, properties, and applications in chemical and biosensing are extensively surveyed. Graphene-based chemical and biosensors may similarly be classified into three main groups including chemoresistive, electrochemical, and other sensing platforms. Chemoresistive graphene-based chemical sensors have been widely developed for ultrasensitive gas-phase chemical sensing with single molecule detection capability. Graphene-based electrochemical sensors for chemical and biosensing have shown excellent performances toward various non-bio and bio-analytes compared to most other carbon-based electrodes due to its very high electron transfer rate of highly dense edge-plane-like defective active sites, excellent direct electrochemical oxidation of small biomolecules and direct electrochemistry of enzyme while graphene FET chemoresistive biosensors for detections of DNA, protein/DNA mixture, and other antibody-specific biomolecules have been reported with high sensitivity and specificity. In addition, the graphene's performance considerably depends on synthesis method and surface functionalized graphene oxides prepared by chemical, thermal, and particularly electrochemical reductions are demonstrated to be highly promising for both electrochemical and chemoresistive sensing platforms. However, large-scale economical production of graphene is still not generally attainable and graphene-based chemical and biosensors still suffer from poor reproducibility due to difficulty of controlling graphene sensor structures. Therefore, novel methods for well-controlled synthesis and processing of graphene must be further developed. Furthermore, effective

A. Wisitsoraat (✉) and A. Tuantranont
Nanoelectronics and MEMS Laboratory, National Electronics and Computer Technology Center (NECTEC), National Science and Technology Development agency (NSTDA), Pathumthani, Thailand
e-mail: anurat.wisitsoraat@nectec.or.th; adisorn.tuantranont@nectec.or.th

A. Tuantranont (ed.), *Applications of Nanomaterials in Sensors and Diagnostics*,
Springer Series on Chemical Sensors and Biosensors (2013) 14: 103–142
DOI 10.1007/5346_2012_47, © Springer-Verlag Berlin Heidelberg 2013,
Published online: 15 February 2013

doping methods should be developed and applied to enhance its sensing behaviors. Lastly, graphene's chemical and biological interaction and related charge transport mechanisms are not well understood and should be further studied.

Keywords Chemical and biosensors, Chemoresistive sensor, Electrochemical sensor, FET sensors, Graphene

Contents

1 Introduction

Graphene is a two-dimensional nano-carbon material with a honeycomb lattice structure that is the basic building block for all carbon allotropes with other dimensionalities (0D fullerenes, 1D nanotubes, and 3D graphite) [1–3]. Its unique 2D structure leads to exceptional physical, chemical, and electronic properties including huge specific surface area [3, 4], excellent electrical/thermal conductivity [5–11], room-temperature quantum Hall effect (QHE) [12–15], great mechanical strength [16–18], and biochemical compatibility [19–21]. It is thus a promising candidate for a number of applications including field-effect and single-electron transistors [22–30], memory devices [31], supercapacitors [32–35], batteries [36–40], fuel cells [41–43], solar cells [44–47], bioscience/biotechnologies [19–21, 48, 49], and chemical/biosensors [50–61].

Among these, graphene-based chemical/biosensors have been among the most successful applications, which can be attributed to advancement in graphene synthesis and fabrication processes as well as inherent graphene's properties. Graphene-based chemical and biosensors have shown superior performance over other carbon-based materials including its nearest counterpart, carbon nanotubes (CNTs) because of its double surface area compared to CNTs, higher chemical reactivity due to larger number of edge plane per unit mass, higher electron mobility, and conductivity [53, 62, 63]. In addition, it can be inexpensively produced from low-cost graphite with no metallic impurity, which would otherwise be found in most CNTs and can introduce anomalous sensing response [62–65]. Moreover, its ease of processing and low hazardous properties make it more feasible for commercial applications [19–21].

The graphene's chemical and biosensing performance considerably depends on synthesis method. In the next section, main graphene synthesis methods including micromechanical exfoliation, chemical/electrochemical route syntheses, chemical vapor deposition (CVD), and other special techniques will concisely discussed to provide comprehensive understanding of graphene formation processes. More detailed graphene synthesis processes are available in recent review articles [66–69]. Among various methods, chemical/electrochemical route syntheses will particularly be emphasized because they have abundant structural defects [70–72] and functional groups [66, 73], which are advantageous for chemical and biosensing applications [61, 74]. In addition, they can produce graphene nanosheets in aqueous solution at low cost and may easily be scaled up for large-scale production [23, 75]. Moreover, they are most suitable route for formation of functionalized graphene and graphene/polymer composites [76–82], which can be highly useful for chemical and biosensing.

The properties of graphene applied for chemical and biosensors directly dictate the device performance. Thus, key graphene properties especially physical, chemical, and electronic are summarized following the synthesis method. More comprehensive graphene properties can be referred to many graphene's review articles [53, 66–69]. Next, graphene-based chemical and biosensors are successively digested. In chemical and biosensor sections, two main chemical/biosensing platforms, chemoresistive and electrochemical, are extensively covered while other sensing methods are also introduced. In chemoresistive platforms, graphene resistor and field-effect transistor (FET) structures for various gas-phase chemicals and biomolecules are discussed. For electrochemical detections, electrocatalytic properties of graphene, direct electrochemistry of enzyme, and electrochemical performances toward various molecules are elaborated. More aspects of particular chemical or biosensor can be found in technical and review articles devoted for each topic [53, 66–69]. Lastly, concluding remarks on the present results and the future research direction for graphene-based chemical and biosensors are provided.

2 Graphene Synthesis

Since its discovery, a number of graphene synthesis methods have been explored. Important methods are discussed in the following subsections and their graphical mechanism, advantages, and disadvantages are summarized in Table 1.

Table 1 Summary of mechanism, advantages, and disadvantages of various graphene synthesis methods

Method	Mechanisms	Advantages	Disadvantages
Cleavage by adhesive tape	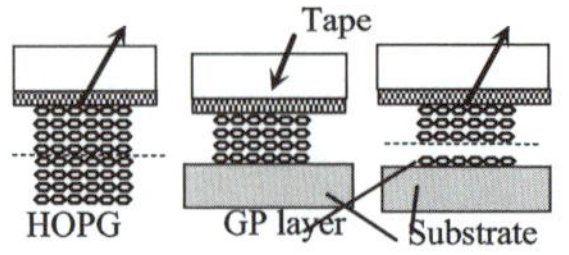 Tape HOPG GP layer Substrate	High-quality monolayer graphene sheets with large size	Require high skill operator, slow and difficult to control, unsuitable for mass production
Mechanical exfoliation by ultrasonication in solution	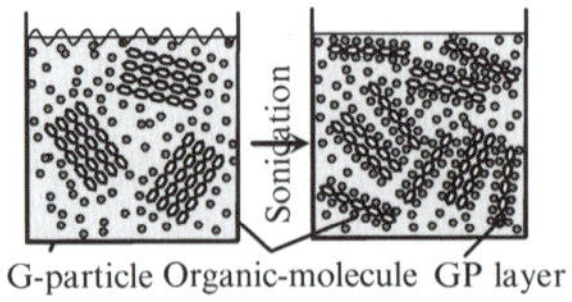 Sonication G-particle Organic-molecule GP layer	Scalable for mass production and medium graphene quality	Random number of graphene layers, low production yield
Mechanical exfoliation with intercalation of small molecules	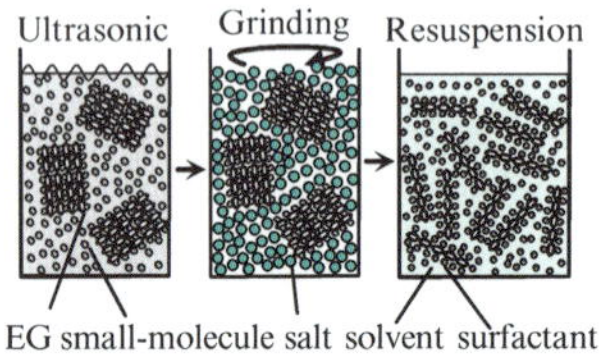 Ultrasonic Grinding Resuspension EG small-molecule salt solvent surfactant	High-quality monolayer graphene, high yield, scalable for mass production	Complicated and time-consuming process
Chemical oxidation of graphite and subsequent reduction	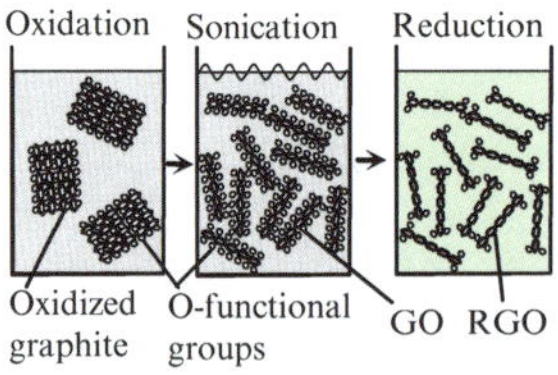 Oxidation Sonication Reduction Oxidized graphite O-functional groups GO RGO	Suitable for mass production, high yield of monolayer graphene, low cost, and simple	High defect density, poor electronic properties
Surface functionalization of graphene oxide	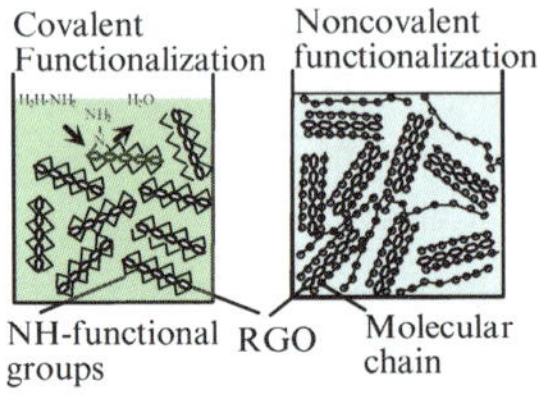 Covalent Functionalization Noncovalent functionalization NH-functional groups RGO Molecular chain	Suitable for mass production, monolayer graphene, low cost, and simple process	High defect density, poor electronic properties
Electrolytic exfoliation	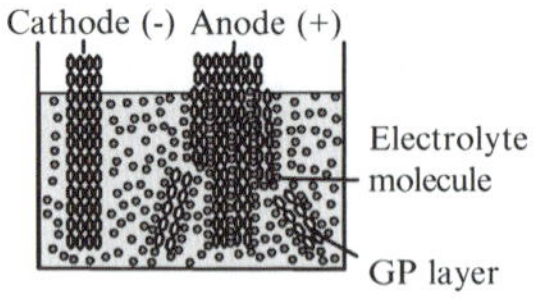 Cathode (-) Anode (+) Electrolyte molecule GP layer	Single step process, scalable for mass production, and high yield	Few-layer graphene structure, slow process

(continued)

Table 1 (continued)

Method	Mechanisms	Advantages	Disadvantages
Thermal CVD	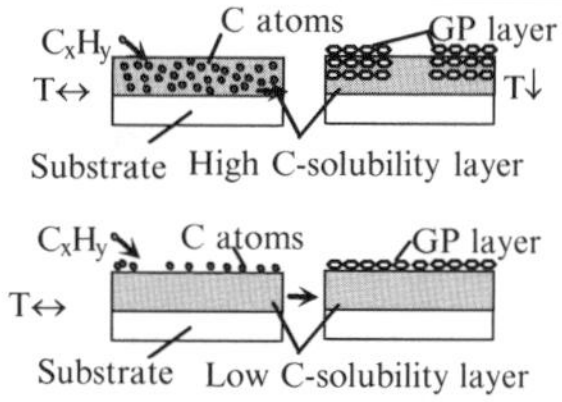 C_xH_y, C atoms, GP layer, T↔, T↓, Substrate, High C-solubility layer; C_xH_y, C atoms, GP layer, T↔, Substrate, Low C-solubility layer	Large-area, uniform and high-quality graphene with well-controlled number of layers and doping	Require high temperature, metal substrate, layer transfer process
PECVD	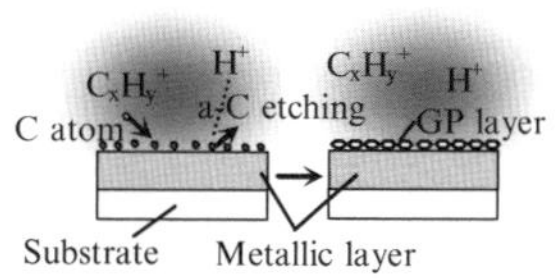 $C_xH_y^+$, H^+, a-C etching, C atom, GP layer, Substrate, Metallic layer	Relatively low temperature, directional growth, and ease of doping	Relatively low quality, uniformity, and high cost
Ultrahigh vacuum annealing of SiC and other C-compound	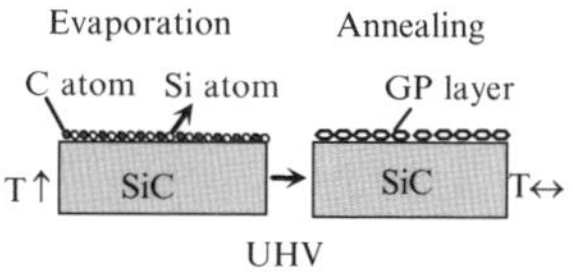 Evaporation, Annealing, C atom, Si atom, GP layer, T↑, SiC, T↔, UHV	Good quality, no need of layer transfer to other substrate	Require very high temperature and ultrahigh vacuum, difficult to control uniformity and growth pattern
Total organic synthesis	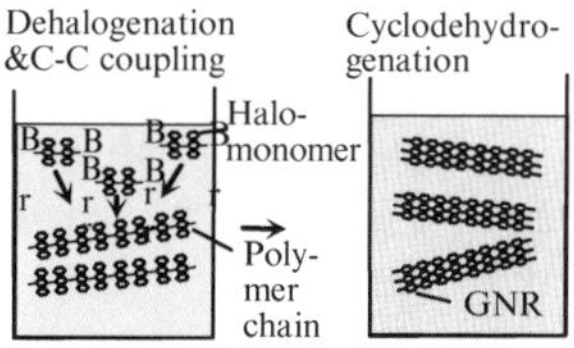 Dehalogenation &C-C coupling, Cyclodehydro-genation, Halo-monomer, Poly-mer chain, GNR	GNR structure with precise control of the composition, structure, and properties	Difficult to control dispersibility and a planar geometry for large GNR
Unzipping CNTs	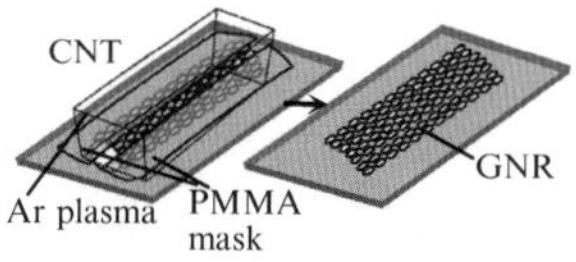 CNT, Ar plasma, PMMA mask, GNR	Scalable for mass production, GNR structure with well-controlled size and structure	High cost and relatively poor electronic properties

2.1 Exfoliation and Cleavage

Exfoliation and cleavage are the most fundamental methods for graphene production. The approach is based on physical extraction of graphene from crystalline graphite, which can be done in three following means.

2.1.1 Cleavage by Adhesive Tape

Mechanical exfoliation by adhesive tape is the first method that led to the discovery of monolayer graphene [1, 2]. In its most basic form, common cellophane tape is used to successively peel off layers from highly oriented pyrolytic graphite (HOPG) sheet and then pressed down against a clean substrate to deposit a sample. The graphitic film present on the tape is typically much thicker than one layer. When tape is lifted away, substrate-layer van der Waals attraction will delaminate the bottommost layer from others as shown in Table 1, leaving monolayer graphene on the substrate. The technique can produce high-quality graphene crystallites with size of more than 100 μm^2 that is electrically isolated for fundamental studies of transport physics and other properties but is not yet scalable to large area. However, there have been attempts to achieve patterned graphene by micromechanical exfoliation from mesa-structure graphite that can be formed by photolithographic patterning and oxygen plasma etching [83].

2.1.2 Mechanical Exfoliation by Ultrasonication in Solution

Mechanical exfoliation in solutions typically employs high intensity ultrasound to physically exfoliate layers from graphite powders dispersed in organic solvents. The quality of graphene in the dispersion prepared by this method is considerably depending on the solvent. Solvents suitable for graphene exfoliation include *N*-methyl-pyrrolidone (NMP) [84], sodium dodecylbenzene sulfonate (SDBS) [85], and sodium cholate (SC) [86]. NMP is reported to be highly effective for defect-free monolayer graphene production because NMP has similar surface energy to graphene but it suffers from high cost and high boiling point while SDBS and SC are aqueous surfactants that are coated on graphene, making them more stable dispersion by preventing aggregation via Coulomb repulsion. These methods produce dispersion that contains graphene with mixed numbers of layer. Separation techniques such as density gradient ultracentrifugation have thus been employed to select graphene sheets in the dispersion with controlled thickness. Although this technique may be implemented in a large scale, its production yield is still considered relatively low.

2.1.3 Mechanical Exfoliation with Intercalation of Small Molecules

In this approach, small non-covalently attaching molecules or polymers is incorporated between the layers of graphite by subjecting graphite to shear intensive mechanical stirring with ultrasonic solvent at room temperature or chemical/electrochemical oxidation in an intercalating acid to prepare graphite intercalation compounds (GICs). Acetic acid, acetic acid anhydride, concentrated sulfuric acid, and hydrogen peroxide are common ultrasonic solvents that are intercalated between graphite layers [87, 88]. The treated graphite is now expandable and it is then expanded to become expanded graphite (EG) by brief heating (60 s) at high temperature (900–1,000°C) in forming gas [89]. Next, EG is mechanically exfoliated to yield graphene sheet by grinding with salt crystals such as NaCl and reintercalated with organic molecules such as oleum. The exfoliated graphite is then dispersed in solvent such as *N,N*-dimethylformamide (DMF) and treated with additive such as tetrabutylammonium (TBA) to further isolate graphene in the solution [90]. The advantages of this method include high-quality monolayer graphene sheets, high production yield, and stable graphene suspension in organic solvent with very low agglomeration.

2.2 Chemical Route Synthesis

Chemical synthesis of graphene is mainly based on graphite oxidation and reduction. It is considered one of the most practical routes for mass production of graphene at low cost. Graphene can be chemically synthesized with and without surface functionalization schemes. In addition, the oxidation and reduction of graphene can also be done electrochemically.

2.2.1 Chemical Oxidation of Graphite and Subsequent Reduction

This method is based on chemical conversion of graphite to graphite oxide (GO) and subsequent reduction of GO to graphene [4, 91]. Generally, GO is synthesized by the method of Brodie [92], Staudenmaier [93], or Hummers [94] or their variation, which involve oxidation of graphite to various levels. A combination of potassium chlorate ($KClO_3$) with nitric acid (HNO_3) and potassium permanganate ($KMnO_4$) with sulfuric acid (H_2SO_4) are used to oxidize graphite in Brodie/Staudenmaier and Hummers methods, respectively. The oxidation is normally conducted at room temperature under constant stirring for a long time (>72 h). The GO product is then filtered, thoroughly washed, and dried before redispersed into final solvent. An important variant of these methods is to use graphite salts made by intercalating graphite with strong acids such as H_2SO_4, HNO_3, or $HClO_4$ as precursors for oxidation to GO. GO is heavily oxygenated bearing hydroxyl and epoxy groups on sp^3 hybridized carbon on the basal plane. Hence, GO is highly hydrophilic and readily exfoliated under moderate sonication in water and other solvents such as DMF, tetrahydrofuran (THF), *N*-methyl-2-pyrrolidone (NMP), and ethylene glycol, yielding stable dispersion consisting mostly of single layered sheets (graphene oxide). The thickness of graphene oxide sheets is typically 1 nm, which is more than 0.34 nm of pristine graphene sheet due to the displacement of sp^3 carbon atoms above and below the original graphene plane and the presence of covalently bound oxygen atoms. Sufficiently diluted colloidal suspension of graphene oxide prepared by this mean is clear, homogeneous, and stable indefinitely due to its hydrophilicity and electrostatic repulsion of negative charge on graphene oxide surface.

Graphene can be obtained from graphene oxide by various reduction means including chemical, thermal, solvothermal, and electrochemical reduction methods. Chemical reduction of graphene oxide sheets may be achieved using several reducing agents including hydrazine [95, 96], sodium borohydrate ($NaBH_4$) [97, 98], gaseous hydrogen [99], hydroquinone [100], and alkaline solutions [101]. Hydrazine does not react with water and is found to be the best to produce stable aqueous suspension of reduced graphene oxide (RGO) nanosheets at a suitable pH without any stabilizer. $NaBH_4$ is a more effective reductant for graphene oxide in aqueous solution compared to hydrazine because $NaBH_4$ can eliminate most oxygen-containing groups, resulting in graphene with higher C:O ratio and lower sheet resistance. Gaseous hydrogen can also produce graphene with relatively high C:O ratio while hydroquinone and alkaline solutions are not as effective as hydrazine and $NaBH_4$. It should be noted that reduced graphene in organic solvent tend to agglomerate or restack to form graphite through Van der Waals interaction due to their hydrophobic nature similar to CNTs and other nanomaterials.

Thermal reduction utilizes a short high-temperature (900–1,100°C) heat treatment to remove the oxide functional groups from graphene oxide surfaces [79, 102, 103]. Moreover, exfoliation of monolayer graphene can take place when the decomposition rate of the epoxy and hydroxyl sites of graphite oxide exceeds the diffusion rate of the evolved gases. The thermal reduction/exfoliation can produce 80% single layer RGO and the removal of the oxide groups causes about 30% mass loss, indicating large numbers of vacancies and structural defects. A simple one-step solvothermal reduction is another promising method to produce a stable RGO dispersion in organic solvent [104, 105]. The deoxygenation of GO in solvothermal process involves thermal deoxygenation at 200°C when refluxing GO in NMP and a concomitant reaction of GO with NMP molecules. For electrochemical reduction, GO is reduced in an electrolyte at a small negative potential vs Ag/AgCl and oxygen functional groups can be effectively removed from GO [106].

2.2.2 Surface Functionalization of Graphene Oxide

The surface functionalization of graphene oxide plays an important role in controlling exfoliation behavior of RGO and can also be used to change the surface properties to anchor various type molecules. There are two main surface functionalization approaches, covalent and non-covalent functionalization. In the first approach, oxygen functional groups on graphene oxide surfaces, including carboxylic groups at edges and epoxy/hydroxyl groups on the basal plane can be modified with organic isocyanates to yield various chemically modified GOs. Isocyanates react with the carboxyl and hydroxyl groups of GO to form amide and carbamate esters, respectively. These functional groups reduce the hydrophilicity of GO, resulting in stable dispersion of exfoliated single graphene sheets in polar aprotic solvents and facilitating intimate mixing of the graphene oxide sheets with polymer matrix. Moreover, modified graphene oxide can be chemically reduced in presence of a host polymer to enhance electrical conductivity of graphene-polymer nanocomposites [77, 78, 80, 82, 107–112].

In order to anchor biological molecules on graphene oxide, the carboxylic groups can be activated by various molecules including thionyl chloride ($SOCl_2$) [82, 111, 113], 1-ethyl-3-(3-dimethylaminopropyl)-carbodiimide (EDC) [114], *N*,

N-dicyclohexylcarbodiimide (DCC) [80], or 2-(7-aza-1H-benzotriazole-1-yl)-1,1,3,3-tetramethyluronium hexafluorophosphate (HATU) [59]. Covalently attached functional groups such as amides or esters are then formed by subsequent addition of nucleophilic species such as amines or alcohols [113]. The dispersability of modified graphene oxide in organic solvents can be improved by attachment of hydrophobic long, aliphatic amine groups while interesting nonlinear optical properties is obtained from porphyrin-functionalized primary amines and fullerene-functionalized secondary amines functional groups [110, 111]. Moreover, the amine groups and hydroxyl groups on graphene oxide can be used to attach polymers through either grafting-onto or grafting-from approaches, in which an atom transfer radical polymerization (ATRP) initiator such as α-bromoiso-butyryl-bromide is attached to graphene surfaces to initiate the living polymerization [108, 115]. In addition to carboxylic groups, epoxy groups on graphene oxide can be used to attach to different functional groups through reactions with various amine ending chemicals such as octadecylamine [100], an ionic liquid (1-(3-aminopropyl)-3-methylimidazolium bromide) [113].

In the second approach, graphene oxide is non-covalently functionalized via weak interactions including π–π, Van der Waals, and electrostatic attraction between graphene oxide and target molecules. The sp^2 network of graphene oxide interacts with conjugated polymers and aromatic compounds via π–π bonds, resulting in stabilization of chemically RGO (CR-GO). Poly(sodium 4-styrenesulfonate) (PSS) [116], sulfonated polyaniline [117], poly(3-hexylthiophene) (P3HT) [118], 7,7,8,8-tetracyanoquinodimethane anion [119], tetrasulfonate salt of copper phthalocyanine (TSCuPc) [120], porphyrin [121], pyrene/perylenediimide decorated with water-soluble moieties [122], and cellulose derivatives [123] are among conjugated polymers and aromatic compounds employed in this approach. Aromatic molecules can stably anchor onto the RGO surface without disturbing its electronic conjugation by introducing electrostatic charges. Dye-labeled DNA has also been used to functionalize graphene oxide for proteins and DNA detection [58]. The binding between the dye-labeled DNA and target molecule will alter the conformation of dye-labeled DNA and will release the dye-labeled DNA from GO, restoring of dye fluorescence, which was quenched when dye-labeled DNA attached on GO.

Although chemical synthesis of graphene generally provides large quantity of monolayer graphene oxide, high-density structural defects revealed by Raman spectroscopy are also formed due to the invasive chemical treatment [124]. Even after subsequent chemical reduction and thermal annealing, a large fraction of structural defects that significantly degrade graphene electronic properties are found to still remain by XPS analysis [125]. Therefore, physical exfoliation approaches would be more desirable when high-quality graphene structure is required.

2.2.3 Electrolytic Exfoliation

In this method, graphite is oxidized by electrochemical oxidation and exfoliated into graphene sheets. In electrolytic exfoliation process, two graphite rods are placed in an electrolysis cell filled with an electrolyte and a constant potential ranging from 3 to 20 V is applied between electrodes. A group of electrolytes including polymer solutions such as poly-styrene-sulfonate (PSS) has been reported [126]. Graphite oxidation gradually occurs on the positive electrode (anode) and graphene layers are exfoliated as black precipitates in the reactor. The electrolysis is conducted for several hours to obtain stable dispersion with desired graphene concentrations. The dispersed product may be centrifuged to separate large agglomerates from supernatant portion of the dispersion. This method can be used to produce high-quality graphene with varying geometry, thickness, and size that can be controlled by applied voltage, time, and properties of electrolyte. In addition, it can produce stable graphene nanosheets in aqueous solution at low cost and may easily be scaled up for large-scale production. Moreover, it is the most suitable route for formation of graphene in conducting polymer matrix, which can be highly useful for electrochemical transduction and bioreceptor immobilization.

2.3 Chemical Vapor Deposition

CVD is a standard gas-phase synthesis process that has been widely used to grow a variety of thin film material from reaction and condensation of gas source activated by some kinds of energy. Recently, it has been applied for graphene growth and it is now regarded as the most promising technique for large-scale production of mono- or few-layer graphene films.

2.3.1 Thermal CVD

Thermal CVD is the most basic CVD process that utilizes only thermal energy to activate chemical decomposition. The first successful synthesis of few-layer graphene films using thermal CVD was obtained using camphor as the precursor on Ni foils in 2006 [127]. Since then, many CVD studies have been conducted to grow graphene layers on several types of substrates using different precursors [6, 128–138]. The CVD growth mechanisms of graphene can be divided into two main cases depending on carbon solubility of substrates. For high carbon solubility (>0.1 at.%) substrates such as Co and Ni, the growth process involves the diffusion of the carbon into the substrate at the growth temperature and the subsequent precipitation of carbon out of the bulk metal upon the cooling [133–136]. Therefore, the thickness and crystalline ordering of the precipitated graphene layers is controlled by the substrate material, temperature, pressure, time, cooling rate, and the concentration of carbon dissolved in substrate, which depend on the type and concentration of the carbonaceous gas and the thickness of substrate. For example, CVD graphene growth on Ni substrate is typically conducted at a vacuum of 10^{-2} Torr and temperature of 1,000°C using a diluted methane gas. The average number of graphene layers grown on a Ni catalyst ranges from three to eight, depending on the reaction time and cooling rates. To produce graphene monolayer on this type of substrate, thinner Ni layers of thickness less than 300 nm were deposited on SiO_2/Si substrates [6].

For low carbon solubility (<0.001 at.%) substrate such as Cu, the graphene growth entails surface-dominated process [136, 137] beginning with catalytic decomposition of hydrocarbon species (C_xH_y) on Cu. Depending on the temperature, methane pressure, methane flow, and hydrogen partial pressure, the surface is either under-saturated, saturated, or supersaturated with C_xH_y species. Carbon nuclei will then be formed and grown on saturated or supersaturated surface. Graphene islands would grow until neighboring islands connected and fully cover the Cu surface under certain temperature, methane flow rate, and methane partial pressure. Unlike high carbon solubility case, monolayer and bilayer graphene are predominantly formed on low carbon solubility Cu substrate, which is believed to induce the self-limiting graphene growth.

The main advantages of CVD process are large area and uniform graphene with high quality and well-controlled number of layers. In addition, high-quality graphene layers can be transferred to another substrate over large area by simple processes that require no complicated mechanical or chemical treatment [7, 139]. CVD-grown graphene film on a foil is transferred to another substrate via adhesive polymer supports and etching of the copper layers to release the graphene layers. Moreover, CVD approach allows substitutional doping of graphene by simple introduction of gases such as NH_3 [140–142]. The nitrogen atoms can be substituted in graphene as pyridinic, graphitic, and pyrrolic forms that lead to interesting properties. For instance, N-doped graphene electrode enhances oxygen reduction capacity in fuel cells and increases reversible discharge capacitance in lithium ion batteries compared to pristine graphene due possibly to surface defects induced by nitrogen doping [141, 142].

2.3.2 Plasma-Enhanced CVD

Plasma-enhanced CVD (PECVD) employs plasma as an additional energy source to activate chemical decomposition, allowing graphene synthesis at a lower temperature compared to thermal CVD. The first reported few and monolayer PECVD graphene was obtained from a radio frequency (rf) PECVD system using a gas mixture of 5–100% CH_4 in H_2 at 12 Pa operating pressure, 900 W rf power and 680°C substrate temperature [143, 144]. More PECVD graphene studies have been conducted to understand the graphene growth mechanism and optimize experimental conditions [145–148]. PECVD growth graphene generally requires shorter deposition time (<5 min) and lower temperature (600–700°C) compared to thermal CVD growth. The PECVD growth mechanism entails a balance between the surface diffusion of C-bearing growth species from precursor gas and graphitic etching by atomic hydrogen. In addition, graphene sheets erected vertically can be produced through this method and believed to be attributed to the plasma electric field direction [147, 148].

2.3.3 Ultrahigh Vacuum Annealing of SiC and Other Substrates

In this method, graphene layers are obtained by rearrangement of surface carbon atoms that remain upon sublimation of silicon atoms from SiC surface due to ultrahigh vacuum (UHV) annealing at around 1,200°C for a few minutes [149–155]. The thickness of graphene layers depends on the annealing time and temperature. More recently, annealing in gas at a higher temperature (~400°C above UHV temperature) has been used instead of UHV to produce graphene on SiC with improved thickness homogeneity [156, 157]. This method has been attractive especially for semiconductor industry because graphene on SiC requires no transfer for further processing [24, 27, 157, 158]. However, it suffers from relatively high cost, challenging large-area thickness control problem, different growth patterns on different SiC polar faces (i.e., Si-face or C-face), and the lack of understanding of the structure and electronic properties of the interface layer between graphene and substrate [154, 155, 159–161]. This approach can be applied to other carbon-contained metallic substrates including Ru, Ir, Ni, Co, and Pt [162–164]. For example, UHV annealing of (0 0 0 1) Ru crystals produces a very sparse free-standing graphene structure with a linear growth of macroscopic single-crystalline domains.

2.4 *Other Synthesis Approaches*

2.4.1 Total Organic Synthesis

This method is based on chemical transformation of organic monomer precursor into graphene-like polyacyclic hydrocarbons (PAHs) via thermal activation steps [165]. Firstly, monomers are thermally deposited onto a gold surface such that the halogen substituents from the precursors is removed, leaving molecular building blocks for targeted graphene nanoribbons (GNRs) in the form of surface-stabilized biradical species. The biradical species diffuse across the surface and undergo radical addition reactions to form linear polymer chains as imprinted by the specific chemical functionality pattern of the monomers. Next, a second thermal activation is performed to establish a surface-assisted cyclodehydrogenation, leading to a fully aromatic extended system. This process can produce graphene products with atomic precision. For example, $N = 7$ GNRs and the chevron-type $N = 6/N = 9$ GNRs (N = number of fully hydrogen-terminated armchair edges) have been sequentially synthesized on a Ag (111) surface from 1,4-diiodo-2,3,5,6-tetraphenylbenzene coupling with 4-bromophenylboronic acid [166] or 10,10′-dibromo-9,9′-bianthryl precursor monomers [167]. Thus, the total organic synthesis is a versatile approach that can make GNR with precise control of the composition, structure, and properties but problems may arise from poor dispersability of a nonplanar large PAHs structure.

2.4.2 Unzipping CNTs

This method is based on the concept of unzipping or breaking CNTs walls to unroll them into GNRs, which can be done by three different means. Firstly, CNTs can be unzipped by plasma etching [168–170]. A polymer mask such as polymethyl methacrylate (PMMA) is used to coat CNTs dispersed on a Si substrate and PMMA coats MWCNTs everywhere except on MWCNT sidewalls attached to substrate. The sidewalls are then exposed to Ar or H_2 plasma etching after peeling PMMA-MWCNT film off from the substrate, leading to unzipped CNTs-GNRs. The PMMA residue can be removed using acetone vapor and calcination at 300°C. The produced GNRs typically have smooth edges and uniform width of half of the circumference of starting MWCNTs. The second CNT unzipping approach is through the oxidization and longitudinal unzipping of MWCNTs [171–174] in concentrated sulfuric acid, followed by treatment with $KMnO_4$. The oxidative unzipping generated the oxidized GNRs, which can be effectively reduced and doped by a high-temperature (700–800°C) annealing treatment. The last method utilizes longitudinal cutting using metal clusters as nanoscalpels that induce local oxidation for controlled cutting of CNTs [175]. The main advantage of CNT unzipping approach is the ease of large-scale production of GNRs with well-controlled size, structure, and properties. However, GNRs obtained by various unzipping methods tend to have inferior electronic properties compared to mechanically exfoliated graphene.

3 Graphene Properties

Graphene has recently been demonstrated to have extraordinary structural, electronic, chemical, optical, mechanical, and thermal properties. The following includes its important properties for device applications.

3.1 Structure and Morphology

The ideal graphene structure is an infinite planar two-dimensional sheet of honeycomb sp^2 lattice that contributes to a delocalized network of electrons. However, real graphene structure has finite size and is not perfectly planar. In general, freely suspended graphene has microscopic corrugations or “intrinsic” ripples with a lateral periodicity of about 8–10 nm and a height displacement of 0.7–1 nm [176]. In addition, Graphene can be produced in two main forms including nanosheet and nanoribbon with dimensions largely varied from a few nanometers to hundreds of microns depending on synthesis methods. In each graphene structure, there are always a variety of defects including topological defects (pentagons, heptagons, or their combination), vacancies, adatoms, edges/cracks, adsorbed impurities, and so on [70, 71, 177]. Mechanical exfoliation is the method that produces minimal defects while CVD produces moderately low defect density and chemical routes tend to generate highest defect density compared to other methods. The large numbers of lattice defects of graphene synthesized by CVD and chemical routes lead to wrinkled structures that are different from the rippled structure of mechanically exfoliated graphene [134, 178].

3.2 Electronic Properties

For a monolayer graphene, three of the four valence electrons on each carbon form the σ (single) bonds while the fourth electron forms one-third of a π bond with each of its neighbors producing a carbon–carbon bond order in graphite of one and one-third. With no chemical bonding in the *out-of-plane* direction, extremely weak interactions in this direction lead to nearly free propagation of charge and thermal carriers, resulting in large out-of-plane electrical and thermal conductivities of more than three orders of magnitude compared to their in-plane analogues. Thus, graphene's charge carriers may be considered as relativistic particles, which are electrons those have lost their rest mass that can be well described with (2+1) dimensional Dirac equation. As a result, the electronic band structure of monolayer graphene exhibits zero-band gap property, in which valence and conduction bands intersect in the reciprocal space at the same point, which is referred to as Dirac points. Evidence for the existence of massless Dirac quasiparticles in graphene is experimentally proven by the cyclotron mass measurement [179]. Due to massless nature of Dirac electron, graphene has very high electron mobility and electronic conductivity. The mobility of more than 200,000 cm^2/V s at carrier density of 2×10^{11} cm^{-2} for mechanically exfoliated suspended layer of graphene above a Si/SiO_2 gate has been reported [180, 181].

Nevertheless, zero band gap property of graphene considerably limits its use for nanoelectronics because charge carriers in graphene cannot be fully depleted for electronic switching. The band gap of graphene can be induced by reducing the connectivity of the π electron network based on two main approaches. Firstly, controlled oxidation of a few layers of graphene can result in a small band gap graphene [182, 183]. However, this approach is currently impractical because it is very difficult to control graphene oxidation of the first few layers. Secondly, band gap of the graphene can be modified by lateral quantum confinement using graphene nano ribbon (GNR) [171, 172, 184–189] or graphene quantum dot structure [189] and by biasing bilayer graphene [190, 191]. Doping and edge functionalization also change the band gap in nanoribbons [192]. The band gap property of GNRs greatly depends on lateral dimensions determined by preparation methods. Total organic synthesis and unzipping CNTs are promising for fabrication of GNR can produce GNR with low defects, high mobility, and well-controlled dimensions. Nevertheless, the mobility of GNRs is at least an order of magnitude lower than those of large 2D graphene sheets due to edge scattering [186].

Another practical problem of graphene is that its massless electronic structure can be greatly disrupted with the presence of defects that act as scattering sites. Defects can be introduced from intrinsic and extrinsic sources. Intrinsic defects are those that have been generated by synthesis process. Thus, mechanical exfoliation and CVD are normally preferred over chemical methods when electronic properties are critical. Extrinsic defects [193–195] including surface charge traps, interfacial phonons, substrate ripples, and absorbed impurity depend largely on interface between graphene and other materials in the system.

Ambipolar electric field effect at room temperature is another important electronic property of monolayer graphene. With this effect, charge carriers can be tuned between electrons and holes by applying a required gate voltage in concentrations as high as 10^{13} cm^{-2} [1, 3, 31]. In positive gate bias the Fermi level rises above the Dirac point which promotes electrons populating into conduction band, whereas, in negative gate bias the Fermi level drops below the Dirac point promoting the holes in valence band.

Anomalous QHE is another spectacular property of graphene. With this effect, hall conductivity of graphene is discretized at half integer positions according to $\sigma_{xy} = \pm(N + \frac{1}{2})(4e^2/h)$ where N, e, and h are Landau level index, electron charge, and Plank's constant, respectively. The effect occurs due to energy quantization of Dirac fermions in graphene under an applied magnetic field (B), which is given by $E_N = \pm v_F\sqrt{2e\hbar BN}$ where v_F is Fermi energy [12–15]. QHE in monolayer graphene can occur at room temperature unlike other materials whose QHE happens at temperatures below the boiling point of Helium.

3.3 Chemical Properties

Graphene has very high chemical activity like other carbon-based materials as it can participate in reactions as either a reducing agent (electron donor) or an oxidizer (electron acceptor). This is a direct consequence of its electronic structure, which results in both an electron affinity and an ionization potential of 4.6 eV [196]. From the chemistry of CNTs, there are differences in chemical reactivity between the different crystallographic directions (zigzag or armchair) [197–199] and unrolled or flattened planar graphene with different basal plane and edge (zigzag or armchair) of graphene would thus exhibit similarly different chemical activity [200]. However, graphene would provide double activity because of its twice specific surface area compared to CNTs. Substitutional doping by the replacement of carbon with other elements including boron (p-type) and nitrogen (n-type) is another important method to modify its chemical activity and electronic properties [43, 201].

3.4 Optical Properties

The most important optical property of graphene is its high optical transmittance. Monolayer graphene has low absorbance of 2.3% over a broad wavelength range from 300 to 2,500 nm [202]. The absorption linearly increases as the number of graphene layers increases with absorption fine structure constant (α) of ~1/37 ($\pi\alpha = 2.3\%$) [203]. Its UV absorption edge occurs at ~250 nm due to the inter band electronic transition from the unoccupied π^* states. Since Fermi energy and band gap of graphene can be tuned by applied electric field, the optical transition should also electrically tunable and this will be highly useful for development of high performance graphene-based optoelectronics devices [204].

High-quality graphene does not exhibit photoluminescence (PL) due to its zero band gap. However, PL can be attained from band gap-induced graphene including GO, GNRs, and graphene quantum dot structures [182, 183]. Broad PL from solid GO and liquid GO suspension has been observed and the progressive chemical reduction quenched the PL of GO. In addition, blue PL has been observed from GO thin films deposited by thoroughly exfoliated suspensions and is believed to originate from the recombination of electron–hole pairs, localized within small sp^2 carbon clusters embedded within the GO sp^3 matrix [205].

3.5 *Mechanical Properties*

Free-standing monolayer graphene is reported to have excellent mechanical properties with the highest elastic modulus of 1 TPa, intrinsic tensile strength, and stiffness [206]. In addition, graphene exhibits an interesting electromechanical behavior. The electronic band structure of graphene can be tuned by the applied strain due to atomic elongation. The band gap of 0.25 eV is found to provoke under the highest uniaxial strain (0.78%) applied to a single layer graphene [207].

3.6 *Thermal Properties*

Free-standing monolayer graphene is also reported to have excellent thermal properties with the highest thermal conductivity of 5,000 W/mK, almost twice higher than single-walled CNTs [208]. However, thermal conductivity of the graphene on various supports is considerably lower due to phonon scattering on substrate. Thermal conductivity of typical graphene on support is about 600 W/mK. The thermal conductivity is also affected by structural factors including number of graphene layers, defects edge, and isotopic doping due to phonon scattering between layers, at defects and at dopants [209, 210].

4 Graphene-Based Chemical Sensors

Graphene-based chemical sensors may be broadly classified into three main groups including chemoresistive, electrochemical, and other chemical-sensing platforms. Table 2 provides up-to-date summary of materials, synthesis method, structures, analytes, and minimum reported concentration for graphene-based chemical sensors.

Table 2 Summary of up-to-date reported graphene-based chemical sensors

Active material	Synthesis method	Analytes	Structure	Measurement	Minimum reported concentration	Refs.
RGO	CO-CR	NH_3	FET	*I* vs *t*	N/A	[67]
RGO	CO-CR	NO_2, NH_3, DNT	Resistor	*R* vs *t*	~ppm	[74]
RGO	CO-TR/CO-CR	H_2O (vapor)	Resistor	*R* vs *t*	~1%	[211]
RGO	CO-TR	NO_2, NH_3	FET	*I* vs *t*	~ppm, ~0.1%	[212, 213]
RGO	CO-TR	HCN, CEES DMMP, DNT	FET	*G* vs *t*	~10 ppb	[214]
RGO	CO-TR	H_2O vapor, NO_2, Cl_2	Resistor	*R* vs *t*	~ppm	[215]
G	ME	H_2O vapor, octanoic acid, TMA	FET	*I* vs *t*	~ppm	[216]
RGO	CO-TR	NO_2	FET	*I* vs *t*	~ppm	[217]
RGO + Pd	CO-ER	H_2	Resistor	*R* vs *t*	N/A	[218]
RGO	CO-CR	NH_3	FET	*I* vs *t*	~ppm	[219]
RGO	CO-CR	$Fe(CN)_6^{3-/4-}$	EC	CV, LSV	~mM	[61]
RGO	CO-CR	$Ru(NH_3)_6^{3+/2+}$, $Fe(CN)_6^{3-/4-}$	EC	CV	~mM	[200]
RGO	CO-CR	$Ru(NH_3)_6^{3+/2+}$, $Fe(CN)_6^{3-/4-}$, $Fe^{3+/2+}$	EC	CV	~mM	[220]
RGO	CO-TR	PC, AA	EC	CV, SWV	~10 nM, 10 μM	[221]
RGO	CO-TR	$Fe(CN)_6^{3-/4-}$, CPZ	EC	CV, DPV	~0.1 mM, ~10 nM	[222]
RGO	CO-TR	$Fe(CN)_6^{3-/4-}$, AA	EC	CV, *I* vs *t*	~1 mM, ~0.1 μM	[223]
RGO	CO-CR, CO-ER	$Fe(CN)_6^{3-/4-}$	EC	CV	~mM	[224]
N-doped RGO	CO-TR	$Fe(CN)_6^{3-/4-}$	EC	CV	~mM	[53]
RGO-Nafion	CO-CR	Pb^{2+}, Cd^{2+}	EC	DPASV	~nM	[225, 226]
G	ME	H^+	ISFET	*I* vs *t*	~0.01 pH	[227]
G	ME	Na^+	ISFET	*I* vs *t*	~nM	[228]
RGO-CaM/ MT-II	CO-TR	Na^+, K^+, Ca^{2+}, Mg^{2+}, Hg^{2+}	ISFET	*I* vs *t*	~nM	[229]
Pt/RGO/SiC	CO-TR	H_2	Diode	*I* vs *t*	~0.1%	[230]
RGO	CO-CR	H_2, CO	SAW	*f* vs *t*	~0.01%, ~10 ppm	[231]

EC electrochemical, *CO* chemical oxidation, *CR* chemical reduction, *TR* thermal reduction, *ER* electrochemical reduction, *ME* mechanical exfoliation, *CV* cyclic voltammetry, *LSV* linear sweep voltammetry, *SWV* square wave voltammetry, *DPSAV* differential pulse anodic stripping voltammetry, *PC* paracetamol, *CPZ* chlorpromazine, *AA* ascorbic acid, *TMA* trimethylamine, *DNT* dinitrotoulene, *HCN* hydrogen cyanide, *CEES* chloroethyl ethyl sulfide, *DMMP* dimethyl methyl phosphonate, *CaM* calmodulin, *MT-II* metallothionein, *R* resistance, *I* current, *f* resonance frequency

4.1 Graphene-Based Chemoresistive Chemical Sensor

Graphene-based chemoresistive chemical sensor exploits the resistance changes of graphene layer due to molecular adsorption on graphene. As molecules adsorb on graphene surface, graphene transfers charges to or from molecules, causing its Fermi level, carrier density, and electrical resistance to change. Graphene is a promising candidate for chemoresistive sensor due to its high carrier mobility, large specific surface area, and low Johnson noise [3, 15, 179, 232, 233]. The graphene-based chemoresistive sensors have been mostly employed for detection of gas-phase chemical or vapor such as H_2 NO_2, NH_3, H_2O, CO, dinitrotoluene (DNT), iodine, and ethanol and hydrazine hydrate [74, 211–219, 234, 235]. The sensitivity to gas or vapor of graphene-based sensor considerably depends on synthesis process that determines quality and conductivity of graphene as well as active chemical functional groups on graphene edges that interact with adsorbed molecules, resulting in charge carrier transfer. Among graphene prepared by various methods, RGO exhibits the most promising performance due to its diverse functionalities formed on graphene edges [211–213]. It has been demonstrated that unRGO exhibits very low response to gases such as NO_2 and the response become much higher after reduction due to the recovery of active sp^2 bond and enhanced carrier concentration [74]. Additional studies further suggest that gases interact with graphene structural defects such as vacancies as well as surface functional groups, which are highly dependent on reduction process [211, 214]. RGO chemical sensors for highly sensitive NO_2 and Cl_2 detection by chemical reduction with various agents including hydrazine and ascorbic acid have recently been demonstrated [215]. Moreover, it is found that RGO prepared by high-temperature reduction exhibits faster and higher chemoresistive response to vapor than chemically reduced GO and it is attributed to the formation of larger number of defects by the high-temperature process [211]. Moreover, RGO has been applied for detection of chemical warfare agents and explosives, achieving sensitivities at parts per billion levels [214].

The graphene-based chemoresistive sensors generally employ two basic device structures including graphene thin film resistor and graphene FET. Graphene thin film resistors comprise a graphene network layer and two metal contacts while graphene FET has similar structure but with additional underlying back gate dielectric and gate electrode [67]. The additional gate electrode allows more effective carrier control in graphene layer by ambipolar electric field effect, in which gate voltage can tune charge carriers between electrons and holes. In addition, it allows study of charge-transfer mechanisms between molecules and graphene by monitoring gate voltage that corresponds to Dirac point position in graphene. For instance, graphene FET exhibit Dirac point shift from 0 to −30 V upon exposure to NH_3 gas, suggesting that NH_3 molecules transfer electrons to graphene and make graphene surface become n-type [67]. From similar studies, CO molecules act as electron donors while H_2O and NO_2 molecules behave like electron acceptors for graphene [233]. In other reports, p-type RGO FET shows high response to NO_2 with large resistance decrease upon exposure to low concentration NO_2 while RGO FET becomes n-type and its conductivity decrease when reacted to NH_3 due to reducing reaction between NH_3 and oxygen groups in RGO [74, 213, 217, 235].

4.2 Graphene-Based Electrochemical Chemical Sensor

Carbon-based materials have been most widely used in electroanalysis and electrocatalysis [65, 236, 237] due to their high redox potential and electrochemical stability. Graphene is a better platform for the electrocatalytic study of carbon materials than CNTs due to its absence of metallic impurity [53, 238]. From recent studies, graphene exhibits a wide electrochemical potential window of ~2.5 V for 0.1 M PBS (pH 7.0) similar to other carbon-based electrodes including graphite, GC, and boron-doped diamond electrodes but the charge-transfer resistance of graphene measured by AC impedance spectra is much lower [61, 236, 239, 240]. In addition, graphene-based electrodes exhibits reversible redox peaks toward standard redox couples including $[Fe(CN)_6]^{3-/4-}$ and $[Ru(NH_3)_6]^{3+/2+}$ and $Fe^{3+/2+}$ with predominantly diffusion limited mechanisms [220]. It should be noted that these group of analytes are standard systems for electrode comparison because $[Ru(NH3)_6]^{3+/2+}$ has nearly ideal outer-sphere redox system that is insensitive to most surface defects or impurities on electrodes while $[Fe(CN)_6]^{3-/4-}$ is surface-sensitive but not oxide-sensitive and $Fe^{3+/2+}$ is both surface-sensitive and oxide-sensitive [236]. The peak-to-peak potential separations (ΔE_p) in CVs of graphene are ~65 mV (10 mV/s) for $[Fe(CN)_6]^{3-/4-}$ and 60–65 mV (100 mV/s) for $[Ru(NH_3)_6]^{3+/2+}$, which are much smaller than those of GC electrode and very close to the ideal value of 59 mV [220]. The low ΔE_p value infers a fast ET on graphene electrode for a single-electron electrochemical reaction [241]. The direct electron transfer rate constants (k_0) for $[Ru(NH_3)_6]^{3+/2+}$ of graphene and GC are 0.18 and 0.055 cm/s, respectively. Similarly, k_0 for $[Fe(CN)_6]^{3-/4-}$ of graphene and GC are found to be 0.49 and 0.029 cm/s, respectively while k_0 for $Fe^{3+/2+}$ of graphene electrode are few orders of magnitude higher than that of GC electrode. The fast electron transfer of graphene can be attributed to the active surface of graphene having high density of edge-plane-like defective active sites for electron transfer and unique electronic structure of graphene, especially the high density of the electronic states over a wide energy range [220, 236, 242].

The performance of graphene-based electrochemical sensors significantly depends on synthesis method. From many reports, chemically and thermally RGOs are highly promising for electrochemical sensing of many electroactive analytes particularly drugs such as paracetamol (PC), chlorpromazine (CPZ), and ascorbic acid (AA) [61, 220–223]. For instance, RGO prepared by thermal reduction exhibits excellent PC detection performance with very high sensitivity and low detection limit [221]. Similarly, RGO prepared by chemical reduction shows excellent sensitivity as well as low detection limit for CPZ and AA [222, 223]. Moreover, RGO produced through the electrochemical reduction exhibits relatively better electrochemical performance for $[Fe(CN)_6]^{3-/4-}$ detection than chemically and thermally reduced one due possibly to larger density of active edge-plane defective sites produced by electrochemical reduction [224, 243–245]. Furthermore, nitrogen doping has recently been demonstrated to be a promising mean to enhance chemical and biosensing performances of graphene. Nitrogen-doped graphene prepared by thermal reduction and doping exhibits relatively better electrochemical performance for $[Fe(CN)_6]^{3-/4-}$ detection than undoped one [53].

Graphene-based electrochemical sensors have also been developed for the environmental analysis of heavy metal ions (Pb^{2+} and Cd^{2+}) [225, 226]. Graphene is incorporated into Nafion layer, which is ion-selective molecule typically used for these heavy metal detections, to improve the detection sensitivity. It is demonstrated that Nafion/RGO composite electrode offers considerably enhanced sensitivity as well as improved selectivity for the metal ion (Pb^{2+} and Cd^{2+}) detections. The electrochemical signal is greatly enhanced on graphene electrodes and Pb^{2+} and Cd^{2+} peaks are well distinguished. The linear range and detection limits for both Cd^{2+} and Pb^{2+} are much better than those of Nafion film-modified bismuth electrode and ordered mesoporous carbon-coated GCE [246, 247] and they are comparable to Nafion/CNT-coated bismuth film electrode [248]. The improved sensitivity and selectivity can be credited to high specific surface area and electron transfer rate of graphene that produce strongly adsorbed target ions, enhancing the charge transfer and reducing fouling effect of nonconductive Nafion [225, 226]. Graphene ion-sensitive (IS) FET, another electrochemical sensor type that use G-FET structure with external reference electrode, has also recently been demonstrated for detection of ions such as H^+ (pH sensor) [227], Na^+, K^+, Ca^{2+}, Mg^{2+}, and Hg^{2+} [228, 229]. Reference electrode acts as an external top gate and electrochemical reaction from ions in electrolytes affect the electrical potential of graphene channels resulting in drain current modulation. Reported G-ISFET sensors exhibit high sensitivity and low detection limit.

4.3 Other Graphene-Based Chemical Sensors

Recently, there have also been few reports of graphene chemical sensors based on other platforms such as multilayer graphene diode and surface acoustic wave (SAW) device. In the first platform, Pt/RGO/SiC diode has been demonstrated for hydrogen (H_2) sensing [230]. This structure exploits the change of barrier height at Pt/graphene interface as a result of hydrogen adsorption on Pt and charge carrier transfer through graphene/SiC junction. The diode current is rapidly increased upon hydrogen adsorption. In the other platform, thin RGO nanosheets are deposited on 36° YX lithium tantalate ($LiTaO_3$) SAW interdigitated transducers (IDTs) [231]. The SAW structure utilizes the change of acoustic impedance as a result of gas adsorption on graphene sensing layer that results in the change of resonance frequency. The graphene-based SAW sensors exhibit promising sensing performance towards H_2 and CO at low (40°C) and room-operating temperatures.

5 Graphene-Based Biosensors

Graphene-based biosensors are defined here as sensors that employ graphene/ biological receptors as an active sensing material and/or apply for sensing biological analytes. They may be similarly classified into three groups including electrochemical, FET, and other biosensing platforms. Table 3 provides up-to-date summary of materials, synthesis method, structures, analytes, and minimum reported concentration for graphene-based chemical sensors.

5.1 Graphene-Based Electrochemical Biosensor

Graphene-based electrochemical biosensor may be divided into two main types including non-enzyme-based and enzyme-based electrochemical biosensors.

5.1.1 Non-enzyme-Based Electrochemical Biosensor

Non-enzyme-based electrochemical biosensors rely on direct oxidation of biomolecules by electrode. Detection of various biomolecules including β-nicotinamide adenine dinucleotide redox couple (NAD^+/NADH), dopamine (DA), and deoxy-nucleic acid (DNAs) using nonenzymatic graphene-based electrochemical sensors have been studied. NAD^+/NADH are coenzyme transferring electrons in redox reactions for cell metabolisms of biological substrates such as lactate, alcohol, or glucose catalyzed by dehydrogenases [65, 263]. CR-GO-modified GC [220] and edge-plane pyrolytic graphite electrodes (EPPGEs) [249] exhibit high electron transfer rate and low oxidation potentials (0.40 V) toward NAD^+/NADH, which are much better than graphite/GC, bare GC, and bare EPPGEs but comparable to CNTs-based electrodes [226, 264, 265]. In addition, surface fouling associated with the accumulation of reaction products are effectively suppressed. High-resolution X-ray photoelectron spectroscopy and ab initio molecular dynamics NAD^+/NADH adsorption studies suggest that oxygen-containing groups such as carboxylic on graphene edges enhance NAD^+ adsorption and hence electrochemical activity [263]. Hence, the obtained performances should be attributed to high density of active edge-plane-like defective sites on CR-GO for electron transfer to biological species [266]. Subsequently, non-covalent functionalization of graphene with methylene green (MG) has been employed to further reduce NADH oxidation potential to 0.14 V [250]. The superior NADH oxidation behaviors may partly be attributed to much less graphene agglomeration due to surface functionalization.

Multilayer graphene nanoflake films (MGNFs)-based electrode synthesized by catalyst-free microwave PECVD is reported for DA detection [241]. DA is a vital neurotransmitter for central nervous, renal, hormonal, and cardiovascular systems. The MGNFs electrodes exhibits high sensitivity, low detection limit (0.17 μM), and

Table 3 Summary of up-to-date reported graphene-based biosensors

Active material	Synthesis method	Analytes	Structure	Measurement	Minimum reported concentration	Ref.
RGO-Chit-GOx	CO-TR	Glucose	EC	CV	~10 μM	[51]
RGO-AuNPs-Chit	CO-CR	H_2O_2, O_2, glucose	EC	CV	~0.1 mM, ~0.1 mM	[52]
RGO, RGO/Nafion, fatigued-RGON	CO-CR	OPH	EC	CV and *I* vs *t*	~μM, ~0.1 μM, ~0.1 μM	[55]
SDBS functionalized RGO-HRP	CO-CR	H_2O_2	EC	*I* vs *t*	~10 μM	[60]
RGO	CO-CR	DNA base, DA, AA, UA, AP, H_2O_2,	EC	DPV	~10 μg/mL, ~mM, ~10 μM	[61]
RGO-ADH	CO-CR	NADH, ethanol	EC	DPV, *I* vs *t*	~10 μM, ~0.1 mM	[61]
RGO-GOx	CO-CR	Glucose	EC	*I* vs *t*	~μM	[61]
RGO	CO-CR	DA	EC	DPV, *I* vs *t*	~μM	[198]
RGO	CO-CR	DA, AA, ST	EC	CV, DPV	~mM	[200]
RGO	CO-CR	DA	EC	CV	~mM	[220]
G	MWPECVD	DA, AA, UA	EC	CV	~mM	[241]
RGO	CO-TR	DA, AA	EC	CV, SWV	~10 μM	[221]
RGO	CO-CR, CO-ER	NADH	EC	CV	~mM	[224]
G	ME	BSA	ISFET	*I* vs *t*	~10 nM	[227]
RGO	CO-CR	H_2O_2, NADH	EC	CV	~0.1 mM, ~mM	[249]
RGO-MG	CO-CR	NADH	EC	CV	~mM	[250]
RGO-AuNPs	CO-ER	DNA	EC	CV	~μM	[251]
IL-RGO	CO-CR	DNA bases, DA	EC	CV	~50 nm, ~50 nM	[252]
GO-MnO_2	CO	H_2O_2	EC	CV and *I* vs *t*	~μM	[253]
G-TPA-HRP	ME-SL	H_2O_2	EC	CV and *I* vs *t*	~0.1 μM	[254]
PtNPs-RGO/Chit-GOx	CO-CR	Glucose	EC	CV and *I* vs *t*	~0.1 μM	[255]
G/Nafion-GOx	ME	Glucose	EC	CV, *I* vs *t*	~10 μM	[256]
RGO-GOx	CO-ER	Glucose	EC	CV	~10 μM	[257]
RGO-DNA	CO-CR	DNA, bacterium	FET	*I* vs *t*	~ss-DNA, ~single molecule, ~single bacterium	[59]

(continued)

Table 3 (continued)

Active material	Synthesis method	Analytes	Structure	Measurement	Minimum reported concentration	Ref.
RGO-AuNPs-antibody	CO-TR	Protein	FET	*I* vs *t*	~ng/ml	[258]
G-DNA	CVD	DNA	FET	*I* vs *t*	~ss-DNA	[259]
G-Aptamer	ME	IgE	FET	*I* vs *t*	~nM	[260, 261]
GO-DNA	CO	DNA, protein	PL	Fluorescence intensity	~nM	[58]
GO-CdTe (NPs or QDs)-TCC	CO	Glutathione	CL	Luminescent intensity	~10 μM	[262]

IL Ionic liquid, *NPs* nanoparticles, *EC* electrochemical, *DPV* differential pulse voltammetry, *ME-SL* mechanical exfoliation in solution, *MWPECVD* microwave plasma CVD, *TPA* tetrasodium1,3,6,8-pyrenetetrasulfonic acid, *Chit* chitosan, *DA* dopamine, *AA* ascorbic acid, *UA* uric acid, *AP* acetaminophen, OPH organo phosphate, TCC thiole-containing compound, *IgE* immunoglobulin E, *BSA* bovine serum albumin, *PL* photoluminescence, *CL* chemiluminescence

well-separated oxidation peaks for important interfering species including ascorbic acid (AA) and uric acid (UA). MGNFs provide not only better sensitivity but also selectivity than traditional solid electrodes whose DA, AA, and UA oxidation peaks are overlapping. In other reports, graphene-based electrodes have shown high sensitivity with a linear range of 5–200 mM and high selectivity for dopamine against AA and serotonin (ST) [198, 200], which are superior to CNTs-based electrodes. In addition to edge-plane sites/defects, the superior graphene performance may be attributed to the higher conductivity, better electron transport to substrate via nanoconnector formed by graphene [241], more sp^2-like plane surface area and p–p stacking interaction of DA and graphene [198].

Nonenzymatic graphene-based electrochemical DNA sensors have been developed based on direct oxidation DNA capability of graphene [61]. DNA detection using CR-GO-modified GC electrode is demonstrated using differential pulse voltammetry (DPV). DPV signals from CR-GO/GC of the four free DNA bases including guanine (G), adenine (A), thymine (T), and cytosine (C) exhibit well-separated peak potentials while those from GC electrode are largely overlapping. Thus, CR-GO/GC can concurrently detect four free DNA bases. In addition, CR-GO/GC electrode can separately perceive all four DNA bases in single-stranded (ss)-DNAs as well as double-stranded (ds)-DNAs at physiological pH (7.3) with no prehydrolysis step. This allows non-hybridized and unlabeled detection of a single-nucleotide polymorphism (SNP) site for short oligomers with a specific sequence. These excellent performances may be attributed to CR-GO's antifouling properties and high electron transfer kinetics, resulting from high density of edge-plane defective sites of monolayer graphene and many active sites of oxygen-containing functional groups beneficial for electron transfer between the electrode and DNA species [266]. Recently, graphene-gold nanoparticle composite prepared by electrochemical synthesis is reported to provide even superior DNA detection performance due to synergistic effect between graphene and gold nanoparticles [251]. In addition, ionic liquid (IL)-functionalized graphene has been reported very recently to provide better electrochemical performance for DNA bases as well as DA detections than unfunctionalized one due to high ionic conductivity of IL and dispersibility of IL-functionalized graphene [252]. Moreover, graphene ISFET, another electrochemical sensor type that use G-FET structure with external reference electrode, has also recently been developed for sensing of protein such as bovine serum albumin (BSA) [227]. The detection of BSA is based on induced positive ions on BSA in phthalate solution that can be sensed electrochemically with high sensitivity.

5.1.2 Enzyme-Based Electrochemical Biosensor

Enzyme-based biosensors utilize enzyme immobilized on functionalized electrode surface to react with target biomolecules to generate products that can be electrochemically oxidized or reduced, resulting in electrochemical response. Hydrogen peroxide (H_2O_2) is a fundamental mediator for electroanalysis of food, medicine, and other chemical products and is the most common enzymatic product of oxidase widely used for enzyme-based biosensors. The electrochemical detection performance toward H_2O_2 of graphene (CR-GO)-based electrode has recently been studied. CR-GO-modified GC electrode exhibits high electron transfer rate, wide dynamic range, and lower oxidation/reduction potentials (0.20/0.10 V), which are much better than graphite/GC and GC electrodes [61]. Recently, GO-MnO_2 nanocomposite has been shown to have superior H_2O_2 sensitivity and selectivity [253]. The excellent H_2O_2 detection performance of graphene-based electrodes make them promising for enzyme-based electrochemical sensors.

Although an electrode exhibits the high electrocatalytic activity toward an enzymatic product, it may not be effective for enzyme-based electrochemical detection if it cannot provide efficient electron transfer between the active center of enzyme and electrode. This is often the case for common electrodes because the active centers of most redox enzymes are located deeply in a hydrophobic cavity of the molecule [267] and electron mediators are typically required to assist the electron transfer. Such process is quite slow and undesirable in some systems and hence the direct electron communication between the electrode and the active center of the enzyme, referred to as direct electrochemistry of enzyme, is preferred for biosensor and related systems [268–271]. Direct electrochemistry of redox enzymes may be realized on advanced electrodes comprising nanostructured materials that have excellent electron transfer property. CNTs, metal nanoparticles, and some nanoporous structures are found to have such property and the direct electron transfers between enzymes and these electrodes are demonstrated to be highly effective [65, 272–275].

Graphene is expected to exhibit better direct electrochemistry of enzymes due to its higher electron transfer rate and large specific surface area than CNTs [224]. The direct electrochemistry of horseradish peroxidase (HRP) on graphene prepared by mechanical exfoliation in solution has recently been demonstrated to be highly effective for selective H_2O_2 detection [254]. While most attention has been paid on the direct electrochemistry of glucose oxidase (GOD) on graphene prepared by different methods for development of glucose biosensors [51, 61, 255–257, 276]. From these studies, graphene-GOD-modified carbon electrodes exhibit a pair of well-defined redox CV peaks corresponding to reversible electron transfer of redox active center (flavin adenine dinucleotide, FAD) in GOD while carbon electrodes only modified with GOD do not show the redox peaks. In addition, the formal potential (the average of cathodic and anodic peak potentials) is close to the standard electrode potential of $FAD/FADH_2$ of ~0.43 V vs Ag/AgCl [51, 275]. These indicate that graphene significantly enhances a direct electron transfer of GOD. Moreover, low redox peak-to-peak separation of ~69 mV, high electron transfer rate constant (k_s) of ~2.8 s/1, nearly unity ratio of cathodic to anodic current intensity and the linear current density relationship with scan rates indicates that the redox process of GOD on graphene electrode is reversible and surface-limited [51, 61]. The direct electron transfer of GOD on graphene is also stable with no changes after tens of CV cycles and good response retention of more than 95% after 1 week storage. Furthermore, the GOD sensing performances of graphene-based electrodes are considerably much better than most of reported CNTs-based electrodes [277–279] due possibly to its high enzyme loading on its large surface area [51].

The performances of graphene-based biosensor are considerably affected by graphene preparation, surface functionalization, and enzyme immobilization methods. Glucose sensor based on polyethylenimine-functionalized ionic liquid/graphene nanocomposite is reported to exhibit wide glucose dynamic range (2–14 mM), good repeatability, and high stability [276] while glucose biosensor based on CR-GO [61] exhibits better performances with wide linear range, high sensitivity, and low detection limit of 2 μM (S/N ~ 3), which are also better than those of other carbon materials-based electrodes including CNTs [280–282], carbon nanofibers [283], exfoliated graphite nanoplatelets [256], and highly ordered mesoporous carbon [284]. In addition, GOD/CR-GO/GC sensor provides very fast glucose response (9 s response time) and high stability. In another report, glucose biosensors based on GOD/chitosan functionalized graphene composite show excellent sensitivity and long-term stability [51]. The role of chitosan is to improve graphene dispersion and immobilization of enzyme molecules.

Recently, graphene/metal nanoparticles (NP) systems have been employed for biosensing. Biosensors based on graphene/AuNPs/chitosan composites [52] show excellent electrocatalytical activity toward H_2O_2 and O_2 while GOD/graphene/PtNPs/chitosan glucose biosensor [255] exhibit high sensitivity with a low detection limit of 0.6 μM. The enhanced performance can be attributed to the synergistic surface area and conductivity effect of graphene and metal nanoparticles [52, 255]. Dehydrogenase biosensors based on functionalized graphene have also been developed. For instance, ethanol dehydrogenase biosensor based on graphene-ADH exhibits relatively fast response, wide linear range, and low detection limit compared to ADH-graphite/GC and ADH/GC electrodes [61]. This enhanced performance can be attributed to excellent catalytic activity of graphene toward NADH and the effective transport of ethanol and reaction products through enzyme immobilized graphene matrixes [285].

5.2 Graphene-Based Field-Effect Transistor Biosensor

In graphene FET biosensors, the graphene layer of FET structure is functionalized and immobilized with biological receptors, which can be at both biocellular and the biomolecular scale. The binding between target biomolecules and bioreceptor results in charge transfer to/from graphene layer and change of conductivity. Graphene FET biosensors have been developed for detection of a wide variety of biological target including bacteria, DNA, protein/DNA mixture, and other antibody-specific biomolecules [59, 258, 260, 261]. The graphene-based bacteria sensor utilizing a p-type graphene FET immobilized with bacteria antibody is shown to be ultra sensitive with a single bacterium detection capability [59]. For DNA sensor, single-stranded DNA is tethered on graphene p-type FET. Upon hybridization with its complementary DNA strand, the surface hole density is found to significantly increase. Thus, the DNA graphene sensor provides a label-free, reversible, and highly sensitive DNA detection. In another development, a polarity-specific polyelectrolyte molecular transistor is employed for protein/DNA detection. The DNA/protein mixture is immobilized on RGO transistor to simultaneously detect DNA as well as protein from the presence and dynamic cellular secretion of biomolecules with good specificity. The conductivity/mobility of RGO transistor would be increased upon DNA hybridization due to the electron transfer from charged amine group to RGO while protein binding causes the change in the opposite direction [258]. In another report, CVD-grown graphene FETs configured as solution-gated transistors can label freely and electrically detect DNA hybridization with single-base specificity [259]. The DNA detection is attributed to the electronic n-doping from DNA to graphene. The immobilization-specific antibodies on RGO FETs yield graphene-based immunosensor. For example, a highly sensitive and selective FET biosensor using Au NP-antibody conjugates decorated with GO sheets have been reported for detection of a pathogen rotavirus [258]. As another instance, graphene FET is modified with aptamer for sensitive and selective detection of immunoglobulin E [260, 261].

5.3 Other Graphene-Based Biosensors

Recently, there have also been few reports of graphene biosensors based on other platforms such as photoluminescence and electrochemiluminescence. For instance, GO has been used as binding medium for labeled protein and DNA detection with photoluminescence, achieving very low minimum detection range on the order of nM [58]. CdTe/RGO composite also exhibited the chemical–biological sensing where graphene worked as an amplified electrogenerated chemiluminescence (ECL) of QDs platform [262] for glutathione drug detection with graphene-based electronics glutathione drug sensor.

6 Concluding Remarks

In conclusion, graphene's extraordinary structural, electronic, and physiochemical properties make it highly attractive for chemical and biosensing applications. Graphene-based chemical and biosensors may similarly be classified into three main groups including chemoresistive, electrochemical, and other sensing platforms. Chemoresistive graphene-based chemical sensors using resistor and FET structures have been widely developed for ultrasensitive gas-phase chemical sensing with single molecule detection capability. Similarly, graphene-based electrochemical sensors have shown excellent sensitivity and selectivity toward various electroactive analytes compared to most other carbon-based electrodes due to its very high electron transfer rate, which can be attributed to high density of edge-plane-like defective active sites and high density of the electronic states over a wide energy range. For electrochemical biosensing, graphene has shown superior performance in both non-enzyme- and enzyme-based electrochemistry due to its excellent direct electrochemical oxidation of small biomolecules such as NADH, dopamine, and DNA, and direct electrochemistry of enzyme such as glucose oxidase and dehydrogenase. Graphene FET chemoresistive biosensors for detection of a wide variety of biological target including bacteria, DNA, protein/DNA mixture, and other antibody-specific biomolecules have been reported with high sensitivity and specificity.

In addition, the performance of graphene-based chemical and biosensors considerably depends on synthesis method. From most reports, surface functionalized graphene oxides prepared by chemical, thermal, and particularly electrochemical reductions are highly promising for both electrochemical and chemoresistive sensing platforms. Recently, nitrogen doping has been shown to be a promising mean to enhance chemical and biosensing performances of graphene. However, there are still very few chemical or biosensing studies on doped graphene and hence it is a potential subject to be further explored. Furthermore, large-scale economical production of graphene is still not generally attainable and graphene-based chemical and biosensors still suffer from poor reproducibility because the exact structures of graphene-based sensors are difficult to be controlled. Therefore, novel methods for well-controlled synthesis and processing of graphene must be further developed. Lastly, physics and chemistry of interaction between graphene surface and chemicals or biomolecules as well as charge transport mechanisms are not yet well understood and they should be further studied for effective application of graphene in chemical and biosensing.

References

1. Novoselov KS, Geim AK, Morozov SV et al (2004) Electric field in atomically thin carbon films. Science 306(5696):666–669
2. Novoselov KS, Jiang D, Schedin F et al (2005) Two-dimensional atomic crystals. Proc Natl Acad Sci USA 102(30):10451–10453
3. Geim AK, Novoselov KS (2007) The rise of graphene. Nat Mater 6(3):183–191
4. Park S, Ruoff RS (2009) Chemical methods for the production of graphenes. Nat Nanotechnol 4(4):217–224
5. Service RF (2009) Carbon sheets an atom thick give rise to graphene dreams. Science 324(5929):875–877
6. Kim KS, Zhao Y, Jang H et al (2009) Large-scale pattern growth of graphene films for stretchable transparent electrodes. Nature 457(7230):706–710
7. Bae S, Kim H, Lee Y et al (2010) Roll-to-roll production of 30-inch graphene films for transparent electrodes. Nat Nanotechnol 5(8):574–578
8. Eda G, Lin YY, Miller S et al (2008) Transparent and conducting electrodes for organic electronics from reduced graphene oxide. Appl Phys Lett 92(23):233305
9. Mattevi C, Eda G, Agnoli S et al (2009) Evolution of electrical, chemical, and structural properties of transparent and conducting chemically derived craphene thin films. Adv Funct Mater 19(16):2577–2583
10. Balandin AA, Ghosh S, Bao W et al (2008) Superior thermal conductivity of single-layer graphene. Nano Lett 8(3):902–907
11. Yu A, Ramesh P, Sun X et al (2008) Enhanced thermal conductivity in a hybrid graphite nanoplatelet – carbon nanotube filler for epoxy composites. Adv Mater 20(24):4740–4744
12. Nomura K, MacDonald AH (2006) Quantum hall ferromagnetism in graphene. Phys Rev Lett 96(25):256602
13. Novoselov KS, Jiang Z, Zhang Y et al (2007) Room-temperature quantum hall effect in graphene. Science 315(5817):1379
14. Novoselov KS, McCann E, Morozov SV et al (2006) Unconventional quantum hall effect and berry's phase of 2pi in bilayer graphene. Nat Phys 2(3):177–180
15. Zhang Y, Tan YW, Stormer HL et al (2005) Experimental observation of the quantum hall effect and berry's phase in graphene. Nature 438(7065):201–204
16. Cai D, Yusoh K, Song M (2009) The mechanical properties and morphology of a graphite oxide nanoplatelet/polyurethane composite. Nanotechnology 20(8):085712
17. Rafiee MA, Rafiee J, Wang Z et al (2009) Enhanced mechanical properties of nanocomposites at low graphene content. ACS Nano 3(12):3884–3890
18. Vadukumpully S, Paul J, Mahanta N et al (2011) Flexible conductive graphene/poly(vinyl chloride) composite thin films with high mechanical strength and thermal stability. Carbon 49(1):198–205
19. Park S, Mohanty N, Suk JW et al (2010) Biocompatible, robust free-standing paper composed of a tween/graphene composite. Adv Mater 22(15):1736–1740
20. Wang K, Ruan J, Song H et al (2010) Biocompatibility of graphene oxide. Nanoscale Res Lett 6(1):1–8
21. Yan X, Chen J, Yang J et al (2010) Fabrication of free-standing, electrochemically active, and biocompatible graphene oxide-polyaniline and graphene-polyaniline hybrid papers. ACS Appl Mater Interfaces 2(9):2521–2529
22. Chang H, Sun Z, Yuan Q et al (2010) Thin film field-effect phototransistors from bandgap-tunable, solution-processed, few-layer reduced graphene oxide films. Adv Mater 22(43):4872–4876
23. Joung D, Chunder A, Zhai L et al (2010) High yield fabrication of chemically reduced graphene oxide field effect transistors by dielectrophoresis. Nanotechnology 21(16):165202
24. Kedzierski J, Hsu PL, Healey P et al (2008) Epitaxial graphene transistors on SiC substrates. IEEE Trans Electron Dev 55(8):2078–2085

25. Lee CG, Park S, Ruoff RS et al (2009) Integration of reduced graphene oxide into organic field-effect transistors as conducting electrodes and as a metal modification layer. Appl Phys Lett 95(2):023304
26. Liao L, Bai J, Cheng R et al (2010) Top-gated graphene nanoribbon transistors with ultrathin high-k dielectrics. Nano Lett 10(5):1917–1921
27. Lin YM, Dimitrakopoulos C, Jenkins KA et al (2010) 100-ghz transistors from wafer-scale epitaxial graphene. Science 327(5966):662
28. Lin YM, Jenkins KA, Alberto VG et al (2009) Operation of graphene transistors at giqahertz frequencies. Nano Lett 9(1):422–426
29. Xia F, Farmer DB, Lin YM et al (2010) Graphene field-effect transistors with high on/off current ratio and large transport band gap at room temperature. Nano Lett 10(2):715–718
30. Ponomarenko LA, Schedin F, Katsnelson MI et al (2008) Chaotic dirac billiard in graphene quantum dots. Science 320(5874):356–358
31. Myung S, Park J, Lee H et al (2010) Ambipolar memory devices based on reduced graphene oxide and nanoparticles. Adv Mater 22(18):2045–2049
32. Stoller MD, Park S, Yanwu Z et al (2008) Graphene-based ultracapacitors. Nano Lett 8(10):3498–3502
33. Vivekchand SRC, Rout CS, Subrahmanyam KS et al (2008) Graphene-based electrochemical supercapacitors. J Chem Sci 120(1):9–13
34. Wang H, Hao Q, Yang X et al (2009) Graphene oxide doped polyaniline for supercapacitors. Electrochem Commun 11(6):1158–1161
35. Wang Y, Shi Z, Huang Y et al (2009) Supercapacitor devices based on graphene materials. J Phys Chem C 113(30):13103–13107
36. Guo P, Song H, Chen X (2009) Electrochemical performance of graphene nanosheets as anode material for lithium-ion batteries. Electrochem Commun 11(6):1320–1324
37. Liang M, Zhi L (2009) Graphene-based electrode materials for rechargeable lithium batteries. J Mater Chem 19(33):5871–5878
38. Wang C, Li D, Too CO et al (2009) Electrochemical properties of graphene paper electrodes used in lithium batteries. Chem Mater 21(13):2604–2606
39. Yang S, Cui G, Pang S et al (2009) Fabrication of cobalt and cobalt oxide/graphene composites: towards high-performance anode materials for lithium ion batteries. ChemSusChem 3(2):236–239
40. Yoo EJ, Kim J, Hosono E et al (2008) Large reversible Li storage of graphene nanosheet families for use in rechargeable lithium ion batteries. Nano Lett 8(8):2277–2282
41. Seger B, Kamat PV (2009) Electrocatalytically active graphene-platinum nanocomposites. Role of 2-d carbon support in pem fuel cells. J Phys Chem C 113(19):7990–7995
42. Shao Y, Liu J, Wang Y et al (2009) Novel catalyst support materials for pem fuel cells: current status and future prospects. J Mater Chem 19(1):46–59
43. Shao Y, Sui J, Yin G et al (2008) Nitrogen-doped carbon nanostructures and their composites as catalytic materials for proton exchange membrane fuel cell. Appl Catal Environ 79(1):89–99
44. Wu J, Becerril HA, Bao Z et al (2008) Organic solar cells with solution-processed graphene transparent electrodes. Appl Phys Lett 92(26):263302
45. Liu Q, Liu Z, Zhang X et al (2009) Polymer photovoltaic cells based on solution-processable graphene and p3ht. Adv Funct Mater 19(6):894–904
46. Liu Q, Liu Z, Zhang X et al (2008) Organic photovoltaic cells based on an acceptor of soluble graphene. Appl Phys Lett 92(22):223303
47. Yin Z, Sun S, Salim T et al (2010) Organic photovoltaic devices using highly flexible reduced graphene oxide films as transparent electrodes. ACS Nano 4(9):5263–5268
48. Yang K, Wan J, Zhang S et al (2011) In vivo pharmacokinetics, long-term biodistribution, and toxicology of pegylated graphene in mice. ACS Nano 5(1):516–522

49. Nair RR, Blake P, Blake JR et al (2010) Graphene as a transparent conductive support for studying biological molecules by transmission electron microscopy. Appl Phys Lett 97 (15):153102
50. Ang PK, Chen W, Wee ATS et al (2008) Solution-gated epitaxial graphene as pH sensor. J Am Chem Soc 130(44):14392–14393
51. Kang X, Wang J, Wu H et al (2009) Glucose oxidase-graphene-chitosan modified electrode for direct electrochemistry and glucose sensing. Biosens Bioelectron 25(4):901–905
52. Shan C, Yang H, Han D et al (2009) Graphene/aunps/chitosan nanocomposites film for glucose biosensing. Biosens Bioelectron 25(5):1070–1074
53. Shao Y, Wang J, Wu H et al (2010) Graphene based electrochemical sensors and biosensors: a review. Electroanalysis 22(10):1027–1036
54. Baby TT, Aravind SSJ, Arockiadoss T et al (2010) Metal decorated graphene nanosheets as immobilization matrix for amperometric glucose biosensor. Sens Actuators B Chem 145(1):71–77
55. Choi BG, Park H, Park TJ et al (2010) Solution chemistry of self-assembled graphene nanohybrids for high-performance flexible biosensors. ACS Nano 4(5):2910–2918
56. Dong H, Gao W, Yan F et al (2010) Fluorescence resonance energy transfer between quantum dots and graphene oxide for sensing biomolecules. Anal Chem 82(13):5511–5517
57. Feng L, Chen Y, Ren J et al (2011) A graphene functionalized electrochemical aptasensor for selective label-free detection of cancer cells. Biomaterials 32(11):2930–2937
58. Lu CH, Yang HH, Zhu CL et al (2009) A graphene platform for sensing biomolecules. Angew Chem Int Ed 48(26):4785–4787
59. Mohanty N, Berry V (2008) Graphene-based single-bacterium resolution biodevice and DNA transistor: interfacing graphene derivatives with nanoscale and microscale biocomponents. Nano Lett 8(12):4469–4476
60. Zeng Q, Cheng J, Tang L et al (2010) Self-assembled graphene-enzyme hierarchical nanostructures for electrochemical biosensing. Adv Funct Mater 20(19):3366–3372
61. Zhou M, Zhai Y, Dong S (2009) Electrochemical sensing and biosensing platform based on chemically reduced graphene oxide. Anal Chem 81(14):5603–5613
62. Pumera M, Ambrosi A, Bonanni A et al (2011) Graphene for electrochemical sensing and biosensing. Trends Analyt Chem 29(9):954–965
63. Ratinac KR, Yang W, Gooding JJ et al (2011) Graphene and related materials in electrochemical sensing. Electroanalysis 23(4):803–826
64. Balasubramanian K, Burghard M (2006) Biosensors based on carbon nanotubes. Anal Bioanal Chem 385(3):452–468
65. Wang J (2005) Carbon-nanotube based electrochemical biosensors: a review. Electroanalysis 17(1):7–14
66. Singh V, Joung D, Zhai L et al (2011) Graphene based materials: past, present and future. Prog Mater Sci 56(8):1178–1271
67. Zhu Y, Murali S, Cai W et al (2010) Graphene and graphene oxide: synthesis, properties, and applications. Adv Mater 22(35):3906–3924
68. Allen MJ, Tung VC, Kaner RB (2009) Honeycomb carbon: a review of graphene. Chem Rev 110(1):132–145
69. Soldano C, Mahmood A, Dujardin E (2010) Production, properties and potential of graphene. Carbon 48(8):2127–2150
70. Chen JH, Cullen WG, Jang C et al (2009) Defect scattering in graphene. Phys Rev Lett 102 (23):236805
71. Meyer JC, Kisielowski C, Erni R et al (2008) Direct imaging of lattice atoms and topological defects in graphene membranes. Nano Lett 8(11):3582–3586
72. Bagri A, Mattevi C, Acik M et al (2010) Structural evolution during the reduction of chemically derived graphene oxide. Nat Chem 2(7):581–587
73. Szabo T, Berkesi O, Forgo P et al (2006) Evolution of surface functional groups in a series of progressively oxidized graphite oxides. Chem Mater 18(11):2740–2749

74. Fowler JD, Allen MJ, Tung VC et al (2009) Practical chemical sensors from chemically derived graphene. ACS Nano 3(2):301–306
75. Eda G, Chhowalla M (2010) Chemically derived graphene oxide: towards large-area thin-film electronics and optoelectronics. Adv Mater 22(22):2392–2415
76. Kou R, Shao Y, Wang D et al (2009) Enhanced activity and stability of pt catalysts on functionalized graphene sheets for electrocatalytic oxygen reduction. Electrochem Commun 11(5):954–957
77. Qi X, Pu KY, Zhou X et al (2010) Conjugated-polyelectrolyte-functionalized reduced graphene oxide with excellent solubility and stability in polar solvents. Small 6(5):663–669
78. Ramanathan T, Abdala AA, Stankovich S et al (2008) Functionalized graphene sheets for polymer nanocomposites. Nat Nanotechnol 3(6):327–331
79. Schniepp HC, Li JL, McAllister MJ et al (2006) Functionalized single graphene sheets derived from splitting graphite oxide. J Phys Chem B 110(17):8535–8539
80. Veca LM, Lu F, Meziani MJ et al (2009) Polymer functionalization and solubilization of carbon nanosheets. Chem Commun 18:2565–2567
81. Xu LQ, Yang WJ, Neoh KG et al (2010) Dopamine-induced reduction and functionalization of graphene oxide nanosheets. Macromolecules 43(20):8336–8339
82. Xu Y, Liu Z, Zhang X et al (2009) A graphene hybrid material covalently functionalized with porphyrin: synthesis and optical limiting property. Adv Mater 21(12):1275–1279
83. Zhang Y, Small JP, Pontius WV et al (2005) Fabrication and electric-field-dependent transport measurements of mesoscopic graphite devices. Appl Phys Lett 86(7):1–3
84. Hernandez Y, Nicolosi V, Lotya M et al (2008) High-yield production of graphene by liquid-phase exfoliation of graphite. Nat Nanotechnol 3(9):563–568
85. Lotya M, Hernandez Y, King PJ et al (2009) Liquid phase production of graphene by exfoliation of graphite in surfactant/water solutions. J Am Chem Soc 131(10):3611–3620
86. Green AA, Hersam MC (2009) Solution phase production of graphene with controlled thickness via density differentiation. Nano Lett 9(12):4031–4036
87. Kang F, Leng Y, Zhang TY (1996) Influences of H_2O_2 on synthesis of H_2SO_4-gics. J Phys Chem Solid 57(6–8):889–892
88. Kang F, Zhang TY, Leng Y (1997) Electrochemical behavior of graphite in electrolyte of sulfuric and acetic acid. Carbon 35(8):1167–1173
89. Pan YX, Yu ZZ, Ou YC et al (2000) New process of fabricating electrically conducting nylon 6/graphite nanocomposites via intercalation polymerization. J Polym Sci B Polym Phys 38(12):1626–1633
90. Li X, Zhang G, Bai X et al (2008) Highly conducting graphene sheets and langmuir-blodgett films. Nat Nanotechnol 3(9):538–542
91. Dreyer DR, Park S, Bielawski CW et al (2010) The chemistry of graphene oxide. Chem Soc Rev 39(1):228–240
92. Brodie BC (1860) Sur le poids atomique du graphite. Ann Chim Phys 59:466–472
93. Staudenmaier L (1898) Verfahren zur darstellung der graphitsaure. Ber Dtsch Chem Ges 31:1481–1487
94. Hummers WS Jr, Offeman RE (1958) Preparation of graphitic oxide. J Am Chem Soc 80(6):1339
95. Parades JI, Villar-Rodil S, Martínez-Alonso A et al (2008) Graphene oxide dispersions in organic solvents. Langmuir 24(19):10560–10564
96. Eda G, Fanchini G, Chhowalla M (2008) Large-area ultrathin films of reduced graphene oxide as a transparent and flexible electronic material. Nat Nanotechnol 3(5):270–274
97. Si Y, Samulski ET (2008) Synthesis of water soluble graphene. Nano Lett 8(6):1679–1682
98. Shin HJ, Kim KK, Benayad A et al (2009) Efficient reduction of graphite oxide by sodium borohydride and its effect on electrical conductance. Adv Funct Mater 19(12):1987–1992
99. Wu ZS, Ren W, Gao L et al (2009) Synthesis of high-quality graphene with a pre-determined number of layers. Carbon 47(2):493–499

100. Wang S, Chia PJ, Chua LL et al (2008) Band-like transport in surface-functionalized highly solution-processable graphene nanosheets. Adv Mater 20(18):3440–3446
101. Fan X, Peng W, Li Y et al (2008) Deoxygenation of exfoliated graphite oxide under alkaline conditions: a green route to graphene preparation. Adv Mater 20(23):4490–4493
102. McAllister MJ, Li JL, Adamson DH et al (2007) Single sheet functionalized graphene by oxidation and thermal expansion of graphite. Chem Mater 19(18):4396–4404
103. Zhu Y, Stoller MD, Cai W et al (2010) Exfoliation of graphite oxide in propylene carbonate and thermal reduction of the resulting graphene oxide platelets. ACS Nano 4(2):1227–1233
104. Murugan AV, Muraliganth T, Manthiram A (2009) Rapid, facile microwave-solvothermal synthesis of graphene nanosheets and their polyaniline nanocomposites for energy strorage. Chem Mater 21(21):5004–5006
105. Dubin S, Gilje S, Wang K et al (2010) A one-step, solvothermal reduction method for producing reduced graphene oxide dispersions in organic solvents. ACS Nano 4(7):3845–3852
106. Zhou M, Wang Y, Zhai Y et al (2009) Controlled synthesis of large-area and patterned electrochemically reduced graphene oxide films. Chem Eur J 15(25):6116–6120
107. Ansari S, Giannelis EP (2009) Functionalized graphene sheet-poly(vinylidene fluoride) conductive nanocomposites. J Polym Sci B Polym Phys 47(9):888–897
108. Fang M, Wang K, Lu H et al (2009) Covalent polymer functionalization of graphene nanosheets and mechanical properties of composites. J Mater Chem 19(38):7098–7105
109. Ganguli S, Roy AK, Anderson DP (2008) Improved thermal conductivity for chemically functionalized exfoliated graphite/epoxy composites. Carbon 46(5):806–817
110. Geng J, Jung HT (2010) Porphyrin functionalized graphene sheets in aqueous suspensions: from the preparation of graphene sheets to highly conductive graphene films. J Phys Chem C 114(18):8227–8234
111. Liu ZB, Xu YF, Zhang XY et al (2009) Porphyrin and fullerene covalently functionalized graphene hybrid materials with large nonlinear optical properties. J Phys Chem B 113(29):9681–9686
112. Nguyen DA, Lee YR, Raghu AV et al (2009) Morphological and physical properties of a thermoplastic polyurethane reinforced with functionalized graphene sheet. Polym Int 58(4):412–417
113. Yang H, Shan C, Li F et al (2009) Covalent functionalization of polydisperse chemically-converted graphene sheets with amine-terminated ionic liquid. Chem Commun 26:3880–3882
114. Liu Z, Robinson JT, Sun X et al (2008) Pegylated nanographene oxide for delivery of water-insoluble cancer drugs. J Am Chem Soc 130(33):10876–10877
115. Lee SH, Dreyer DR, An J et al (2009) Polymer brushes via controlled, surface-initiated atom transfer radical polymerization (atrp) from graphene oxide. Macromol Rapid Commun 31(3):281–288
116. Stankovich S, Piner RD, Chen X et al (2006) Stable aqueous dispersions of graphitic nanoplatelets via the reduction of exfoliated graphite oxide in the presence of poly(sodium 4-styrenesulfonate). J Mater Chem 16(2):155–158
117. Bai H, Xu Y, Zhao L et al (2009) Non-covalent functionalization of graphene sheets by sulfonated polyaniline. Chem Commun 13:1667–1669
118. Chunder A, Liu J, Zhai L (2010) Reduced graphene oxide/poly(3-hexylthiophene) supramolecular composites. Macromol Rapid Commun 31(4):380–384
119. Hao R, Qian W, Zhang L et al (2008) Aqueous dispersions of TCNQ-anion-stabilized graphene sheets. Chem Commun 48:6576–6578
120. Chunder A, Pal T, Khondaker SI et al (2010) Reduced graphene oxide/copper phthalocyanine composite and its optoelectrical properties. J Phys Chem C 114(35):15129–15135
121. Wojcik A, Kamat PV (2010) Reduced graphene oxide and porphyrin. An interactive affair in 2-D. ACS Nano 4(11):6697–6706

122. Su Q, Pang S, Alijani V et al (2009) Composites of craphene with large aromatic molecules. Adv Mater 21(31):3191–3195
123. Yang Q, Pan X, Huang F et al (2010) Fabrication of high-concentration and stable aqueous suspensions of graphene nanosheets by noncovalent functionalization with lignin and cellulose derivatives. J Phys Chem C 114(9):3811–3816
124. Stankovich S, Dikin DA, Piner RD et al (2007) Synthesis of graphene-based nanosheets via chemical reduction of exfoliated graphite oxide. Carbon 45(7):1558–1565
125. Becerril HA, Mao J, Liu Z et al (2008) Evaluation of solution-processed reduced graphene oxide films as transparent conductors. ACS Nano 2(3):463–470
126. Wang G, Wang B, Park J et al (2009) Highly efficient and large-scale synthesis of graphene by electrolytic exfoliation. Carbon 47(14):3242–3246
127. Somani PR, Somani SP, Umeno M (2006) Planer nano-graphenes from camphor by CVD. Chem Phys Lett 430(1–3):56–59
128. Kotov NA (2006) Materials science: carbon sheet solutions. Nature 442(7100):254–255
129. Cao H, Yu Q, Colby R et al (2010) Large-scale graphitic thin films synthesized on Ni and transferred to insulators: structural and electronic properties. J Appl Phys 107(4):044310
130. Lee S, Lee K, Zhong Z (2010) Wafer scale homogeneous bilayer graphene films by chemical vapor deposition. Nano Lett 10(11):4702–4707
131. Bhaviripudi S, Jia X, Dresselhaus MS et al (2010) Role of kinetic factors in chemical vapor deposition synthesis of uniform large area graphene using copper catalyst. Nano Lett 10(10):4128–4133
132. Gomez De Arco L, Zhang Y, Kumar A et al (2009) Synthesis, transfer, and devices of single- and few-layer graphene by chemical vapor deposition. IEEE Trans Nanotechnol 8(2):135–138
133. Reina A, Jia X, Ho J et al (2009) Large area, few-layer graphene films on arbitrary substrates by chemical vapor deposition. Nano Lett 9(1):30–35
134. Chae SJ, Güneş F, Kim KK et al (2009) Synthesis of large-area graphene layers on poly-nickel substrate by chemical vapor deposition: wrinkle formation. Adv Mater 21(22):2328–2333
135. Yu Q, Lian J, Siriponglert S et al (2008) Graphene segregated on Ni surfaces and transferred to insulators. Appl Phys Lett 93(11):113103
136. Li X, Cai W, Colombo L et al (2009) Evolution of graphene growth on Ni and Cu by carbon isotope labeling. Nano Lett 9(12):4268–4272
137. Li X, Cai W, An J et al (2009) Large-area synthesis of high-quality and uniform graphene films on copper foils. Science 324(5932):1312–1314
138. Cai W, Moore AL, Zhu Y et al (2010) Thermal transport in suspended and supported monolayer graphene grown by chemical vapor deposition. Nano Lett 10(5):1645–1651
139. Lee Y, Bae S, Jang H et al (2010) Wafer-scale synthesis and transfer of graphene films. Nano Lett 10(2):490–493
140. Wei D, Liu Y, Wang Y et al (2009) Synthesis of N-doped graphene by chemical vapor deposition and its electrical properties. Nano Lett 9(5):1752–1758
141. Qu L, Liu Y, Baek JB et al (2010) Nitrogen-doped graphene as efficient metal-free electrocatalyst for oxygen reduction in fuel cells. ACS Nano 4(3):1321–1326
142. Reddy ALM, Srivastava A, Gowda SR et al (2010) Synthesis of nitrogen-doped graphene films for lithium battery application. ACS Nano 4(11):6337–6342
143. Wang JJ, Zhu MY, Outlaw RA et al (2004) Free-standing subnanometer graphite sheets. Appl Phys Lett 85(7):1265–1267
144. Wang J, Zhu M, Outlaw RA et al (2004) Synthesis of carbon nanosheets by inductively coupled radio-frequency plasma enhanced chemical vapor deposition. Carbon 42(14):2867–2872
145. Dato A, Radmilovic V, Lee Z et al (2008) Substrate-free gas-phase synthesis of graphene sheets. Nano Lett 8(7):2012–2016

146. Vitchev R, Malesevic A, Petrov RH et al (2010) Initial stages of few-layer graphene growth by microwave plasma-enhanced chemical vapour deposition. Nanotechnology 21(9):095602
147. Malesevic A, Vitchev R, Schouteden K et al (2008) Synthesis of few-layer graphene via microwave plasma-enhanced chemical vapour deposition. Nanotechnology 19(30):305604
148. Zhu M, Wang J, Holloway BC et al (2007) A mechanism for carbon nanosheet formation. Carbon 45(11):2229–2234
149. Penuelas J, Ouerghi A, Lucot D et al (2009) Surface morphology and characterization of thin graphene films on SiC vicinal substrate. Phys Rev B Condens Matter Mater Phys 79 (3):033408
150. Berger C, Song Z, Li T et al (2004) Ultrathin epitaxial graphite: 2D electron gas properties and a route toward graphene-based nanoelectronics. J Phys Chem B 108(52):19912–19916
151. de Heer WA, Berger C, Wu X et al (2007) Epitaxial graphene. Solid State Commun 143(1–2):92–100
152. Hass J, De Heer WA, Conrad EH (2008) The growth and morphology of epitaxial multilayer graphene. J Phys Condens Matter 20(32):016602
153. Ni ZH, Chen W, Fan XF et al (2008) Raman spectroscopy of epitaxial graphene on a SiC substrate. Phys Rev B Condens Matter Mater Phys 77(11):115416
154. Peng X, Ahuja R (2008) Symmetry breaking induced bandgap in epitaxial graphene layers on SiC. Nano Lett 8(12):4464–4468
155. Zhou SY, Gweon GH, Fedorov AV et al (2007) Substrate-induced bandgap opening in epitaxial graphene. Nat Mater 6(10):770–775
156. Emtsev KV, Bostwick A, Horn K et al (2009) Towards wafer-size graphene layers by atmospheric pressure graphitization of silicon carbide. Nat Mater 8(3):203–207
157. Tedesco JL, Jernigan GG, Culbertson JC et al (2010) Morphology characterization of argon-mediated epitaxial graphene on C-face SiC. Appl Phys Lett 96(22):222103
158. Wu X, Sprinkle M, Li X et al (2008) Epitaxial-graphene/graphene-oxide junction: an essential step towards epitaxial graphene electronics. Phys Rev Lett 101(2):026801
159. Berger C, Song Z, Li X et al (2006) Electronic confinement and coherence in patterned epitaxial graphene. Science 312(5777):1191–1196
160. Kim S, Ihm J, Choi HJ et al (2008) Origin of anomalous electronic structures of epitaxial graphene on silicon carbide. Phys Rev Lett 100(17):176802
161. Varchon F, Feng R, Hass J et al (2007) Electronic structure of epitaxial graphene layers on SiC: effect of the substrate. Phys Rev Lett 99(12):125007
162. Sutter PW, Flege JI, Sutter EA (2008) Epitaxial graphene on ruthenium. Nat Mater 7(5):406–411
163. Vozquez De Parga AL, Calleja F, Borca B et al (2008) Periodically rippled graphene: growth and spatially resolved electronic structure. Phys Rev Lett 100(5):056807
164. Wintterlin J, Bocquet ML (2009) Graphene on metal surfaces. Surf Sci 603(10–12):1841–1852
165. Wu J, Pisula W, Müllen K (2007) Graphenes as potential material for electronics. Chem Rev 107(3):718–747
166. Yang X, Dou X, Rouhanipour A et al (2008) Two-dimensional graphene nanoribbons. J Am Chem Soc 130(13):4216–4217
167. Cai J, Ruffieux P, Jaafar R et al (2010) Atomically precise bottom-up fabrication of graphene nanoribbons. Nature 466(7305):470–473
168. Jiao L, Zhang L, Wang X et al (2009) Narrow graphene nanoribbons from carbon nanotubes. Nature 458(7240):877–880
169. Xie L, Jiao L, Dai H (2010) Selective etching of graphene edges by hydrogen plasma. J Am Chem Soc 132(42):14751–14753
170. Jiao L, Wang X, Diankov G et al (2010) Facile synthesis of high-quality graphene nanoribbons. Nat Nanotechnol 5(5):321–325
171. Shimizu T, Haruyama J, Marcano DC et al (2010) Large intrinsic energy bandgaps in annealed nanotube-derived graphene nanoribbons. Nat Nanotechnol 6(1):45–50

172. Sinitskii A, Dimiev A, Corley DA et al (2010) Kinetics of diazonium functionalization of chemically converted graphene nanoribbons. ACS Nano 4(4):1949–1954
173. Sinitskii A, Dimiev A, Kosynkin DV et al (2010) Graphene nanoribbon devices produced by oxidative unzipping of carbon nanotubes. ACS Nano 4(9):5405–5413
174. Kosynkin DV, Higginbotham AL, Sinitskii A et al (2009) Longitudinal unzipping of carbon nanotubes to form graphene nanoribbons. Nature 458(7240):872–876
175. Elias AL, Botello-Méndez AR, Meneses-Rodríguez D et al (2010) Longitudinal cutting of pure and doped carbon nanotubes to form graphitic nanoribbons using metal clusters as nanoscalpels. Nano Lett 10(2):366–372
176. Stolyarova E, Kwang TR, Ryu S et al (2007) High-resolution scanning tunneling microscopy imaging of mesoscopic graphene sheets on an insulating surface. Proc Natl Acad Sci USA 104(22):9209–9212
177. Hashimoto A, Suenaga K, Gloter A et al (2004) Direct evidence for atomic defects in graphene layers. Nature 430(7002):870–873
178. Xu K, Cao P, Heath JR (2009) Scanning tunneling microscopy characterization of the electrical properties of wrinkles in exfoliated graphene monolayers. Nano Lett 9(12):4446–4451
179. Novoselov KS, Geim AK, Morozov SV et al (2005) Two-dimensional gas of massless dirac fermions in graphene. Nature 438(7065):197–200
180. Bolotin KI, Sikes KJ, Jiang Z et al (2008) Ultrahigh electron mobility in suspended graphene. Solid State Commun 146(9–10):351–355
181. Morozov SV, Novoselov KS, Katsnelson MI et al (2008) Giant intrinsic carrier mobilities in graphene and its bilayer. Phys Rev Lett 100(1):016602
182. Gokus T, Nair RR, Bonetti A et al (2009) Making graphene luminescent by oxygen plasma treatment. ACS Nano 3(12):3963–3968
183. Luo Z, Vora PM, Mele EJ et al (2009) Photoluminescence and band gap modulation in graphene oxide. Appl Phys Lett 94(11):111909
184. Han MY, Ozyilmaz B, Zhang Y et al (2007) Energy band-gap engineering of graphene nanoribbons. Phys Rev Lett 98(20):206805
185. Barone V, Hod O, Scuseria GE (2006) Electronic structure and stability of semiconducting graphene nanoribbons. Nano Lett 6(12):2748–2754
186. Evaldsson M, Zozoulenko IV, Xu H et al (2008) Edge-disorder-induced Anderson localization and conduction gap in graphene nanoribbons. Phys Rev B Condens Matter Mater Phys 78(16):161407
187. Son YW, Cohen ML, Louie SG (2006) Energy gaps in graphene nanoribbons. Phys Rev Lett 97(21)
188. Yan Q, Huang B, Yu J et al (2007) Intrinsic current–voltage characteristics of graphene nanoribbon transistors and effect of edge doping. Nano Lett 7(6):1469–1473
189. Ritter KA, Lyding JW (2009) The influence of edge structure on the electronic properties of graphene quantum dots and nanoribbons. Nat Mater 8(3):235–242
190. Nilsson J, Castro Neto AH, Guinea F et al (2008) Electronic properties of bilayer and multilayer graphene. Phys Rev B Condens Matter Mater Phys 78(4):045405
191. Zhang Y, Tang TT, Girit C et al (2009) Direct observation of a widely tunable bandgap in bilayer graphene. Nature 459(7248):820–823
192. Cervantes-Sodi F, Csányi G, Piscanec S et al (2008) Edge-functionalized and substitutionally doped graphene nanoribbons: electronic and spin properties. Phys Rev B Condens Matter Mater Phys 77(16):165427
193. Meyer JC, Geim AK, Katsnelson MI et al (2007) The structure of suspended graphene sheets. Nature 446(7131):60–63
194. Hwang EH, Adam S, Sarma SD (2007) Carrier transport in two-dimensional graphene layers. Phys Rev Lett 98(18):186806
195. Chen JH, Jang C, Xiao S et al (2008) Intrinsic and extrinsic performance limits of graphene devices on SiO_2. Nat Nanotechnol 3(4):206–209

196. Zhao Q, Gan Z, Zhuang Q (2002) Electrochemical sensors based on carbon nanotubes. Electroanalysis 14(23):1609–1613
197. Shao Y, Yin G, Gao Y et al (2006) Durability study of PtC and PtCNTs catalysts under simulated pem fuel cell conditions. J Electrochem Soc 153(6):A1093–A1097
198. Wang Y, Li Y, Tang L et al (2009) Application of graphene-modified electrode for selective detection of dopamine. Electrochem Commun 11(4):889–892
199. Lin Y, Cui X, Ye X (2005) Electrocatalytic reactivity for oxygen reduction of palladium-modified carbon nanotubes synthesized in supercritical fluid. Electrochem Commun 7(3):267–274
200. Alwarappan S, Erdem A, Liu C et al (2009) Probing the electrochemical properties of graphene nanosheets for biosensing applications. J Phys Chem C 113(20):8853–8857
201. Groves MN, Chan ASW, Malardier-Jugroot C et al (2009) Improving platinum catalyst binding energy to graphene through nitrogen doping. Chem Phys Lett 481(4–6):214–219
202. Nair RR, Blake P, Grigorenko AN et al (2008) Fine structure constant defines visual transparency of graphene. Science 320(5881):1308
203. Kravets VG, Grigorenko AN, Nair RR et al (2010) Spectroscopic ellipsometry of graphene and an exciton-shifted van hove peak in absorption. Phys Rev B Condens Matter Mater Phys 81(15):155413
204. Bonaccorso F, Sun Z, Hasan T et al (2010) Graphene photonics and optoelectronics. Nat Photonics 4(9):611–622
205. Eda G, Lin YY, Mattevi C et al (2010) Blue photoluminescence from chemically derived graphene oxide. Adv Mater 22(4):505–509
206. Tsoukleri G, Parthenios J, Papagelis K et al (2009) Subjecting a graphene monolayer to tension and compression. Small 5(21):2397–2402
207. Ni ZH, Yu T, Lu YH et al (2008) Uniaxial strain on graphene: Raman spectroscopy study and band-gap opening. ACS Nano 2(11):2301–2305
208. Pop E, Mann D, Wang Q et al (2006) Thermal conductance of an individual single-wall carbon nanotube above room temperature. Nano Lett 6(1):96–100
209. Nika DL, Pokatilov EP, Askerov AS et al (2009) Phonon thermal conduction in graphene: role of umklapp and edge roughness scattering. Phys Rev B Condens Matter Mater Phys 79(15):155413
210. Jiang JW, Lan J, Wang JS et al (2011) Isotopic effects on the thermal conductivity of graphene nanoribbons: localization mechanism. J Appl Phys 107(5):054314
211. Jung I, Dikin D, Park S et al (2008) Effect of water vapor on electrical properties of individual reduced graphene oxide sheets. J Phys Chem C 112(51):20264–20268
212. Lu G, Ocola LE, Chen J (2009) Gas detection using low-temperature reduced graphene oxide sheets. Appl Phys Lett 94(8):083111
213. Lu G, Ocola LE, Chen J (2009) Reduced graphene oxide for room-temperature gas sensors. Nanotechnology 20(44):445502–445510
214. Robinson JT, Perkins FK, Snow ES et al (2008) Reduced graphene oxide molecular sensors. Nano Lett 8(10):3137–3140
215. Dua V, Surwade SP, Ammu S et al (2010) All-organic vapor sensor using inkjet-printed reduced graphene oxide. Angew Chem Int Ed 49(12):2154–2157
216. Dan Y, Lu Y, Kybert NJ et al (2009) Intrinsic response of graphene vapor sensors. Nano Lett 9(4):1472–1475
217. Ganhua L, Leonidas EO, Junhong C (2009) Gas detection using low-temperature reduced graphene oxide sheets. Appl Phys Lett 94:083111
218. Sundaram RS, Gomez-Navarro C, Balasubramanian K et al (2008) Electrochemical modification of grapheme. Adv Mater 20(16):3050–3053
219. Lu G, Yu K, Ocola LE et al (2011) Ultrafast room temperature NH_3 sensing with positively gated reduced graphene oxide field-effect transistors. Chem Commun 47(27):7761–7763
220. Tang L, Wang Y, Li Y et al (2009) Preparation, structure, and electrochemical properties of reduced graphene sheet films. Adv Funct Mater 19(17):2782–2789

221. Kang X, Wang J, Wu H et al (2010) A graphene-based electrochemical sensor for sensitive detection of paracetamol. Talanta 81(3):754–759
222. Parvin MH (2011) Graphene paste electrode for detection of chlorpromazine. Electrochem Commun 13(4):366–369
223. Li F, Li J, Feng Y et al (2011) Electrochemical behavior of graphene doped carbon paste electrode and its application for sensitive determination of ascorbic acid. Sens Actuators B Chem 157(1):110–114
224. Wang J, Yang S, Guo D et al (2009) Comparative studies on electrochemical activity of graphene nanosheets and carbon nanotubes. Electrochem Commun 11(10):1892–1895
225. Li J, Guo S, Zhai Y et al (2009) Nafion-graphene nanocomposite film as enhanced sensing platform for ultrasensitive determination of cadmium. Electrochem Commun 11(5):1085–1088
226. Li J, Guo S, Zhai Y et al (2009) High-sensitivity determination of lead and cadmium based on the nafion-graphene composite film. Anal Chim Acta 649(2):196–201
227. Ohno Y, Maehashi K, Matsumoto K (2010) Chemical and biological sensing applications based on graphene field-effect transistors. Biosens Bioelectron 26(4):1727–1730
228. Sofue Y, Ohno Y, Maehashi K et al (2011) Highly sensitive electrical detection of sodium ions based on graphene field-effect transistors. Jpn J Appl Phys 50(6 Pt 2):06GE07
229. Sudibya HG, He Q, Zhang H et al (2011) Electrical detection of metal ions using field-effect transistors based on micropatterned reduced graphene oxide films. ACS Nano 5(3):1990–1994
230. Shafiei M, Spizzirri PG, Arsat R et al (2010) Platinum/graphene nanosheet/SiC contacts and their application for hydrogen gas sensing. J Phys Chem C 114(32):13796–13801
231. Arsat R, Breedon M, Shafiei M et al (2009) Graphene-like nano-sheets for surface acoustic wave gas sensor applications. Chem Phys Lett 467(4–6):344–347
232. Huang B, Li Z, Liu Z et al (2008) Adsorption of gas molecules on graphene nanoribbons and its implication for nanoscale molecule sensor. J Phys Chem C 112(35):13442–13446
233. Leenaerts O, Partoens B, Peeters FM (2008) Adsorption of H_2O, NH_3, CO, NO_2, and NO on graphene: a first-principles study. Phys Rev B Condens Matter Mater Phys 77(12):125416
234. Schedin F, Geim AK, Morozov SV et al (2007) Detection of individual gas molecules adsorbed on graphene. Nat Mater 6(9):652–655
235. Li X, Wang H, Robinson JT et al (2009) Simultaneous nitrogen doping and reduction of graphene oxide. J Am Chem Soc 131(43):15939–15944
236. McCreery RL (2008) Advanced carbon electrode materials for molecular electrochemistry. Chem Rev 108(7):2646–2687
237. Kim SN, Rusling JF, Papadimitrakopoulos F (2007) Carbon nanotubes for electronic and electrochemical detection of biomolecules. Adv Mater 19(20):3214–3228
238. Martin P (2009) Electrochemistry of gaphene: new horizons for sensing and energy storage. Chem Rec 9(4):211–223
239. Jia J, Kato D, Kurita R et al (2007) Structure and electrochemical properties of carbon films prepared by a electron cyclotron resonance sputtering method. Anal Chem 79(1):98–105
240. Niwa O, Jia J, Sato Y et al (2006) Electrochemical performance of angstrom level flat sputtered carbon film consisting of sp2 and sp3 mixed bonds. J Am Chem Soc 128(22):7144–7145
241. Shang NG, Papakonstantinou P, McMullan M et al (2008) Catalyst-free efficient growth, orientation and biosensing properties of multilayer graphene nanoflake films with sharp edge planes. Adv Funct Mater 18(21):3506–3514
242. Fischer AE, Show Y, Swain GM (2004) Electrochemical performance of diamond thin-film electrodes from different commercial sources. Anal Chem 76(9):2553–2560
243. Guo HL, Wang XF, Qian QY et al (2009) A green approach to the synthesis of graphene nanosheets. ACS Nano 3(9):2653–2659
244. Ramesha GK, Sampath NS (2009) Electrochemical reduction of oriented graphene oxide films: an in situ Raman spectroelectrochemical study. J Phys Chem C 113(19):7985–7989

245. Shao Y, Wang J, Engelhard M et al (2009) Facile and controllable electrochemical reduction of graphene oxide and its applications. J Mater Chem 20(4):743–748
246. Kefala G, Economou A, Voulgaropoulos A (2004) A study of nafion-coated bismuth-film electrodes for the determination of trace metals by anodic stripping voltammetry. Analyst 129(11):1082–1090
247. Zhu L, Tian C, Yang R et al (2008) Anodic stripping voltammetric determination of lead in tap water at an ordered mesoporous carbon/nafion composite film electrode. Electroanalysis 20(5):527–533
248. Xu H, Zeng L, Xing S et al (2008) Ultrasensitive voltammetric detection of trace lead(ii) and cadmium(ii) using MWCNTs-nafion/bismuth composite electrodes. Electroanalysis 20(24):2655–2662
249. Lin WJ, Liao CS, Jhang JH et al (2009) Graphene modified basal and edge plane pyrolytic graphite electrodes for electrocatalytic oxidation of hydrogen peroxide and B-nicotinamide adenine dinucleotide. Electrochem Commun 11(11):2153–2156
250. Liu H, Gao J, Xue MQ et al (2009) Processing of graphene for electrochemical application: noncovalently functionalize graphene sheets with water-soluble electroactive methylene green. Langmuir 25:12006–12010
251. Du M, Yang T, Jiao K (2010) Immobilization-free direct electrochemical detection for DNA specific sequences based on electrochemically converted gold nanoparticles/graphene composite film. J Mater Chem 20(41):9253–9260
252. Du Y, Guo S, Dong S et al (2011) An integrated sensing system for detection of DNA using new parallel-motif DNA triplex system and graphene-mesoporous silica-gold nanoparticle hybrids. Biomaterials 32(33):8584–8592
253. Li L, Du Z, Liu S et al (2010) A novel nonenzymatic hydrogen peroxide sensor based on MnO_2/graphene oxide nanocomposite. Talanta 82(5):1637–1641
254. Mao S, Yu K, Lu G et al (2011) Highly sensitive protein sensor based on thermally-reduced graphene oxide field-effect transistor. Nano Res 4(10):921–930
255. Wu H, Wang J, Kang X et al (2009) Glucose biosensor based on immobilization of glucose oxidase in platinum nanoparticles/graphene/chitosan nanocomposite film. Talanta 80(1):403–406
256. Lu J, Drzal LT, Worden RM et al (2007) Simple fabrication of a highly sensitive glucose biosensor using enzymes immobilized in exfoliated graphite nanoplatelets nafion membrane. Chem Mater 19(25):6240–6246
257. Wang Z, Zhou X, Zhang J et al (2009) Direct electrochemical reduction of single-layer graphene oxide and subsequent functionalization with glucose oxidase. J Phys Chem C 113(32):14071–14075
258. Mao S, Lu G, Yu K et al (2010) Specific protein detection using thermally reduced graphene oxide sheet decorated with gold nanoparticle-antibody conjugates. Adv Mater 22(32):3521–3526
259. Dong X, Shi Y, Huang W et al (2010) Electrical detection of DNA hybridization with single-base specificity using transistors based on CVD-grown graphene sheets. Adv Mater 22(14):1649–1653
260. Ohno Y, Maehashi K, Matsumoto K (2010) Label-free biosensors based on aptamer-modified graphene field-effect transistors. J Am Chem Soc 132(51):18012–18013
261. Ohno Y, Maehashi K, Inoue K et al (2011) Label-free aptamer-based immunoglobulin sensors using graphene field-effect transistors. Jpn J Appl Phys 50(7 Pt 1):070120
262. Wang Y, Lu J, Tang L et al (2009) Graphene oxide amplified electrogenerated chemiluminescence of quantum dots and its selective sensing for glutathione from thiol-containing compounds. Anal Chem 81(23):9710–9715
263. Pumera M, Scipioni R, Iwai H et al (2009) A mechanism of adsorption of b-nicotinamide adenine dinucleotide on graphene sheets: Experiment and theory. Chem Eur J 15(41):10851–10856

264. Musameh M, Wang J, Merkoci A et al (2002) Low-potential stable NADH detection at carbon-nanotube-modified glassy carbon electrodes. Electrochem Commun 4(10):743–746
265. Valentini F, Amine A, Orlanducci S et al (2003) Carbon nanotube purification: preparation and characterization of carbon nanotube paste electrodes. Anal Chem 75(20):5413–5421
266. Banks CE, Compton RG (2005) Exploring the electrocatalytic sites of carbon nanotubes for NADH detection: an edge plane pyrolytic graphite electrode study. Analyst 130(9):1232–1239
267. Ghindilis AL, Atanasov P, Wilkins E (1997) Enzyme-catalyzed direct electron transfer: fundamentals and analytical applications. Electroanalysis 9(9):661–674
268. Leger C, Bertrand P (2008) Direct electrochemistry of redox enzymes as a tool for mechanistic studies. Chem Rev 108(7):2379–2438
269. Armstrong FA, Hill HAO, Walton NJ (1988) Direct electrochemistry of redox proteins. Acc Chem Res 21(11):407–413
270. Zhang W, Li G (2004) Third-generation biosensors based on the direct electron transfer of proteins. Anal Sci 20(4):603–609
271. Wu Y, Hu S (2007) Biosensors based on direct electron transfer in redox proteins. Microchim Acta 159(1–2):1–17
272. Yao YL, Shiu KK (2008) Direct electrochemistry of glucose oxidase at carbon nanotube-gold colloid modified electrode with poly(diallyldimethylammonium chloride) coating. Electroanalysis 20(14):1542–1548
273. Katz E, Willner I (2004) Integrated nanoparticle-biomolecule hybrid systems: synthesis, properties, and applications. Angew Chem Int Ed Engl 43:6042–6108
274. Sarma AK, Vatsyayan P, Goswami P et al (2009) Recent advances in material science for developing enzyme electrodes. Biosens Bioelectron 24(8):2313–2322
275. Dai ZH, Ni J, Huang XH et al (2007) Direct electrochemistry of glucose oxidase immobilized on a hexagonal mesoporous silica-mcm-41 matrix. Bioelectrochemistry 70(2):250–256
276. Shan C, Yang H, Song J et al (2009) Direct electrochemistry of glucose oxidase and biosensing for glucose based on graphene. Anal Chem 81(6):2378–2382
277. Guiseppi-Elie A, Lei C, Baughman RH (2002) Direct electron transfer of glucose oxidase on carbon nanotubes. Nanotechnology 13(5):559–564
278. Deng C, Chen J, Chen X et al (2008) Direct electrochemistry of glucose oxidase and biosensing for glucose based on boron-doped carbon nanotubes modified electrode. Biosens Bioelectron 23(8):1272–1277
279. Cai C, Chen J (2004) Direct electron transfer of glucose oxidase promoted by carbon nanotubes. Anal Biochem 332(1):75–83
280. Liu G, Lin Y (2006) Amperometric glucose biosensor based on self-assembling glucose oxidase on carbon nanotubes. Electrochem Commun 8(2):251–256
281. Lin Y, Lu F, Tu Y et al (2004) Glucose biosensors based on carbon nanotube nanoelectrode ensembles. Nano Lett 4(2):191–195
282. Rubianes MD, Rivas GA (2003) Carbon nanotubes paste electrode. Electrochem Commun 5(8):689–694
283. Wu L, Zhang X, Ju H (2007) Amperometric glucose sensor based on catalytic reduction of dissolved oxygen at soluble carbon nanofiber. Biosens Bioelectron 23(4):479–484
284. Zhou M, Shang L, Li B et al (2008) Highly ordered mesoporous carbons as electrode material for the construction of electrochemical dehydrogenase- and oxidase-based biosensors. Biosens Bioelectron 24(3):442–447
285. Chen H, Müller MB, Gilmore KJ et al (2008) Mechanically strong, electrically conductive, and biocompatible graphene paper. Adv Mater 20(18):3557–3561

Molecular Imprinting Technique for Biosensing and Diagnostics

Nenad Gajovic-Eichelmann, Umporn Athikomrattanakul, Decha Dechtrirat, and Frieder W. Scheller

Abstract An introduction into the growing field of molecular imprinting is given, and some principle questions for the design of novel artificial molecular recognition polymers (MIPs) are raised. The limitations of the classical non-covalent imprinting approach are discussed in a brief form. Some novel strategies for the molecular imprinting of macromolecules, especially proteins, are reviewed, as well as new concepts for the integration with transducers and sensors. Two case studies from our own laboratory highlight the question of improving the performance of MIPs by the use of complementary functional monomers and demonstrate a new electrochemical approach to the imprinting of peptides and proteins.

Keywords Electropolymerisation, Functional monomers, Molecular imprinting, Polymers, Protein imprinting

Contents

N. Gajovic-Eichelmann (✉), U. Athikomrattanakul, D. Dechtrirat and F.W. Scheller
Fraunhofer Institute of Biomedical Technology, Am Mühlenberg 13, 14476 Potsdam-Golm, Germany
e-mail: nenad.gajovic@ibmt.fraunhofer.de; nenad.gajovic@ibmt.fhg.de

A. Tuantranont (ed.), *Applications of Nanomaterials in Sensors and Diagnostics*, Springer Series on Chemical Sensors and Biosensors (2013) 14: 143–170
DOI 10.1007/5346_2012_44,
Published online: 8 December 2012

1 General Concept of Molecular Imprinting

Biomimetic, artificial receptor materials with molecular recognition capacity like molecularly imprinted polymers (MIP) have greatly broadened the scope of biosensors and bioanalysis. Novel applications comprise the specific recognition of proteins and peptides [1], nucleotides [2] and catalytically active MIPs [3] besides the well-established imprinting of low-molecular-weight metabolites and drugs [4]. The classical, non-covalent bulk imprinting approach, pioneered by Mosbach [5] as a generalisation of the earlier work by Wulff [6], has turned out the most versatile strategy to the imprinting of small molecules like pesticides, metabolites and drugs [4]. The common basic principle is the mixing of the target molecule (serving as the template) with one or more functional monomers and one or more cross-linking monomers, allowing for the formation of molecular complexes or adducts in solution (pre-arrangement step), prior to the polymerisation step. It is important to realise that the major constituent of nearly all MIPs is the cross-linking monomer, typically making up for 80–90% of the polymer weight. In the resulting porous polymers, being highly cross-linked and rigid, a fraction of the molecular complexes present in the pre-polymerisation mix will be preserved in a quasi-frozen state. In the classical approach, the polymer monoliths are crushed, ground to particles and wet-sieved to achieve a size distribution of typically 25–50 μm. The ground particles may typically have a large surface area of 100–500 $m^2\ g^{-1}$. Upon extraction of the template molecules from the polymer particles, it is believed that specific binding cavities are left behind at the surface, which allow for the selective rebinding of the target [5] (Fig. 1).

1.1 First Considerations When Designing a New MIP

For obvious reasons, the strategy for MIP synthesis will depend on the target to be imprinted.

Is it soluble in common solvents used for MIP preparation? Is it chemically stable? Is it polar or does it bear a charge? Depending on the target, one or few functional monomers will be selected which can form a complex with it. The interactions to be exploited may include ionic interaction, hydrogen bonds, van der Waals forces, hydrophobic interaction, π–π bonds and, as a special case, reversible covalent bonds. The predominant requirement for a proper functional

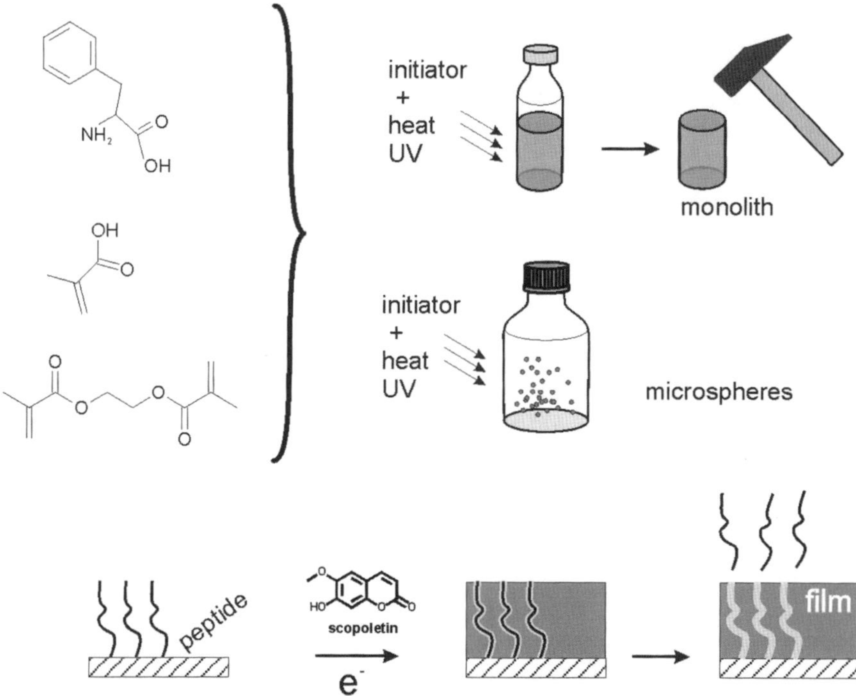

Fig. 1 Schematic of molecular imprinting process, here for the target DL-phenylalanine using methacrylic acid as the functional monomer and EGDMA as cross-linking monomer (*top*: bulk polymer imprinting, intermediate: micro/nanoparticle imprinting). *Lower cartoon* illustrates our novel concept for electrochemical surface imprinting of peptides and proteins

monomer is a high binding constant with the target. It can be determined by methods like NMR titration, isothermal titration calorimetry (ITC) or titration in combination with UV–vis or fluorescence spectroscopy. It is very advisable to determine the binding constant under the same conditions as applied for synthesis, i.e. using the same solvent and the same temperature. The binding constant will provide us with a first hint on the expected performance of the MIP. A few hundreds of monomers are commercially available, with some of them, e.g. methacrylic acid, being used in many applications [4]. If none of these can fulfil the requirements, a synthesis of a custom monomer or a monomer library may be justified. Sellergren's group, amongst others, has recently introduced novel classes of strong binding monomers, which may form stoichiometric complexes even in polar solvents and water [7]. It has become common practice among some MIP researchers to follow a rational design approach [1]. This is often synonymous with selecting the best monomer out of a library of commercial or custom monomers based on molecular modelling results [8]. Several software packages are commercially available for that purpose. Combinatorial libraries of monomers have been developed in analogy to the strategies used in pharma-screening [9].

Once the functional monomer(s) has been found, a cross-linking monomer has to be selected. Although constituting 80–90% of the polymer mass, the cross-linking monomer seems to be less critical for the MIP performance and the vast majority of MIPs use one of these: ethylene glycole dimethacrylate (EGDMA, a polar, divalent cross linker), trimethylolpropane trimethacrylate (TRIM, trivalent cross linker of intermediate polarity), pentaerythritol triacrylate (PETRA, a polar, trivalent cross linker) or divinyl benzene (DVB, an apolar, divalent cross linker). The molar ratios of target:function monomer:cross linker are seldom optimised and most researchers stick to an established recipe (e.g. 1:1:12 or 1:4:20), which seem to give satisfactory results in most cases.

The solvent (porogen) used for polymerisation is critical to the imprinting process: it should be a good solvent for the monomers as well as the target and it should provide a porous polymer. Toluene, acetonitrile and DMSO have often been used. Some studies have been conducted to show that polar solvents will enhance the water compatibility of the MIP [10]. Finally, the cost of the solvent (and its disposal) may dominate the cost of manufacture of an MIP, especially if microspheres or nanoparticles are prepared.

Besides the classical formulation as a bulk monolith, MIPs can be prepared as microspheres [11], nanoparticles or nanogels [12], nanofibres or -filaments [13], inorganic/organic composites [14, 15], thin films [16] or as imprinted hydrogels [17, 18]. The chosen formulation will depend on the application: while ground bulk MIPs lend themselves particularly for solid phase extraction (SPE), thin films or hydrogels may be more easily integrated with biosensors. Inorganic/organic composites may show improved performance in water and nanoparticles or nanofibres may offer higher binding site homogeneity and an improved mass transport, resulting in faster equilibration times. Furthermore, micro-or nanoparticles may be the only option for many "plastic antibody" assay formats, because they will not sediment [19, 20].

Several thousands of MIP-related papers have been published in the last four decades [4], most of them targeting small molecule drugs and metabolites, e.g. antibiotics, steroid hormones, glucose [21] or pesticides like 2,4-D [22] in an organic solvent. The MIP database, run by MJ Whitcombe, is a valuable source of papers and patents on MIPs and can be screened by the target, author or year of publication (www.mipdatabase.com).

1.2 Performance Assessment and Characterisation of MIPs

MIPs are meant to be highly selective adsorbents, so their adsorption properties have to be investigated. All adsorbents are characterised by their adsorption isotherms (for their target and related compounds), specific surface area, porosity and pore size distribution, their solvent compatibility, swelling behaviour, regeneration ability, etc. Adsorption isotherms are most easily determined in batch binding assays [23]. A limited amount of the MIP is added to a dilute solution of the target at a known starting concentration, and, after the binding equilibrium has been reached, the equilibrium concentration of the target, i.e. the non-bound fraction, is determined by UV–vis spectroscopy, radiolabelled ligands, ^{1}H-NMR titration, FTIR, fluorimetry, ITC or other suitable methods. The adsorbed amount is calculated, and plotted vs. the equilibrium concentration (not the starting concentration!) giving the adsorption isotherm. Adsorption isotherms of MIPs are typically non-linear and complex and will require the use of a scientific software program for quantitative analysis. By fitting the data to a simple (e.g. Langmuir or Freundlich) or more complex isotherm (Bi-Langmuir, Langmuir-Freundlich, etc.), the binding constant and binding capacity can be determined. Batch binding experiments are time and material consuming, however. Alternative methods for the determination of the adsorption parameters have been derived from chromatography. These include zonal analysis (a small amount of the target is loaded on the column and the eluting peak shape is evaluated) and frontal analysis (the column is slightly overloaded with sample, and the breakthrough curve is detected) [24]. These methods typically last less than an hour for one experiment, but it may be expensive to prepare a sufficient amount of MIP to fill the column, and the homogeneous packing of a chromatography column with an irregularly shaped material like ground MIP is very challenging and tedious.

The dominant binding forces in the prepolymer mix and in the polymer were investigated by O'Mahony et al. using ^{1}H-NMR, FTIR and Raman spectroscopy [25, 26]. By these methods they could discern the influence of monomers and solvents on the ionic and hydrophobic interactions, which may eventually lead to a true rational design of MIPs. With the advent of MIP nanomaterials, analytical techniques like Raman spectroscopy, AFM and SPR are commonly used to investigate the release of templates and the rebinding [27].

1.3 MIP-Based Nanomaterials

A large variety of alternative formulations for MIPs have been published in recent years. Protocols for the synthesis of monodisperse microspheres (ca. 1–100 μm in diameter) have been developed to improve the use of MIPs as chromatography materials or as analogues for immunological binding assays. Haginaka has recently reviewed the topic of monodisperse imprinted microparticles as affinity-based chromatography media [28]. Non-porous microspheres will typically have specific surface areas in the order of 10–100 $m^2\ g^{-1}$, which is less than a porous bulk MIP (100–500 $m^2\ g^{-1}$) and may not be sufficient for an adsorbent. Nanospheres (ca. 1–100 nm) will have a much larger surface area and they may be used in binding assays similar to immunoassays, because they do not sediment. The recent developments in the field of micro- and nanosized MIP materials and 2-3D patterned MIP structures have been covered in the review by Tokonami et al. [29]. The recent developments in nanostructured MIPs and core-shell nanoparticle approaches are covered in the review by Guan et al. [12]. Imprinted core-shell nanoparticles are composed of two types of materials, often a silica nanoparticle core, coated by a polymer shell, and they can combine material properties that are difficult to achieve with straight polymer nanoparticles. Nanofibres or nanofilaments offer a high surface area, can sometimes be prepared directly at sensor surfaces and may be easier to process and handle than nanoparticles [13].

Novel polymerisation strategies, like living free radical polymerisation, including reversible addition-fragmentation chain transfer (RAFT), metal-catalysed atom transfer radical polymerisation (ATRP), nitroxide-mediated polymerisation (NMP) for the preparation of defined MIP nanomaterials, have been reviewed by Chen et al. [30]. In a recent study, Oxelbark et al. have compared the performance of bupivacaine-imprinted polymers, which were prepared as crushed monoliths, microspheres, silica-based composites and capillary monolith formats, by chromatographic methods [14]. Interestingly, the imprinting factors were the highest and the aqueous compatibility was best with the classical ground monolith particles (i.e. bulk imprinting) approach. From a commercialisation perspective, ground monolith particles may be very interesting, because they can be produced at low cost, with little starting materials and solvents, and inexpensive processing (grinding, wet-sieving, etc.).

1.4 Classical and Novel Applications of MIPs

Molecularly imprinted solid phase extraction materials (MISPE) have been in the focus of MIP technology from the beginning and are still the dominating application area in the literature. Extraction of drugs from complex mixtures, metal ions from wastewater, toxins from raw materials, pesticides from soil and water are a few examples of useful applications for MIP sorbents. In their recent review, Bui and Haupt offer an introduction into the design of MISPE protocols, discerning the different strategies of selective loading and selective washing, and cover the most recent developments in this field [10].

Modern sensor applications using MIP nanomaterials are reviewed elsewhere [31]. Recently, a number of papers have addressed separation processes with imprinted membranes [32]. Extraction of toxic heavy metal ions like nickel, lead, iron or copper has been demonstrated.

Medical applications of MIPs have been the exception, but a recent paper by Hoshino et al. raised much interest [33]. Melittin, the toxic peptide in bee venom, was imprinted in hydrogel nanoparticles. These MIP particles, when administered to living mice, cleared previously injected melittin and scavenged the immunological reaction, like a real antibody. This groundbreaking paper may pave the way for other medical applications of MIPs.

Commercial MIP applications. The first commercial MISPE solutions are offered by Biotage AB, Sweden, who has recently acquired Lund-based MIP Technologies AB. Their MIP library is marketed as a screening platform (ExploraSep™) for the chemical and pharmaceutical industry and for sample preparation by solid phase extraction. Group-selective MIPs for acidic, neutral and basic compounds, diols and aromats, are offered in a 96-well plate format, as columns or bulk materials for scale-up.

Semorex Inc., a US- and Israel-based start-up company, has announced the launch of a medical MIP product, a phosphate binding MIP for the treatment of patients with end-stage renal disease, based on the company's proprietary technology.

Polyintell is a French start-up company specialising in MIP-based solid-phase extraction products for various applications in food & feed safety (e.g. mycotoxins), environmental analysis, diagnostics and drugs of abuse detection.

1.5 Limitations of the Non-covalent Imprinting Approach

Although being most versatile, the non-covalent imprinting approach has its limitations: typically only a small fraction of the exposed surface will be imprinted, while the bulk material, which is mainly constituted of the cross-linking monomers, will permit the adsorption of various hydrophobic and polar species, leading to a significant unspecific binding. For a proper assessment of the imprinting effect, it is necessary to determine the binding affinities and binding capacities of the specific as well as the unspecific binding sites, e.g. by batch binding experiments and proper evaluation of the resulting binding isotherms.

The second problem is the heterogeneity of specific binding sites, resulting in a wide distribution of adsorption energies and, hence, binding affinities among the population of binding sites. It is a consequence of the template-monomer complexes in the pre-polymerisation mix being more or less ordered, resulting in a variation in the number and orientation of weak interactions between them [34]. This inherent heterogeneity leads to complex binding isotherms, and ultimately to a decreased selectivity in extraction, sensing or chromatography applications. This characteristic can be improved by the covalent imprinting approach, pioneered by Wulff [35], which utilises reversible, covalent bonds between templates and functional monomers. Such adducts are generally more stable than their non-covalent counterparts, with negligible dispersion in bond energies. Utilising this approach and using a boronic acid derivative as the covalent receptor group, our group was able to synthesise an MIP with uniform binding sites for D-fructose, a common monosaccharide [36]. The concept of stoichiometric imprinting tries to combine the advantages of covalent imprinting with non-covalent synthesis protocol [37].

Another problem, especially with crushed monolith preparations of MIPs, is the incomplete extraction of the template after polymerisation. A significant portion of the template molecules (up to 50%) will be deeply entrapped inside the bulk material, and may eventually leach out, compromising the performance in many analytical applications like SPE and chromatography. Ellwanger et al. have explored different extraction methods including Soxhlet and microwave-assisted extraction (MAE) for the complete removal of the template from a clenbuterol-imprinted polymer and found MAE to be superior to the commonly used Soxhlet extraction for this particular target [38]. In general, templates can be more easily extracted from MIPs prepared in the form of nanoparticles or thin films [10].

The most severe limitation, however, is the often poor (and virtually unpredictable) water compatibility of most MIPs prepared by the conventional non-covalent technique. Many, if not most, MIPs are meant to be used in purely aqueous solutions or mixtures of organic solvents and buffer. But only few conventional MIPs have been shown to actually work in solvent mixtures with a high water content [7, 14, 24] or in purely aqueous solutions [39]. The reasons for this are diverse: although rarely mentioned in the MIP literature, the combined effect of material hydrophobicity and porosity (i.e. roughness) may result in a superhydrophobic, non-wettable surface [40]. Addition of a small fraction (10%) of a polar solvent like methanol reduces the surface energy of water by ca. 30% and may be sufficient to improve the wetting behaviour [3, 36]. Secondly water may compete for hydrogen bonds and van der Waals bonds at the imprinted binding sites, effectively excluding the target molecules from the interaction sites [10].

1.6 Covalent Imprinting Technique for Saccharide Detection

From a historical perspective, covalent molecular imprinting is an even older technique than non-covalent imprinting. The pioneering work of Wulff [37] has paved the way for this technology. Conceptually the only difference is the use of template-functional monomer pairs, which can form one or more reversible covalent bonds. The vast majority of papers have centred around saccharides and other chiral molecules. Boronic acids are well known for their ability to form cyclic diesters with diols and saccharides in highly alkaline solutions. Vinylphenylboronic acid (VBA) and amino-derivatives thereof have been used as functional monomers for the synthesis of MIPs for saccharides such as D-glucose, D-fructose, D-mannose and derivatives thereof [41]. Wulff et al. have improved the binding constants by using a class of amino-derivatives of VBA as functional monomers, because they have proven to offer higher binding constants at a lower pH than pH 11 [42]. Rajkumar et al. have used this functional monomer for the imprinting of fructosylvalin, the characteristically modified amino acid in glycated haemoglobin (HbA1c) [43] and have used this MIP in a thermometric biosensor [44]. HbA1c is an important biomarker for the assessment of diabetic patients. Using an MIP thermistor as a label-free biosensor system, they could detect fructosylvaline in aqueous solutions with 10% MeOH at pH 8. To improve the performance of boronic acid-based MIPs at a neutral pH, Schumacher and Grüneberger have introduced a novel functional monomer, which is based on the benzoboroxole moiety [36]. Benzoboroxole has been shown to offer much higher binding constants neutral pH with saccharides than VBA [45–47]. In their study with a D-fructose-polymer, the authors showed that their new MIP outperforms the formerly used VBA- or amino-VBA-based polymers at a neutral pH both in terms of binding capacity and binding affinity [36]. Interestingly, the MIP was highly selective: saccharose, a disaccharide composed of D-glucose and D-fructose, was only weakly bound, and the binding capacity for the enantiomer, L-fructose, was reduced by more than 50%.

1.7 Imprinting of Proteins and Peptides

The previously described techniques of bulk molecular imprinting using non-covalent and covalent target/monomer interactions (and the modifications resulting in micro- or nanoparticulate materials) were shown to produce materials able to selectively recognise a vast number of small target molecules in organic solvents. Obviously, proteins, and to a minor extent peptides, are an extremely important target class in life sciences. Synthetic polymer-based molecular receptors for peptides and proteins would open up a wide range of possibilities for new applications in biotechnology, diagnostics and chemistry, complementary to and extending beyond the scope of current antibody technology. Protein purification in one step, diagnostic tests in aggressive sample matrices and novel biosensors are some of the novel possibilities. Therapeutic MIPs in analogy to therapeutic antibodies may eventually have the largest commercial potential of all MIP applications. Compared to small molecule imprinting, where the first commercial products have recently entered the market, we have just taken the first steps along the road towards "plastic antibodies" able to recognise any desired protein. Macromolecular targets like proteins pose additional difficulties, and classical imprinting approaches have mostly failed to deliver satisfactory results. Several reasons for this have been identified: proteins are in general not compatible with organic solvents and will quickly denature in prepolymerisation mixtures based on organic solvents as porogens. Furthermore, mass transport of macromolecules may be prohibitively slow in a highly cross-linked polymer matrix. As a consequence, the protein templates may be difficult to remove from the binding pockets, thus reducing the surface sites available for rebinding [1]. Analogously, rebinding may be too slow for practical applications. Furthermore, the rigid structure of an MIP, lacking the segmental motions of conventional linear polymers, may effectively prevent proteins from entering the specific binding pockets, and unspecific adsorption at the polymer surface may be the predominant binding mode. Finally, the mere availability of sufficient amounts of a highly purified protein target is not granted, and only few proteins are commercially available and low priced.

Several alternative methods have been proposed to address these challenges. An overview of this quickly expanding new field of research can be found in the excellent review by Whitcombe et al. [1]. The technological approaches may be divided into three main strategies: the imprinting of hydrogels as highly flexible, hydrated substitutes for rigid, cross-linked bulk polymers; secondly, the surface imprinting approach, which relies on the imprinting of the polymer surface only, usually applying the polymer as a thin film and the target protein immobilised on a solid surface; and, thirdly, the epitope imprinting approach, which relies on a linear peptide epitope as the target, mimicking the whole protein to be imprinted. The latter approach does not require the protein as a physical material, but rather the sequence and tertiary structure of the protein have to be known (the absolute minimum requirement would be the N- or C-terminal amino acid sequence). Which of the different strategies works best for a given protein target cannot be predicted a priori.

By applying the bulk imprinting strategy in an aqueous solution comprising dissolved horse myoglobin as the protein template together with the water-soluble neutral functional monomer acrylamide and the cross linker *N*-*N'*-methylene bisacrylamide at a low concentration, Hjerten was able to prepare a protein-imprinted hydrogel [48]. This highly hydrated, sparingly cross-linked and macroporous material proved to be highly selective for horse myoglobin in buffered solutions, as it could discriminate between the target and whale myoglobin, which is a very similar protein. The mechanical stability of the sparingly cross-linker material was insufficient, however. Moreover, being a hydrogel, its swelling properties, and, consequently, its porosity will inevitably depend on the ionic strength and pH of the buffer. Guo et al. improved this system by using macroporous chitosan beads as a semi-rigid matrix, and polymerising haemoglobin-imprinted soft polyacrylamide gel inside the macropores [49]. By combining these two materials they obtained hydrophilic, chemically and mechanically stable protein-imprinted particles in high yield. The hybrid gel beads were shown to selectively adsorb haemoglobin in the presence of BSA, illustrating a low degree of unspecific binding. Because of their hydrogel structure, the equilibrium adsorption lasts more than 10 h, making these materials unfeasible for sensing purposes. Bergmann et al. have recently reviewed the literature on protein imprinting in hydrogels [17].

Surface imprinting using immobilised protein templates proved to be a feasible alternative to bulk imprinting in many cases. Rather than mixing the protein templates and functional monomers in solution, the proteins are immobilised on a solid surface as a monolayer, either by adsorption or by covalent attachment [49]. The pre-polymerisation mix is applied as a thin film to this surface and polymerisation is initiated. The resulting polymer film, after separation of the templated surface, will bear imprinted surface sites. Photopolymerisation or electropolymerisation has been shown to be particularly suited for this method. Lin et al. have successfully imprinted lysozyme, ribonuclease A and myoglobin by μ-contact printing immobilised of the respective proteins at a glass surface, followed by photopolymerisation of a thin polymer film on the solid-state template [50].

Surface imprinting of proteins was also demonstrated using silica nanoparticles as the support for the template, resulting in protein-imprinted nanospheres [51].

In an analogous, electrochemical polymerisation approach, Menaker et al. have electropolymerised poly-3,4-ethylenedioxythiophene (PEDOT) doped with polystyrene sulphonate (PSS) to prepare thin, surface-imprinted polymer microrods (8 μm diameter) directly on the transducer [52]. The microrod morphology was obtained by immobilising the template protein, avidin, inside the pores of a sacrificial track-etched porous polycarbonate membrane. This membrane was pressed onto an electrode surface and the electropolymerisation (imprinting) step filled up the pores with the PEDOT/PSS polymer. The membrane was dissolved, leaving behind surface-imprinted polypyrrole microrods at the transducer surface. This elegant strategy eliminates the critical step of (mechanically) separating the surface-imprinted film and the templated surface and the need to re-immobilise the surface-imprinted film at the transducer. Rebinding experiments with fluorescence-labelled avidin and BSA as a competitor were conducted in phosphate buffer, and a high selectivity and affinity (ca. 100 nM) of the MIP film was demonstrated.

Whole protein imprinting relies on the availability of a sufficient amount of highly pure protein templates, a precondition that may be difficult to meet. The presence of native proteins during the polymerisation step poses additional constraints: polymerisation can only be achieved in aqueous solution, limiting the choices of functional monomers, cross linkers and initiators to water-soluble species. Thermal initiation is prohibited in most cases and the protein may form covalent bonds with the polymer backbone during the free-radical polymerisation step [1]. Because of these limitations, an alternative imprinting strategy has been proposed that uses a short (4–10 amino acids) peptide epitope, e.g. the C-terminus or N-terminus, as a surrogate for the protein target. This approach, termed epitope imprinting by its inventors, in order to emphasise the similarity to the immunological determinant recognised by an antibody, offers several advantages [53]: the availability of the template protein in a pure form is not necessary, and only the sequence must be known. Peptides are compatible with a number of organic solvents and can be used in conventional MIP pre-polymerisation mixtures. In an evolution of the original concept and combining it with the ideas of surface imprinting, Nishino et al. have applied peptides, immobilised at a glass surface, as a template to achieve an oriented presentation of the epitope [54]. The C-terminal nonapeptide from cytochrome C was imprinted in a 0.5-mm-thick photopolymerised film. After peeling of the film from the surface, the imprinted sites, borne in the surface contacting face of the film, became available for rebinding experiments. The authors demonstrated a stunning selectivity and affinity constant (90 nM) of the binding sites, proving this concept useful for a number of proteins. In a recent paper, the same group has demonstrated an antibody-like in vivo effect of surface-imprinted MIP particles directed against the toxic peptide, melittin [33]. A different approach for the direct imprinting of a peptide or protein in thin, electropolymerised films is developed and described in the following.

2 Non-covalent Imprinting with Two Complementary Functional Monomers: The Case of Nitrofurantoin

2.1 Non-covalent Imprinting of Two Complementary Monomers

The first application of two different functional monomers in the same MIP was reported by Ramström et al. [55]. They revealed that the MIP prepared from a methacrylic acid and 2-vinylpyridine binds the target, an amino acid derivative,

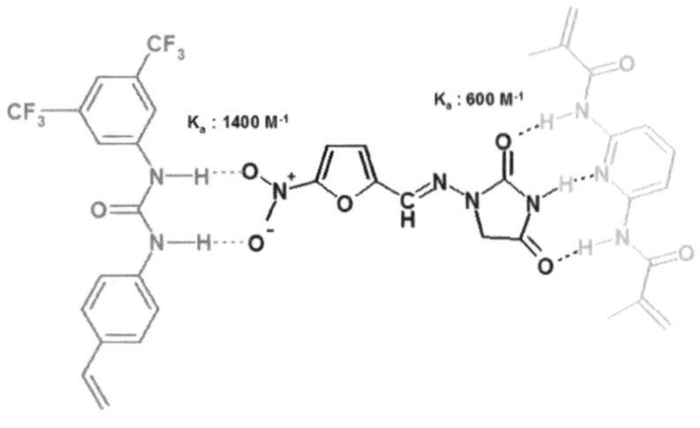

Fig. 2 Ternary complex between nitrofurantoin and two complementary functional monomers (association constants for the binary complexes are indicated)

better than the MIP with only one monomer. In similar works, Takeuchi showed that the selectivity of a triazine-imprinted polymer can be fine-tuned by the relative amounts of MAA and 2-(trifluoromethyl)acrylic acid (TFMAA) [56]. Batch binding experiments showed that a higher proportion of MAA in the polymer favoured the binding of atrazine over ametryn. Several other studies have resorted to commercially available monomers, with the main property being the ionic interaction between the monomer and the target. In a detailed study with nitrofurantoin (NFT), an antibiotic drug, as the model substance, we have explored the potential of combining two complementary, neutral functional monomers in one MIP, starting with custom-synthesised functional monomers [57–59]. NFT belongs to the group of nitrofuran antibiotics and has long been used for the treatment of urinary tract infections in man and husbandry. Because of health risks, it has been banned by the EC for use in husbandry, but it is still commonly used in many food-exporting countries. Because of its quick metabolisation, it was previously not possible to generate antibodies for NFT and to develop sensitive immunoassays. Therefore, an NFT binding MIP is a good option to fill the gap. NFT possesses a neutral nitro-group at the furan ring and a characteristic heteroaromatic hydantoin ring at the opposite site (Fig. 2). Based on a class of monomers originally synthesised for the detection of aromatic carboxy groups [7], a set of three related functional monomers was designed to bind effectively to the aromatic nitro group through two hydrogen bonds, as was demonstrated by ^{1}H-NMR titration of the template-monomer complex [57]. Another monomer was designed to bind the hydantoin moiety of NFT (Fig. 2). Different MIPs were synthesised and characterised using each of the functional monomers alone [58, 59]. The results for an MIP comprising a combination of two complementary functional monomers in one MIP and comparing its performance to the previously reported single functional monomer materials were presented.

2.2 *Material and Template Synthesis*

2.2.1 Materials

Unless otherwise specified, all chemicals were purchased from Sigma-Aldrich, Germany. 1-Hydroxycyclohexyl phenyl ketone (Irgacure 127) was from Wako,

Japan. Acetonitrile, dimethyl sulphoxide (DMSO) and methanol were obtained from Carl Roth, Germany. All chemicals and solvents were of analytical or HPLC grade. Sep-Pack NH2 cartridges (6 mL, 1 g) were purchased from Waters, UK.

2.2.2 Synthesis of an Analogue Template

The analogue template, carboxyphenyl aminohydantoin (CPAH), was synthesised as follows: 1.091 g (7.6 mmol) of 1-aminohydantoin hydrochloride and 1.185 g (7.9 mmol) of 3-carboxybenzaldehyde were dissolved in 140 mL of absolute dioxane. The dioxane was distilled off by using a vigreux column. The end point was determined by the boiling temperature (88°C azeotrop of dioxane/water; 102°C pure dioxane). After 3 h of refluxing, the yellow precipitate was obtained after cooling down. The solvent was evaporated and a yellow precipitate obtained. Dichloromethane was added for recrystallisation. 1.67 g of the yellow end product was obtained by filtration (94% yield).

2.2.3 Synthesis of the Functional Monomer 2,6-Bis(Methaacrylamido) Pyridine (BMP)

2,6-Diaminopyridine was recrystallised in dichloromethane at 70–80°C under an N2 atmosphere, filtrated and dried before use for synthesis. 5.643 g (50 mmol) of 2,6-diaminopyridine and 10 mL of triethylamine were dissolved in 150 mL tetrahydrofuran. 100 mmol of methacryloyl chloride was added drop wise with vigorous stirring at 0°C over 2 h. The reaction was performed under N2 atmosphere. Then the mixture was heated up to 40°C and refluxed for 1 h. White Et3NHCl precipitate occurred in the solution. The reaction was quenched by adding 100 mL chloroform and 200 mL distilled water into the mixture and the aqueous solution was discarded. The organic layer was washed with sodium hydrogen carbonate (3 $\times$ 20 mL), sodium chloride (3 $\times$ 20 mL) and dried over sodium sulphate. After filtration, the solvent was evaporated. The concentrated solution was purified by column chromatography (silica gel 60 (particle size), dichloromethane, and chloroform) to give a pale-yellow solid of diaminopyridine derivative (BMP) (2.36 g, 19.3% yield), mp 151–152°C.

2.2.4 Synthesis of the Functional Monomer 1-(4-Vinylphenyl)-3-(3,5-Bis (Trifluoro-Methyl)Phenyl) Urea (VFU)

VFU was synthesised according to the protocol published earlier by Hall et al. [7]. Briefly, to a stirred solution of 4-vinylaniline (20 mmol) in THF (50 mL), 3,5-bis (trifluoromethyl)phenyl isocyanate (20 mmol) in THF (10 mL) was added. The solution was stirred at room temperature overnight. After solvent evaporation the solid residue was recrystallised from ethanol.

2.2.5 Preparation of MIP Based on Two Functional Monomers

The MIP was prepared using the BMP as the first monomer and the urea-based derivative 1-(4-Vinylphenyl)-3-(3,5-bis(trifluoromethyl)phenyl) urea (VFU) as the second monomer. CPAH was used as the template, PETRA as the cross linker, DMSO: acetonitrile (67:33) as the porogen, with a molar ratio of template: monomer1: monomer2: cross linker as 1:1:1:12. According to its composition, the polymer is called MIP-BMP-VFU. Briefly, 0.5 mmol of CPAH, 0.5 mmol of monomer 1 (BMP) and 0.5 mmol of monomer 2 (VFU) were mixed together in 6 mL of DMSO: acetonitrile (67:33) in 10 mL glass vials and incubated for 4 h at 25°C to allow for self-assembly of the host-guest complexes. Then 6 mmol of PETRA and 0.5 wt% of the photoinitiator (Irgacure 127) were added. Each vial was purged with argon for 5 min. After 2 h pre-incubation at 4°C, polymerisation was initiated under ultraviolet light at 366 nm for 6 h. After the polymer monolith was formed, it was broken to coarse pieces with a mortar, and finely ground in a ball mill (Retsch, type S 100). After sieving and washing with hot methanol in a Soxhlet apparatus overnight, most of the template was removed. The course of template removal was monitored with a UV–vis spectrophotometer at 300 nm. The polymer powder was dried for 24 h at 30°C. The non-imprinted control polymer NIP-BMP-VFU was prepared in the same manner as MIP-BMP-VFU but without the template.

The one-functional-monomer MIPs (MIP-BMP, MIP-VFU), meant as reference materials in this study, were prepared according to the same protocol, with the exception that only one functional monomer was used. The overall composition of the polymer changes only little, since the second monomer makes up for only 8% of the polymer dry weight. The idea to increase the cross linker concentration or to add a "neutral" functional monomer was dismissed, because this would also change the polymer composition and cross-linking degree.

2.2.6 Scanning Electron Microscopy and BET Isotherm Porosimetry

Morphologies of the polymers were recorded as scanning electron microscopy (SEM) images with a Gemini Leo 1550 instrument (Carl Zeiss, Oberkochen, Germany) at an operating voltage of 3 keV. The pore parameters and surface areas of the imprinted polymers were measured using a Quantachrome instrument (Automated Surface Area and Pore Size Analyzer, Quadrasorb SI). 30 mg of dry imprinted polymers were degassed at 60°C for 24 h under nitrogen flow to remove adsorbed gases and moisture. The nitrogen adsorption/desorption analysis were performed at 77 K. The surface areas from multi-point N_2 adsorption isotherms were calculated using the Brunauer, Emmett and Teller (BET) equation.

2.2.7 Characterisation of MIP and NIP by Batch Rebinding Studies

The imprinted and non-imprinted polymers were characterised by batch binding experiments to elucidate their equilibrium binding isotherms. Briefly, CPAH or

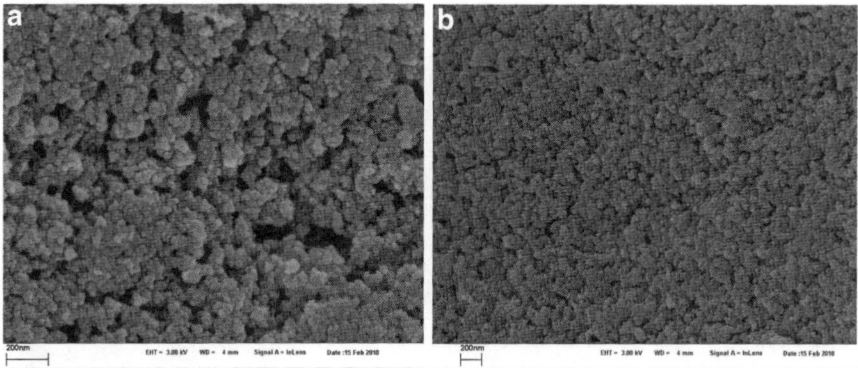

Fig. 3 SEM images of the (**a**) NFT imprinted polymer with two functional monomers (MIP-BMP-VFU) and the (**b**) control polymer NIP-BMP-VFU at 20,000× magnification

NFT solutions (0.001–2.5 mM) were prepared in acetonitrile + 2% DMSO. The DMSO addition was necessary because CPAH and NFT are not soluble in pure acetonitrile. 5 mL of the CPAH or NFT solution was added to 10 mg of polymer in each tube and the mixture was incubated for 24 h at 25°C under stirring. The adsorption isotherms were determined by sedimenting the polymer and measuring the remaining concentration of NFT and CPAH in the supernatant at 300 nm (CPAH) or 370 nm (NFT) using a pre-determined calibration curve for each compound.

2.3 *Nitrofurantoin Detection*

2.3.1 Morphology and Porosity of MIP and NIP

The surface and porosity of MIP-BMP-VFU and NIP-BMP-VFU were qualitatively compared by electron microscopy. Even at the largest magnification of 20,000× no significant difference could be observed (Fig. 3). From these results, comparable BET surface area data can be expected. Indeed, the surface area for the MIP-BMP-VFU exceeded the respective value for the NIP-BMP-VFU by only 16% (Table 1). Interestingly, the polymers with two functional monomers (both MIP and NIP) had five times increased surface area when compared to the polymers prepared with either one of the functional monomers. Taking into account that the stoichiometric ratio of template to monomer was nearly the same in all pre-polymerisation mixtures, namely 1:1 (with only one functional monomer) and 1:1:1 in the case of two functional monomers, this difference is striking. It is clear evidence for an improved pre-arrangement effect caused by the synergistic action of two

Table 1 BET surface area and porosimetry data for three different MIPS (and control polymer NIPs)

Polymers	Surface area[a] (m^2 g^{-1})	Total pore volume[b] (mL g^{-1})	Average pore radius[c] (Å)
MIP-BMP	46.46	0.066	29.42
NIP-BMP	34.03	0.061	26.42
MIP-VFU	36.87	0.047	27.30
NIP-VFU	22.92	0.037	24.72
MIP-BMP-VFU	157.4	0.172	35.94
NIP-BMP-VFU	126.9	0.157	32.40

[a]Determined using multi-point BET method
[b]BJH cumulative desorption pore volume of pores with radius less than 19 Å
[c]Average value of pore radius calculated from several models

complementary functional monomers, which may ultimately increase the density of binding sites, i.e. the binding capacity of the new polymer.

2.3.2 Kinetics of NFT Binding

The time dependence of NFT binding to all imprinted and non-imprinted polymers was investigated. Time constants between 0.8 h and 1.2 h were obtained by fitting the adsorption data to a monoexponential growth function in all cases (i.e. the time to reach 67% saturation), with no clear differences between MIPs and NIPs and irrespective of the quite different saturating binding capacities (data not shown). This result indicates that the adsorption proceeds under diffusion limitation and that the pore morphology should be similar for all polymers under study.

2.3.3 Binding Capacity and Affinity of the Imprinted Polymers

In order to determine the imprinting effect and the overall binding capacity for NFT for the two-functional-monomer MIP and the two one-functional-monomer MIPs in this study, the equilibrium binding isotherms were detected and compared (Fig. 4). A qualitative analysis shows an increased binding capacity for the two-functional-monomer MIP (MIP-BMP-VFU) as compared to the other imprinted (MIP-BMP, MIP-VFU) and non-imprinted polymers (NIP-BMP-VFU, NIP-BMP, NIP-VFU). This was consistent with the significantly higher BET surface of this polymer. Table 2 summarises the imprinting factors for all polymer pairs. Obviously the highest imprinting factor of 3.06 (i.e. the ratio of NFT bound by the MIP divided by the amount bound by the respective NIP) was obtained with the two-functional-monomer polymers. The better imprinting effect was also reflected by the highest figure for the binding site occupancy: 8.1% of the theoretically possible binding sites (under the assumption that each template molecule creates one binding site) were actually occupied at a concentration of 1 mM NFT. This is one of the highest

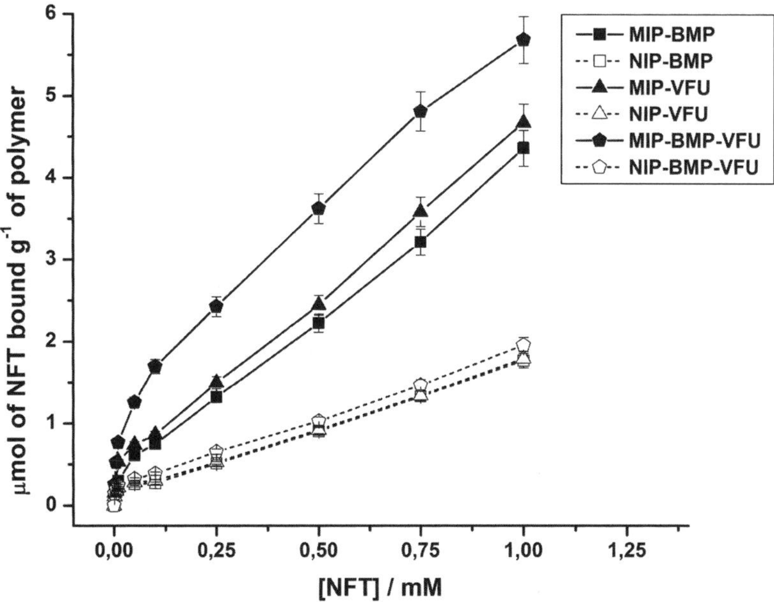

Fig. 4 Binding isotherms for nitrofurantoin at imprinted polymer with two functional monomers (MIP-BMP-VFU) and two MIPs with either one of the two monomers (MIP-BMP, MIP-VFU) and the respective NIP control polymers (condition: 10 mg polymer ad 5 mL solvent (acetonitrile + 2% DMSO))

Table 2 Summary of the imprinting effect for three nitrofurantoin MIPs and control polymers

	Nitrofurantoin		
Polymer	Bound (μmol g^{-1} of polymer)	Imprinting factor	% of binding sites
MIP-BMP	4.380	2.47	5.1
NIP-BMP	1.775		
MIP-VFU	4.670	2.61	5.3
NIP-VFU	1.790		
MIP-2M	5.681	3.06	8.1
NIP-2M	1.854		

$$\text{Imprinting factor} = \frac{\text{Amount of NFT bound by MIP}}{\text{Amount of NFT bound by NIP}}$$

$$\%\ \text{binding sites} = \frac{\text{Amount of NFT bound by MIP} \times 100}{\text{Amount of NFT used for imprinting}}$$

values reported for a bulk imprinted polymer so far, which is even more significant as the actual imprinting has been conducted with the NFT surrogate CPAH, because NFT will deteriorate under the conditions of imprinting.

A Scatchard plot analysis (using a linearised form of the Langmuir isotherm) was performed in order to extract quantitative binding affinity and capacity data,

Table 3 Scatchard analysis of nitrofurantoin binding isotherms at three MIPs

	Nitrofurantoin			
Polymers	K_1 (M^{-1})	B_{max1} (μmol g^{-1})	K_2 (M^{-1})	B_{max2} (μmol g^{-1})
MIP-BMP	79.7	0.82	1.60	5.54
MIP-VFU	121.3	0.92	1.69	5.98
MIP-2M	140.3	1.43	3.16	6.53

under the simplifying assumption of a small population of high-affinity binding sites and a larger population of low-affinity binding sites. The data for both kinds of binding sites are summarised in Table 3, confirming the validity of the hypothesis that the highest-affinity binding sites are present in the two-functional-monomer MIPs. The lower concentration portion of the binding isotherms ($\leq$0.1 mM) was also fitted to a simple Langmuir isotherm using the LMMPro 1.06 software (by C.M. Schulthess) and was in good agreement with the Scatchard analysis (data not shown).

K_1 and B_{max1} are the Langmuir equilibrium constant and the apparent maximum binding capacity for the higher-affinity binding sites (K_2 and B_{max2} are the respective values for the low-affinity binding sites).

In order to improve the affinity of the NFT MIP, a urea-based functional monomer (VFU), capable of donating two hydrogen bonds to the nitro group, and a BMP, capable of donating two and accepting one hydrogen bond from the hydantoin moiety in NFT, was combined in the same polymer. It is thought that the combined actions of the complementary receptor groups are summed up, leading to high-affinity, "multi-point interactions" between the target and the binding pocket inside the MIP. Indeed a synergistic effect was visible as the new polymer outperformed the two reference MIPs in all benchmark tests. Considering the moderate improvement in binding affinity, however, the progress is not satisfactory yet: 15% higher as compared to MIP-VFU, 75% higher as compared to MIP-BMP. We believe that the impact of the porogen (DMSO:acetonitrile in a molar ratio of 67:33) and the cross linker (PETRA) is not yet fully understood with this system and that a rational design approach may help to further improve performance of this polymer. A pronounced group selectivity of the MIPs towards aromatic nitro compounds (VFU) and imido compounds (BMP) has previously been reported by us [59] and a similar selectivity pattern may be expected for the two-functional-monomer MIP.

3 Surface Imprinting of Peptides and Proteins in Ultrathin, Electropolymerised Films

Proteins are among the most interesting targets for MIP sensors: next to the specific extraction of a protein from complex media (e.g. fermentation broth, cell culture supernatant, blood and other biological samples), many immunoassays and other

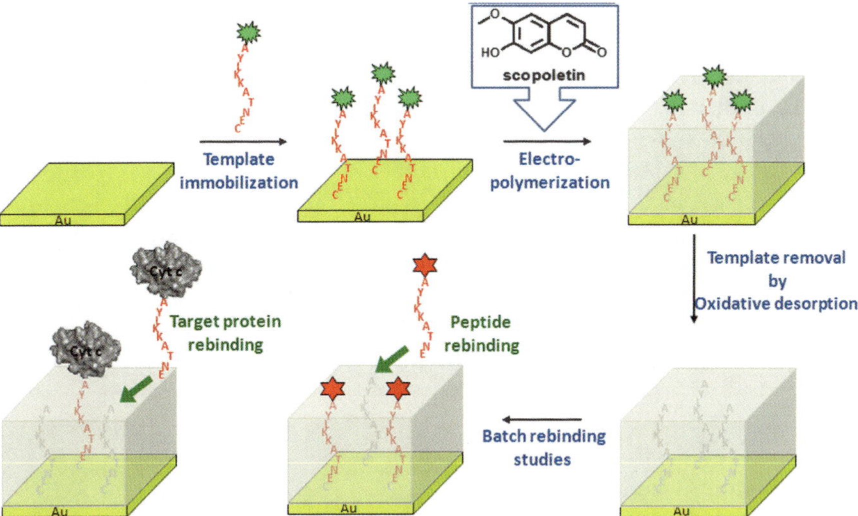

Fig. 5 Electrochemical imprinting of a peptide epitope in an ultrathin electropolymerised poly-scopoletin film (complete workflow, schematic)

diagnostic methods rely on the specific detection of a protein. Being relatively unstable macromolecules, conventional MIP protocols cannot be used for the imprinting of a protein. The aim of this chapter is to demonstrate, for the first time, a new molecular imprinting workflow which combines the concepts of epitope imprinting, surface imprinting and electropolymerisation imprinting. The technique is based on electropolymerisation of a non-conductive hydrophilic film [60] in the presence of a chemisorbed peptide template at a gold surface. The template is removed by electrochemical oxidation after the imprinting, leaving behind binding pockets for the peptide. The principle is illustrated in Fig. 5.

3.1 Electrochemical Imprinting

3.1.1 Materials

Scopoletin, cytochrome C (cyt c) from bovine heart, and all other chemicals were purchased from Sigma-Aldrich, Germany, and were of analytical grade or higher. Dy-633-NHS reactive fluorescence dye was from Dyomics GmbH, Germany. TAMRA-labelled peptide template (TAMRA-AYLKKATNEC) and other peptides used were synthesised by Centic Biotec, Germany. Dy-633-labelled peptide (C(Dy-633-M)-AYLKKATNE) was synthesised by Biosyntan (Germany). Ultrapure water from a water purification system (Sartorius, Germany) was used.

Gold-sputtered glass disks (25 mm diameter) with an Au-film thickness of 200 nm were obtained from SSENS, The Netherlands.

3.1.2 Instrumentation

An electrochemical workstation Autolab PGStat30 equipped with GPES software from Eco Chemie, The Netherlands, was used for all electrochemical works. A three-electrode system with a thin-film gold disk, a Pt wire counter electrode, and an Ag/AgCl (1M KCl) reference electrode were used.

Fluorescence imaging was conducted with a Tecan LS Reloaded microarray scanner (Tecan GmbH, Germany) using Array-pro software. Film-thickness determination was conducted with a Nanofilm EP3 imaging ellipsometer (Accurion, Germany) or with a multimode ellipsometer (Multiskop, Optrel GmbH, Germany).

AFM images were taken with a Veeco DI CP II multi-mode AFM instrument in the tapping mode, with standard cantilevers.

3.1.3 Peptide Chemisorption

A peptide, consisting of the C-terminal nonapeptide from bovine heart cytochrome c (residues 97–104 of Cyt c, AYLKKATNE), extended by an extra cysteine at its C-end and labelled with the TAMRA fluorescence dye at its N-terminal end, was synthesised using standard FMOC chemistry and purified by HPLC. The sequence was verified by its mass spectrum.

The adsorption isotherm at the gold surface of the labelled peptide, diluted in 0.1 M phosphate buffer pH 7, was determined in the concentration range from 0.01 mM to 1 mM by fluorescence imaging using a microarray scanner. All scanner settings were kept constant between experiments to maintain reproducibility. The gold surface was purified with a 30-s treatment in hot piranha solution (30% H_2SO_4, 70% H_2O_2. Take care! piranha solution is extremely harmful!) and dried under N_2 before use. Monolayer coverage was achieved at concentrations above 0.1 mM at a fixed incubation time of 3 h at room temperature (data not shown).

To be used as the template, the peptide was chemisorbed to a gold surface from a 0.05 mM solution in 0.1 M phosphate buffer containing 5 mM TCEP, resulting in ca. 60% surface coverage (data not shown). After chemisorption the surface was incubated in phosphate buffered solution pH 7.4 with 0.1% Tween 20 for 20 min, rinsed with water, dried with N_2 and stored in the dark before fluorescence imaging.

3.1.4 Electropolymerisation Imprinting

An aqueous solution of 0.25 mM scopoletin in 0.1 M NaCl solution with 20 mM EDTA was freshly prepared. Electropolymerisation was conducted using a single potential pulse (0.7 V for 5 s, then 0 V for 15 s) without prior deoxygenation of the solution. The gold disk was rinsed with water, dried with N_2 and stored in the dark before fluorescence measurement.

Non-imprinted control films were prepared in the same manner on bare gold surfaces, omitting the template.

3.1.5 Electrochemical Template Stripping

The peptide template was removed by electrochemical oxidation of the thiol (30 s at 1.4 V) in 0.1 M phosphate buffer pH 7. The gold disk with the imprinted film was rinsed with water and incubated in PBS with 0.1% Tween 20 overnight. It was rinsed with water, dried with N_2 and stored dry until used. The percentage of template removal was calculated by comparing the fluorescence intensity of the surface before and after the stripping step.

3.1.6 Batch Rebinding Experiment

The gold disk with the imprinted film was incubated for 18 h in a solution of the Dy633-labelled peptide or in a solution of Dy633-labelled cytochrome c (labelled with the Dy633 according to the standard protocol provided by Dyomics). All solutions were prepared in 0.1 M phosphate buffer, pH 7. A 5-min rinsing step with PBS containing 0.1% Tween-20 was employed to remove nonspecifically bound peptide or protein.

3.2 Surface Imprinting of Peptides and Proteins

3.2.1 Peptide Chemisorption

Monolayer coverage was achieved at concentrations above 0.1 mM at a fixed incubation time of 3 h at room temperature (data not shown). Chemisorption from a 0.05 mM solution in 0.1 M phosphate buffer containing 5 mM TCEP resulted in ca. 60% surface coverage with a high reproducibility (data not shown).

3.2.2 Film Characterisation

The thickness of the electropolymerised poly-scopoletin film as estimated from SPR spectroscopy using the Optrel Multiskope ellipsometer with a 635-nm laser was 4 ± 0.5 nm for the imprinted and 5 ± 0.5 nm for non-imprinted films. The index of refraction was determined as 1.46 and the extinction coefficient as 0.04 (i.e. the film is transparent).

Our attempts to obtain AFM images for the imprinted 4 nm film failed, because of stickiness and AFM artefacts (data not shown). It was, however, possible to image imprinted and non-imprinted poly-scopoletin films of ca. 20 nm thickness

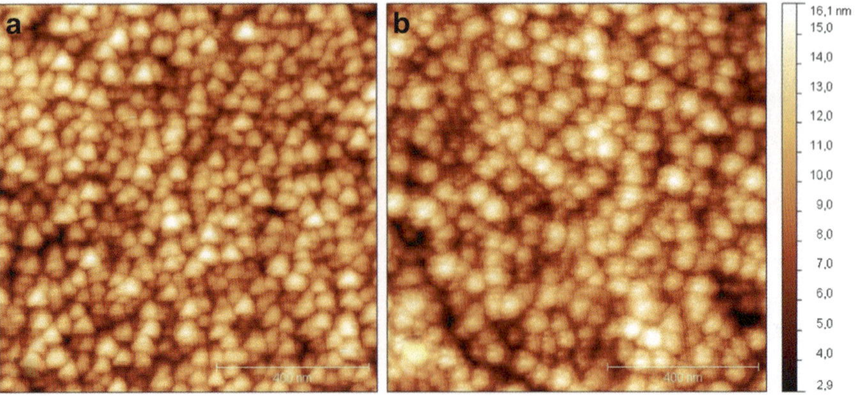

Fig. 6 AFM tapping mode images of a 20-nm-thick (**a**) imprinted and (**b**) non-imprinted poly-scopoletin film

(Fig. 6). Such thick films would not work, because the templates are permanently entrapped in the bulk, and they were prepared merely for the purpose of AFM imaging. These films show a pronounced, grainy structure and a high surface roughness on the same order as the film thickness (ca 16 nm for MIP and NIP). Previous AFM data obtained with 40-nm-thick poly-scopoletin films, prepared by a cyclic voltammetry protocol, consistently showed a continuous, defect-free film with a smooth topography [60]. These preliminary results indicate that the film morphology may be strongly influenced by the deposition protocol favouring a grainy, porous structure with the ultra-thin film obtained by pulse potential deposition.

3.2.3 Template Stripping Efficiency

With the electrochemical oxidation protocol used and based on fluorescence intensity data before and after the template stripping, more than 80% of the chemisorbed template were consistently removed from the 4-nm-thick film (data not shown). This is consistent with the hypothesis that the film thickness is approximately that of the immobilised peptide template.

3.2.4 Peptide Rebinding

The peptide used for the rebinding experiment had the same nonamer sequence as the peptide used as a template, but the C-terminal cystein was omitted in order to avoid chemisorption to the gold, which would confuse the adsorption results. Instead, a cystein was added at the N-terminus. In addition, the peptide was labelled with the deep-red dye Dy633, which could be detected in the red channel of the

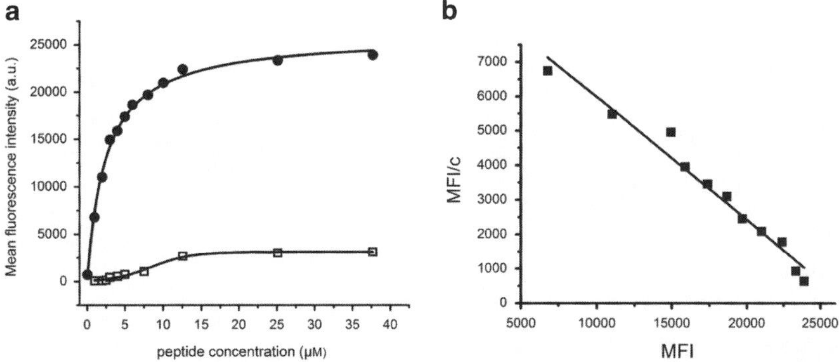

Fig. 7 Peptide rebinding in 0.1 M phosphate buffer at pH 7 to an imprinted poly-scopoletin film (**a**: *filled circles,* **b**: *filled squares*) and a non-imprinted film (**a**: *open squares*). (**a**) Binding isotherm extracted from fluorescence images in arbitrary units (*MFI* mean fluorescence intensity). (**b**) Scatchard plot linearisation of binding isotherm (only for the MIP)

microarray reader. In this way, the fluorescence signals were clearly discernible from the signals due to entrapped template peptides (which were labelled with the green fluorescent dye TAMRA). Figure 7a shows the binding isotherm for the peptide (C-(Dy633-M)-AYLKKATNE) rebinding to the imprinted film and the non-imprinted control film. Binding to the imprinted film was characterised by a 10-times higher binding capacity ($\geq$ imprinting factor: 10) as compared to the non-imprinted film and an ideal Langmuir isotherm behaviour. The linear Scatchard plot (Fig. 7b) supports this observation. Most MIPs show a highly non-linear graph in the Scatchard plot, as a result of the heterogeneity in binding sites. As a consequence, we may assume an almost homogeneous population of binding sites, characterised by a single affinity constant (Kd = 2.5 μM). This is an interesting result, as the rather weak interaction between the film and the peptide should be dominated by weak bonds, like hydrogen bonds and van der Waals forces. The apparent homogeneity of binding sites may be a consequence of the site-oriented immobilisation of the template peptide and the homogeneity of the electropolymerised film itself. In contrast to conventional MIPs, electropolymers are prepared from only one monomer, which may explain the apparent homogeneity of binding sites.

3.2.5 Cytochrome C Rebinding

Similar to the rebinding of the AYLKKATNE-peptide, rebinding of fluorescence-labelled cytochrome c holoprotein was conducted in pure phosphate buffer at pH 7. Again the binding capacity of the imprinted film exceeds the value for the non-imprinted one by a factor of 6 ($\geq$ imprinting factor : 6, Fig. 8). These data clearly show that the epitope imprinting of cytochrome c, as demonstrated by Nishino [54], could be reproduced with our electrochemical polymerisation strategy.

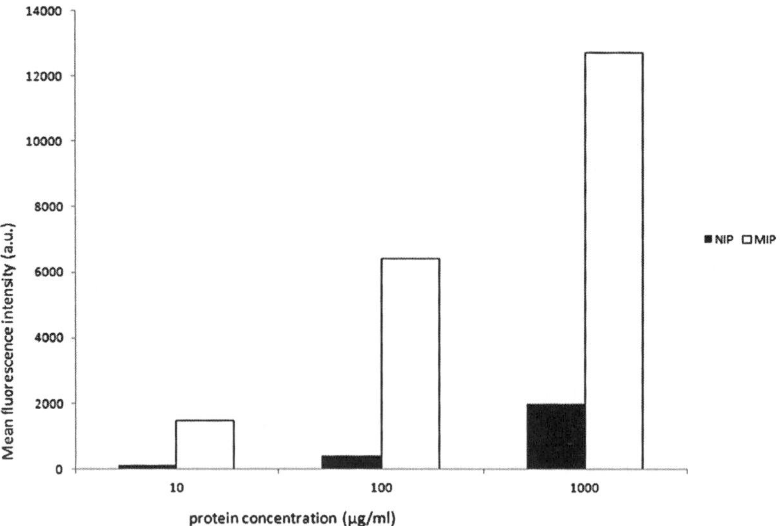

Fig. 8 Cytochrome C rebinding to an imprinted and non-imprinted poly-scopoletin film in 0.1 M phosphate buffer at pH 7, (fluorescence intensity data given in arbitrary units (*MFI* mean fluorescence intensity))

Based on the original Nishino and Menaker approaches, we have developed a workflow for the surface imprinting of thin electropolymerised films with a peptide epitope. Cytochrome c was imprinted by its nonameric C-terminal peptide epitope. The imprinted, hydrophilic poly-scopoletin film possesses a moderately high affinity towards the imprinting peptide (Kd $= 2.5\ \mu M$), and cytochrome c was detected in buffered solution at the micromolar concentration range in purely aqueous solution. The new technique combines the advantages of the epitope imprinting with the excellent film thickness control that is inherent to electropolymerisation. The imprinted polymer film is grown directly at the electrode surface, including all processing steps. This facilitates the integration with electrochemical, SPR or QCM transducers.

4 Outlook

The field of molecularly imprinted materials is growing at a fast pace, as can be judged from the increasing number of publications and the increasing diversity of methods. This is in contrast to the still quite limited number of commercially available MIPs. Only recently MISPE have become commercially available. Sample preparation by group-selective MISPE for chromatographic analysis seems to be an interesting market niche and may work as an opener for MIP technologies in general. The unique versatility of the molecular imprinting concept and the recent advances in the rational design of MIP materials, as well as new and unique MIP nanomaterials, will surely promote new applications in biotechnology, analytical chemistry, biosensors and diagnostics. Moreover, the trend towards "green" chemistry and environmentally friendly processes may give an additional momentum for MIP applications in industry. The design of tailor-made MIPs for the detection of proteins in aqueous solutions and complex biological media like blood has become a realistic option in recent years. Significant progress has been achieved by modifications of the classical bulk imprinting approach, resulting in imprinted hydrogel materials with good aqueous compatibility, and by new strategies like surface imprinting and epitope imprinting using immobilised templates. Protein-imprinted electropolymerised thin films can be prepared directly on transducer surfaces with high thickness control using only minute amounts of the protein template. This technique is particularly suited for the functionalisation of SPR, QCM and electrochemical biosensors.

References

1. Whitcombe MJ, Chianella I, Larcombe L, Piletsky SA, Noble J, Porter R, Horgan A (2011) The rational development of molecularly imprinted polymer-based sensors for protein detection. Chem Soc Rev 40:1547–1571
2. Sreenivasan K (2007) Synthesis and evaluation of molecularly imprinted polymers for nucleic acid bases using aniline as a monomer. React Funct Polym 67:859–864
3. Lettau K, Warsinke A, Katterle M, Danielsson B, Scheller FW (2006) A bi-functional molecularly imprinted polymer (MIP): analysis of binding and catalysis by a thermistor. Angew Chem Int Ed 45:6986–6990
4. Alexander C, Andersson HS, Andersson LI, Ansell RJ, Kirsch N, Nicholls IA, O'Mahony J, Whitcombe MJ (2006) Molecular imprinting science and technology: a survey of the literature for the years up to and including 2003. J Mol Recognit 19:106–180
5. Vlatakis G, Andersson LI, Müller R, Mosbach K (1993) Drug assay using antibody mimics made by molecular imprinting. Nature 361:645–647
6. Wulff G (1995) Molecular imprinting in cross-linked materials with the aid of molecular templates. Angew Chem Int Ed 34:1812–1832
7. Hall AJ, Manesiotis P, Emgenbroich M, Quagilia M, de Lorenzi E, Sellergren BJ (2005) Urea host monomers for stoichiometric molecular imprinting of oxyanions. J Org Chem 70:1732–1736
8. Chianella I, Lotierzo M, Piletsky SA, Tothill IE, Chen B, Karim K, Turner APF (2002) Rational design of a polymer specific for microcystin-LR using a computational approach. Anal Chem 74:1288–1293
9. Chen BN, Piletsky S, Turner APF (2002) Design of "keys". Comb Chem High Throughput Screen 5:409–427
10. Bui BTS, Haupt K (2010) Molecularly imprinted polymers: synthetic receptors in bioanalysis. Anal Bioanal Chem 398:2481–2492
11. Mayes AG, Mosbach K (1996) Molecularly imprinted polymer beads: suspension polymerization using a liquid perfluorocarbon as the dispersing phase. Anal Chem 68:3769–3774
12. Guan G, Liu B, Wang Z, Zhang Z (2008) Imprinting of molecular recognition sites on nanostructures and its applications in chemosensors. Sensors 8:8291–8320

13. Linares AV, Vandevelde F, Pantigny J, Falcimaigne-Cordin A, Haupt K (2009) Polymer films composed of surface-bound nanofilaments with a high aspect ratio, molecularly imprinted with small molecules and proteins. Adv Funct Mater 19:1299–1303
14. Oxelbark J, Legido-Quigley C, Aureliano CSA, Titirici MM, Schillinger E, Sellergren B, Courtois J, Irgum K, Dambies L, Cormack PAG, Sherrington DC, De Lorenzi E (2007) Chromatographic comparison of bupivacaine imprinted polymers prepared in crushed monolith, microsphere, silica-based composite and capillary monolith formats. J Chromatogr A 1160:215–226
15. Yilmaz E, Haupt K, Mosbach K (2000) The use of immobilized templates - a new approach in molecular imprinting. Angew Chem Int Ed 39:2115–2118
16. Weetall HH, Rogers KR (2004) Preparation and characterization of mole-cularly imprinted electropolymerized carbon electrodes. Talanta 62:329–335
17. Bergmann NM, Peppas NA (2008) Molecularly imprinted polymers with specific recognition for macromolecules and proteins. Prog Polym Sci 33:271–288
18. Janiak DS, Kofinas P (2007) Molecular imprinting of peptides and proteins in aqueous media. Anal Bioanal Chem 389:399–404
19. Haupt K, Mayes AG, Mosbach K (1998) Herbicide assay using an imprinted polymer-based system analogous to competitive fluoroimmunoassays. Anal Chem 70:3936–3939
20. Vaihinger D, Landfester K, Krauter I, Brunner H, Tovar GEM (2002) Molecularly imprinted polymer nanospheres as synthetic affinity receptors obtained by miniemulsion polymerization. Macromol Chem Phys 203:1965–1973
21. Piletsky SA, Turner NW, Laitenberger P (2006) Molecularly imprinted polymers in clinical diagnostics. Med Eng Phys 28:971–977
22. Baggiani C, Giraudi G, Giovannoli C, Trotta CF, Vanni A (2000) Chromatographic characterization of molecularly imprinted polymers binding the herbicide 2,4,5-trichlorophenoxyacetic acid. J Chromatogr A 883:119–126
23. Haupt K, Dzgoev A, Mosbach K (1998) Assay system for the herbicide 2,4-dichlorophenoxyacetic acid using a molecularly imprinted polymer as an artificial recognition element. Anal Chem 70:628–631
24. Legido-Quigley C, Oxelbark J, De Lorenzi E, Zurutuza-Elorza A, Cormack PAG (2007) Chromatographic characterisation, under highly aqueous conditions, of a molecularly imprinted polymer binding the herbicide 2,4-dichlorophenoxyacetic acid. Anal Chim Acta 591:22–28
25. O'Mahony J, Molinelli A, Nolan K, Smyth MR, Mizaikoff B (2005) Towards the rational development of molecularly imprinted polymers: ^{1}H NMR studies on hydrophobicity and ion-pair interactions as driving forces for selectivity. Biosens Bioelectron 20:1884–1893
26. O'Mahony J, Molinelli A, Nolan K, Smyth MR, Mizaikoff B (2006) Anatomy of a successful imprint: analysing the recognition mechanisms of a molecularly imprinted polymer for quercetin. Biosens Bioelectron 21:1383–1392
27. Kantarovich K, Tsarfati I, Gheber LA, Haupt K, Bar I (2010) Reading microdots of a molecularly imprinted polymer by surface enhanced Raman spectroscopy. Biosens Bioelectron 26:809–814
28. Haginaka J (2008) Monodispersed, molecularly imprinted polymers as affinity-based chromatography media. J Chromatogr B 866:3–13
29. Tokonami S, Shiigi H, Nagaoka T (2009) Review: micro- and nanosized molecularly imprinted polymers for high-throughput analytical applications. Anal Chim Acta 641:7–13
30. Chen L, Xu S, Li J (2011) Recent advances in molecular imprinting technology: current status, challenges and highlighted applications. Chem Soc Rev 40:2922–2942
31. Shimizu KD, Stephenson CJ (2010) Molecularly imprinted polymer sensor arrays. Curr Opin Chem Biol 14:743–750
32. Park JK, Kim SJ (2004) Separation of phenylalanine by ultrafiltration using D-Phe imprinted polyacrylonitrile-poly(acrylic acid)-poly(acryl amide) terpolymer membrane. Kor J Chem Eng 21:994–998

33. Hoshino Y, Koide H, Urakami T, Kanazawa H, Kodama T, Oku N, Shea KJ (2010) Recognition, neutralization, and clearance of target peptides in the bloodstream of living mice by molecularly imprinted polymer nanoparticles: a plastic antibody. J Am Chem Soc 132:6644–6645
34. Umpleby RJ, Baxter SC, Chen YZ, Shah RN, Shimizu KD (2001) Characterization of the heterogeneous binding site affinity distributions in molecularly imprinted polymers. Anal Chem 73:4584–4591
35. Wulff G, Sarhan H (1972) The use of polymers with enzyme-analogous structures for the resolution of racernates. Angew Chem Int Ed 11:341
36. Schumacher S, Grüneberger F, Katterle M, Hettrich C, Hall DG, Scheller FW, Gajovic-Eichelmann N (2011) Molecular imprinting of fructose using a polymerizable benzoboroxole. Polymer 52:2485–2491
37. Wulff G, Knorr K (2002) Stoichiometric noncovalent interaction in molecular imprinting. Bioseparation 10:257–276
38. Ellwanger A, Karlsson L, Owens PK, Berggren C, Crecenzi C, Ensing K, Bayoudh S, Cormack PAG, Sherrington D, Sellergren B (2001) Evaluation of methods aimed at complete removal of template from molecularly imprinted polymers. Analyst 126:784–792
39. Pan G, Ma Y, Zhang Y, Guo X, Li C, Zhang H (2011) Controlled synthesis of water-compatible molecularly imprinted polymer microspheres with ultrathin hydrophilic polymer shells via surface-initiated reversible addition-fragmentation chain transfer polymerization. Soft Matter 7:8428–8439
40. Lee Y, Ju KY, Lee JK (2010) Stable biomimetic superhydrophobic surfaces fabricated by polymer replication method from hierarchically structured surfaces of Al templates. Langmuir 26:14103–14110
41. Mader HS, Wolfbeis OS (2008) Boronic acid based probes for microde-termination of saccharides and glycosylated biomolecules. Microchim Acta 162:1–34
42. Wulff G, Schauhoff S (1991) Enzyme-analog-built polymers. 27. Racemic resolution of free sugars with macroporous polymers. J Org Chem 56:395–400
43. Rajkumar R, Warsinke A, Möhwald H, Scheller FW, Katterle M (2007) Development of fructosyl valine binding polymer by covalent imprinting. Biosens Bioelectron 22:3318–3325
44. Rajkumar R, Katterle M, Warsinke A, Möhwald H, Scheller FW (2008) Thermometric MIP sensor for fructosyl valine. Biosens Bioelectron 23:1195–1199
45. Hall DG (2005) Boronic acids – preparation and applications in organic synthesis and medicine. Wiley-VCH, Weinheim
46. Rajkumar R, Warsinke A, Möhwald H, Scheller FW, Katterle M (2008) Analysis of recognition of fructose by imprinted polymers. Talanta 76:1119–1123
47. Schumacher S, Katterle M, Hettrich C, Paulke BR, Pal A, Hall DG, Scheller FW, Gajovic-Eichelmann N (2011) Benzoboroxole-modified nanoparticles for the recognition of glucose at neutral pH. Chem Sens 1:1–7
48. Hjerten S, Liao JL, Nakazato K, Wang Y et al (1997) Gels mimicking antibodies in their selective recognition of proteins. Chromatographia 44:227–234
49. Guo TY, Xia YQ, Hao GJ, Song MD, Zhang BH (2004) Adsorptive separation of hemoglobin by molecularly imprinted chitosan beads. Biomaterials 25:5905–5912
50. Lin HY, Hsu CY, Thomas JL, Wang SE, Chen HC, Chou TC (2006) The microcontact imprinting of proteins: the effect of cross-linking monomers for lysozyme, ribonuclease A and myoglobin. Biosens Bioelectron 22:534–543
51. Cutivet A, Schembri C, Kovensky J, Haupt K (2009) Molecularly imprinted microgels as enzyme inhibitors. J Am Chem Soc 131:14699–14702
52. Menaker A, Syritski V, Reut J, Öpik A, Horváth V, Gyurcsányi RE (2009) Electrosynthesized surface-imprinted conducting polymer microrods for selective protein recognition. Adv Mater 21:2271–2275
53. Rachkov A, Minoura N (2001) Towards molecularly imprinted polymers selective to peptides and proteins. The epitope approach. Biochim Biophys Acta 1544:255–266

54. Nishino H, Huang CS, Shea KJ (2006) Selective protein capture by epitope imprinting. Angew Chem Int Ed 45:2392–2396
55. Ramström O, Andersson LI, Mosbach K (1993) Recognition sites incorporating both pyridinyl and carboxy functionalities prepared by molecular imprinting. J Org Chem 58:7562–7564
56. Takeuchi T, Fukuma D, Matsui J (1999) Combinatorial molecular imprinting. Anal Chem 71:285–290
57. Athikomrattanakul U, Promptmas C, Katterle M (2009) Synthetic receptors for neutral nitro derivatives. Tetrahedron Lett 50:359–362
58. Athikomrattanakul U, Katterle M, Gajovic-Eichelmann N, Scheller FW (2009) Development of molecularly imprinted polymers for the binding of nitrofurantoin. Biosens Bioelectron 25:82–87
59. Athikomrattanakul U, Katterle M, Gajovic-Eichelmann N, Scheller FW (2011) Preparation and characterization of novel molecularly imprinted polymers based on thiourea receptors for nitrocompounds recognition. Talanta 84:274–279
60. Gajovic-Eichelmann N, Ehrentreich-Förster E, Bier FF (2003) Directed immobilization of nucleic acids at ultramicroelectrodes using a novel electro-deposited polymer. Biosens Bioelectron 19:417–422

Gold Nanostructure LSPR-Based Biosensors for Biomedical Diagnosis

Mun'delanji C. Vestergaard, Masato Saito, Hiroyuki Yoshikawa, and Eiichi Tamiya

Abstract Progress in nanotechnology has enjoyed exponential growth in the past couple of decades. We have seen design and synthesis of metal nanoparticles (NPs) tailored specifically for biomedical diagnosis. In particular, noble metals have attracted lots of attention. Because of their unique optical and electronic properties, Au and Ag NPs have been exploited in the fabrication of localized surface plasmon resonance (LSPR) chips for detection of biomolecules. They impart increased sensitivity and also allow development of analytical platforms for label-free detection. These metal NPs show specific changes in their absorbance responses in the visible region of the spectrum upon binding with various molecules such as nucleic acids or proteins. In addition, the electronic properties, in particular, of Au and Ag NPs have been employed as labels for detection of proteins and other target molecules. In this chapter, we will focus on the use of Au NPs in LSPR-based biosensor technology. We will discuss the principles and applications of how these NPs have been and can be exploited for medical diagnostics by providing examples, mainly to the work we have conducted in our research group.

Keywords Gold nanoparticles, LSPR, Biosensor, Surface chemistry, Medical diagnosis, Fabrication

---

M.C. Vestergaard
Japan Advanced Institute of Science and Technology, 1-1 Asahidai, Nomi City, Ishikawa 923-1292, Japan

M. Saito, H. Yoshikawa and E. Tamiya (✉)
Department of Applied Physics, Graduate School of Engineering, Osaka University, Suita, Osaka 565-0871, Japan
e-mail: tamiya@ap.eng.osaka-u.ac.jp

A. Tuantranont (ed.), *Applications of Nanomaterials in Sensors and Diagnostics*, Springer Series on Chemical Sensors and Biosensors (2013) 14: 171–188
DOI 10.1007/5346_2012_50,
Published online: 15 March 2013

Contents

1 Introduction

Tremendous progress in nanotechnology in the past few decades has seen the design and fabrication of metal nanoparticles (NPs) with properties suitably geared for biomedical diagnosis. Most notable of these NPs are noble metals such as gold and silver. They have been a subject of immense interest because these metals have unique optical characteristics that are different from the bulk counterparts. They are flexible and of wide applicability due to the fact that the size, composition, shape, assembly, and encapsulation can be easily manipulated [1]. Huang and colleagues in a recent review discussed the photophysical properties of matter, in particular gold NPs. They described what happens when light hits matter. The light can be (1) absorbed, (2) the absorbed light can be reemitted (fluorescence), (3) scattered at the same frequency as the incoming light (Mie or Rayleigh scattering), and (4) the local electromagnetic field of the incoming light can be enhanced. This leads to increase in any spectroscopic signals from the molecules at the material surface (surface-enhanced Raman scattering). These properties are even more enhanced in metal NPs because of unique interaction of light with free electrons in NPs (Fig. 1) [1]. Metal NPs also have unique electronic properties. Generally, they have excellent conductivity and catalytic properties, making them suitable for fabrication of "electronic wires" to enhance the electron transfer between redox centers in proteins and electrode surfaces, and as catalysts to increase electrochemical reactions [2–4].

Because of these optical properties, metal NPs [5–7] are exploited in detection platforms for increased analytical sensitivity. Au and Ag NPs have been exploited in the fabrication of localized surface plasmon resonance (LSPR) chips for detection of biomolecules [8–10]. These metal NPs show specific changes in their absorbance responses in the visible region of the spectrum upon binding with various molecules such as nucleic acids or proteins. The electronic properties, in particular, of Au and Ag NPs have been employed as labels for detection of proteins and other target molecules [11–17]. In this chapter, we will focus on the use of Au NPs in LSPR-based biosensor technology. We will discuss the principles and applications of how these NPs have been and can be exploited for medical diagnostics by providing examples, mainly to the work we have conducted in our research group.

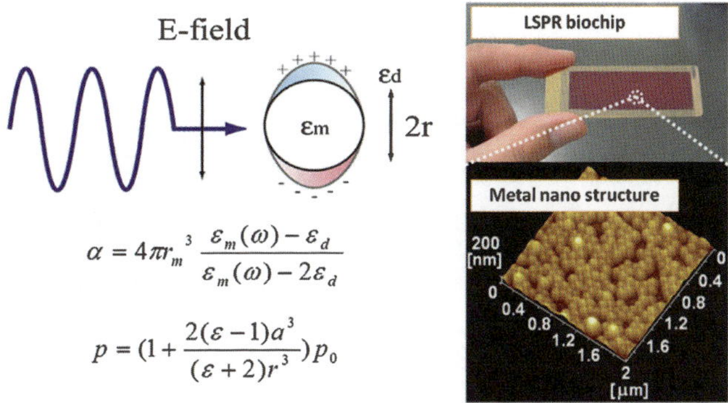

Fig. 1 Excitation of LSPR and gold nanostructures on the LSPR biochip

Biosensor technology has emerged as one of the most promising platforms for studying proteins. Biosensors are devices that combine a biological component (a recognition layer) and a physicochemical detector component (a transducer). The transduction unit can be electrochemical, optical, piezoelectric, magnetic, or calorimetric. The recognition layer can be constructed using enzymes, antibodies, cells, tissues, nucleic acids, peptide nucleic acids, and aptamers [18–21]. Two majority of biosensors are either affinity-based or catalytic-based biosensors. Affinity-based biosensors are used to bind molecular species of interest, irreversibly and non-catalytically. Examples include antibodies, nucleic acids, and hormone receptors. Catalytic-based sensors such as enzymes and microbiological cells recognize and bind a molecule of interest, followed by a catalyzed chemical conversion of that molecule into a product that is then detected. Given the breadth of technologies available, it would be hypocritical to attempt to cover them all. In this chapter we will focus mainly on localized surface plasmon resonance (LSPR), which is an optical transducer biosensor. First, we will discuss the basics in fabrication and/or surface modification of the recognition layers. Then, we will discuss how LSPR was exploited in detecting molecular binding events, by providing concrete examples and data of the different groups of biomolecules that were used as ligands. These include antibodies, nucleic acids, planar membrane bilayers, and aptamers.

2 LSPR-Based Biosensors

LSPR biosensors exploit the collective oscillations of free electrons in metal NPs surrounded by a dielectric media [22–24]. Specifically, the phenomena involve resonant coupling of surface plasmon resonance and photon(s). Briefly, as the size of a metal structure decreases from the bulk scale (m to μm) to the nanoscale (<100 nm), the movement of electrons through the internal metal framework becomes restricted. Consequently, metal NPs display extinction bands in their UV-Visual spectra when the incident light resonates with the conduction band electrons at their surfaces. These charge density oscillations are called LSPR. The excitation of LSPR by light at an

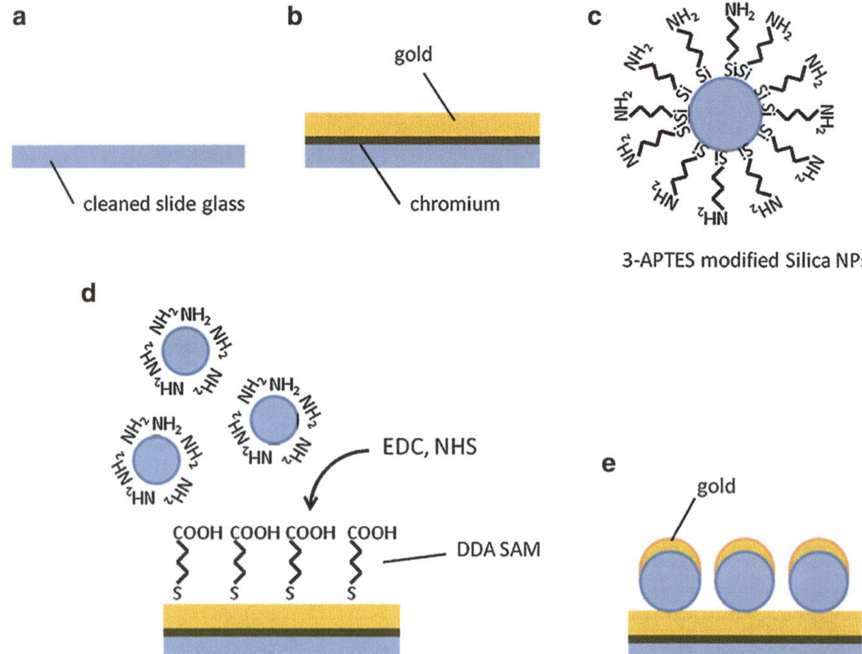

Fig. 2 Fabrication procedure of a gold-capped LSPR nanochip

incident wavelength, where resonance would occur, results in the appearance of intense surface plasmon (SP) absorption bands. The type, size, and shape of the NPs and their distribution affect the intensity and position of the SP absorption. LSPR is also highly sensitive to changes of the surrounding environments [25–28].

The basis of the LSPR detection technique is as follows: upon introducing a ligand and, after, its target molecule, there are changes in the thickness of the metal NP layer surface and this is reflected in an increase in refractive index and absorbance, enabling detection of the molecular species of interest [28]. Because the detection is label free, it can easily be applied to most molecules once the surface chemistry of the sensor surface has been modified accordingly. LSPR biosensors have been used widely to detect biomarkers of several pathologies including Alzheimer's, diabetes, and cancer [29–31]. First, we will discuss the basic design and fabrication procedure for an LSPR chip. The chip is later surface modified with appropriate linkers, recognition molecule(s) and passivated to avoid nonspecific adsorption. We will provide examples mainly from our group on how the fabricated chips have been used for detection of biomedically related molecules.

2.1 LSPR Chip Design and Fabrication

LSPR chips can be designed and fabricated in a few different ways. Here, we provide the fabrication procedure that we have utilized most in our research group [32, 33]. Basically, we used the following process (Fig. 2).

(a) Glass slides are cleaned thoroughly by ultrasonication in ethanol, soaked in piranha solution for 30 min, rinsed thoroughly with ultrapure water and finally, dried under a stream of nitrogen gas.
(b) After, chromium and gold are deposited on the glass slide substrate, using a thermal evaporator in conjunction with a quartz crystal microbalance (QCM), suitably adjusted to deposit the metal elements at the desired thickness. We used ~0.5 Å/s deposition rate, and thickness growth was monitored and manually stopped at 5 nm for chromium and 40 nm for gold. (Warning: Piranha solution is hazardous and highly reactive. It may explode on contact with organic solvents. Extreme care and precaution must be taken at all times).
(c) Silica NPs are surface modified using 1% (v/v) γ-APTES solution in ethanol by continual stirring at RT. After, the suspension is centrifuged and the supernatant decanted to waste. Ultrapure water is used to wash the amino-surface-modified Si NPs, centrifuged and the supernatant discarded as above. This process is repeated at least twice. The NPs are then dried at 120°C for 15 min and stored in a desiccator. Just before use, a Si NP colloidal solution is prepared at 1% (w/v) by dispersing in ultrapure water.
(d) For fabrication of LSPR chip on chromium/gold-deposited glass substrates, first, a multi-spot (~ mm id) silicon sheet is carefully placed on the chip surface. Then, 1 mM 4,4'-Dithiodibutyric acid (DDA) solution is introduced to the gold-layered surface and left for ~1 h to allow formation of self assembled monolayer (SAM) of the DDA. A covalent binding agent, 1-Ethyl-3-(3-dimethylamino-propyl) carbodiimide (EDC, 400 mM) and the previously surface activated Si NPs (10 mg/ml) are mixed 1:1 and introduced to the activated SAM for = 1 h. The EDC serves to activate carboxyl groups of the DDA SAM, which in turn forms esters between the NPs' amino groups and the SAM's carboxyl groups. After addition of each solution, the slides are washed with 50 mM phosphate buffer solution, pH 7.4 (PBS) to remove excess molecular species and subsequently dried at RT.
(e) Deposition of a gold layer (30 nm) to cap the Si NPs (using QCM procedure as reported above) completes the fabrication process.

Gold deposition on the surface of Si NPs enables easy fabrication of a uniform surface with minimum defects. One of the major drawbacks of LSPR chips utilizing the monolayer of Au NPs is the complicated chemistry required to form a self-assembled monolayer (SAM) of Au NPs on a solid surface. During mass-fabrication of numerous chips, small defects in the uniformity of SAMs can cause significant problems in the reproducibility and reliability of the results. Thus, it is important to analyze the surface characteristics of the fabricated chip. In Fig. 1, we show the surface of a fabricated LSPR chip, imaged using atomic force microscopy (AFM). As shown, our robust and highly reproducible fabrication technique circumvented these problems that are commonly faced during the preparation of Au NP monolayers on solid surfaces.

2.2 *Immuno-Chips for Protein Detection*

Immuno sensors exploit the interaction between an antibody (Ab), synthesized in response to the target molecule, an antigen (Ag). Antibodies can be formed when they are attached to an immunogen carrier such as serum albumin. Antibody-binding sites are located at the ends of two arms (Fab units) of the Y-shaped protein. The tail end of the Y (aka Fc unit) contains species-specific structure, commonly used as an antigen for production of species-specific Abs. The antibody is used as the recognition layer in biosensor development. There are a number of immuno sensor formats [32].

2.2.1 Detection of Tau Protein

Using a multi-spot chip fabricated as discussed in Sect. 2.1, we constructed a biosensor for selective detection of tau protein, at room temperature. The study we describe here was conducted and reported by Vestergaard and colleagues [9]. First, the LSPR chip was chemically modified using 1 mM DDA, then EDC (400 mM) and NHS; 100 mM) (1:1 v/v) for SAM formation and functionalization of SAM, respectively, following the procedure discussed in Sect. 2.1. After, we immobilized protein G for ~1 h at concentrations between 0 and 200 μg/mL. We immobilized protein G as our antibody-binding substrate, exploiting the high affinity between the Fc region of antibodies (in our case, anti-mouse IgG) and protein G [20] resulting in "upright" orientation of the antibody for better interaction with the antigen. Then, we blocked unreacted ester groups by immobilizing 0.1 M ethanolamine HCl solution, pH 8.5 for 30 min [33]. Tau-mAb was subsequently immobilized and optimized accordingly. Each immobilization step was followed by stringent washing of the chip surface using PBS and drying at RT. At all stages, we optimized the immobilization of molecules, by carrying out a concentration dependence study. The detection method exploited changes in the thickness of the gold-capped Si NPs chip surface upon introduction of molecular species on the chip surface.

Tau is a 50–65 kDA protein. It is reported that in the brains of AD individuals, tau proteins become abnormally hyperphosphorylated. It is reported to be phosphorylated at more than 20 residues. In healthy individuals, 8–10 of these residues are heterogeneously phosphorylated and lose the capacity to bind to microtubules. Instead, the phosphorylated tau proteins bind to each other inside nerve cells, tying themselves in "knots" known as NFTs [34–36]. Levels of tau are increased in CSF of AD individuals compared to age-matched controls [37, 38], probably due to neuronal and axonal degeneration or accumulation of NFTs [35]. A cutoff value of 195 pg/mL tau in CSF can accurately differentiate clinically diagnosed AD cases from controls with 89% specificity [39].

The fabricated Au-capped NP-modified surface gave high performance, enabling detection of tau protein at 10 pg/mL (Fig. 3). This LoD is much lower than the cutoff value of 195 pg/mL (for AD) for tau protein in cerebral spinal fluid

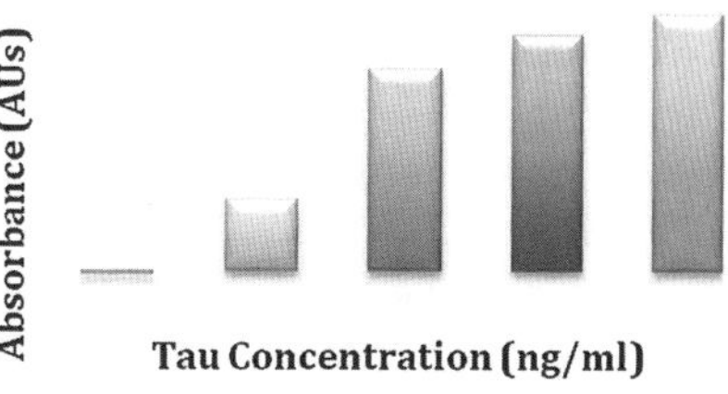

Fig. 3 Detection of tau protein using Au-capped LSPR immuno chip

(CSF). We were also able to demonstrate selectivity of the technique using bovine serum albumin (BSA), perhaps the most-abundant protein component in serum and CSF. The presence of BSA has been shown to interfere with detection of low-abundant biomarkers in biological fluids [40]. In this study, BSA did not interfere with the detection of tau, even at such physiologically low level of concentrations. Such immuno-chips hold huge potential for biomedical diagnosis.

In a separate study, a multi-array LSRP nanochip for simultaneous detection of half a dozen proteins was fabricated based on the design and fabrication of previous chips (Sect. 2.1) [8]. The 300 spot multi-array immuno-sensing platform provided detection limits down to 100 pg/mL.

2.2.2 RNA Apta-Sensor for Detection of Antibody–Antigen Interaction

Aptamers are synthetic oligonucleotides that can be synthesized to selectively bind to low-molecular-weight organic and inorganic substrates and to macromolecules such as proteins, drugs, with high affinity [41–46]. The affinity constants for aptamers are comparable to the binding constants of antibodies to antigens, i.e., it is in the micromolar to nanomolar scale [47]. Here we present work by Ha and colleagues who developed an LSPR-based RNA apta-sensor for detection of antibody–antigen interaction [48].

In their study, the authors focused on immobilizing a thiolated RNA aptamer which is able to catch Fc region of antibody. This would aid in orientating the antibody on the sensor's surface for better exposure to the target antigen. Several truncated and modified aptamers were designed and the optimized aptamer which had the highest affinity to the Fc portion of human IgG1 subclass was selected using surface plasmon resonance [49]. An optimized 23-nucleotide aptamer (Apt8-2, GGA GGUGCUCCGAAAGGAACUCC was prepared and was demonstrated to bind to the Fc domain of human IgG, with high affinity. These aptamers were synthesized and selected from a library of RNA sequences by modifying SELEX (Systematic Evolution of Ligands by EX potential enrichment) with 2-fluoropyrimidines. The affinity of the selected RNA aptamers to Fc portion of antifibrinogen was studied using SPRBIAcore2000. After, the aptamer was first immobilized on an LSPR chip surface (refer to Sect. 2.1 for fabrication). After, the antifibrinogen was immobilized. Using this construct, the authors were able to detect various antifibrinogen:fibrinogen interaction. Once the recognition surface was prepared, detections of antigen–antibody interactions were relatively rapid, taking less than 1.5 h in total [8, 50].

2.3 DNA Biosensors

DNA is particularly well suited for biosensing applications because there is a robust and specific interaction of the base pairs between complementary strands. DNA biosensors convert the Watson–Crick base-pair recognition (hybridization) event into a readable analytical signal. DNA hybridization has an important impact in various fields of life sciences, such as food control, environmental monitoring, and biomedical diagnosis [51]. Here, we discuss a couple of LSPR-based DNA biosensors.

2.3.1 Peptide Nucleic Acid LSPR Sensor for DNA Hybridization

Peptide nucleic acid (PNA) is a synthetic DNA analog. It was originally developed as a gene-targeting drug and hybridized with complementary oligonucleotides. Biosensors based on replacement of the DNA recognition layer with a PNA one offer greatly improved distinction between closely related sequences, as well as several other attractive advantages. The backbone of PNA is composed of repeating N-(2-aminoethyl)glycine units linked by peptide bonds, unlike the DNA or RNA backbone that is composed of deoxyribose and ribose sugar backbones. Since the PNA backbone is neutral (has no charged phosphate groups), its interaction with DNA is stronger than the electrostatic repulsion between DNA duplexes. PNA has therefore been exploited as a ligand in DNA biosensors for sensitive detection of DNA hybridization[51, 52] for applications to detecting single nucleotide polymorphism (SNP), including alcohol dehydrogenase and a mutation implicated in a dominant neurodegenerative dementia known as frontotemporal dementia with parkinsonism linked to chromosome 17 (FTDP-17)[53, 54].

Here we describe detection of DNA hybridization using on a gold-capped NP layer substrate (fabricated as discussed in Sect. 2.1 with slight modification) [55]. A PNA probe, designed to recognize the target DNA which was related to a mutation in tumor necrosis factor (TNF-R) [32], was immobilized on the LSPR chip surface. TNF-R is understood to be the basis of edema in many acute and chronic disease states, including ischemia and reperfusion injury.

For immobilization of the PNA probe (and DNA for comparison), synthetic oligonucleotides were used:

PNA or DNA probe: 5′-biotin- ACC ACC ACT TC -3′
DNA target: 5′- GGT TTC GAA GTG GTG GTC TTG -3′
single-base mismatch target: 5′- GGT TTC GAA GCG GTG GTC TTG -3′

The oligonucleotide stock solutions (100 mg/L) were prepared in 10 mM Tris–HCl, 1 mM EDTA (pH 8.0, TE). Oligonucleotide probe immobilization on the LSPR biosensor was carried out as discussed in Sect. 2.2.1 with a slight modification. DDA solution of 1 mM was introduced to the LSPR-based optical biosensor surface. After SAM formation and functionalization with EDC and then NHS solutions, streptavidin solution (100 μg/mL) was introduced to the NP layer

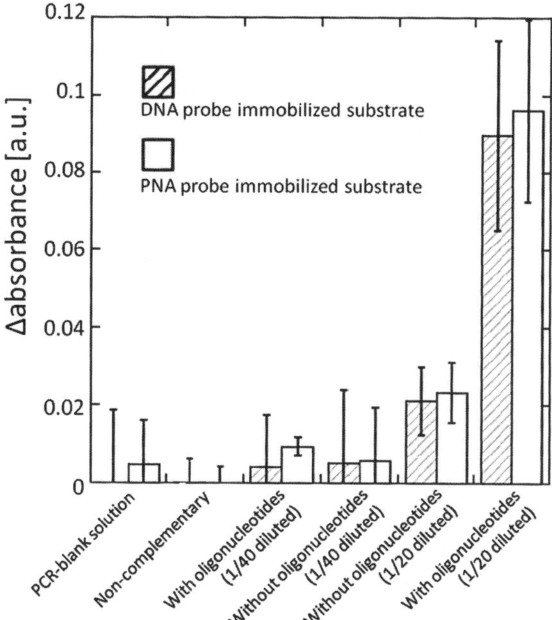

Fig. 4 Absorbance strength change of a PNA or DNA probe immobilized LSPR biosensor for detection of PCR-amplified TNF-R samples

substrate surface for 1 h. Then, biotinylated PNA or DNA probe (1.0 μM) was introduced to the streptavidin-modified LSPR-sensor chip surface for 1 h. Last, the probe-immobilized LSPR-based optical bio sensor surface was rinsed thoroughly with 20 mM phosphate-buffered saline (PBS, pH 7.4) and dried at room temperature.

Mutagenically separated polymerase chain reaction (PCR)-amplified samples of TNF-R were analyzed using the oligonucleotide-immobilized LSPR-based optical biosensor. Two mutagenically separated PCR amplified samples: TNF1 allele (−308G) producing an amplicon of 162 bp in length and TNF2 allele (−308A) producing a 147-bp amplicon was separated using gel electrophoresis. In case of the TNF1 allele, two amplicons were generated. The mutation on TNF2 allele was not detected in any of the donated DNA samples. Thus, the PCR-amplified samples of the TNF1 allele could be assigned to LSPR-based biosensor analysis. The results are shown in Fig. 4. As can be seen, the PNA probe, although not significantly different, gave a slightly higher sensitivity than the DNA probe. This study demonstrates the applicability of the gold-capped nanoparticle chip for diagnosis of mutations. The detection limit for mutations in tumor necrosis factor was as low as 0.677 pM and 0.803 pM for PNA- and DNA-based detection, respectively.

2.3.2 Real-Time Optical Detection of Polymerase Chain Reaction

Real-time monitoring of polymerase chain reaction (PCR) has been established as the main technique for the specific nucleic acid identification of biological samples

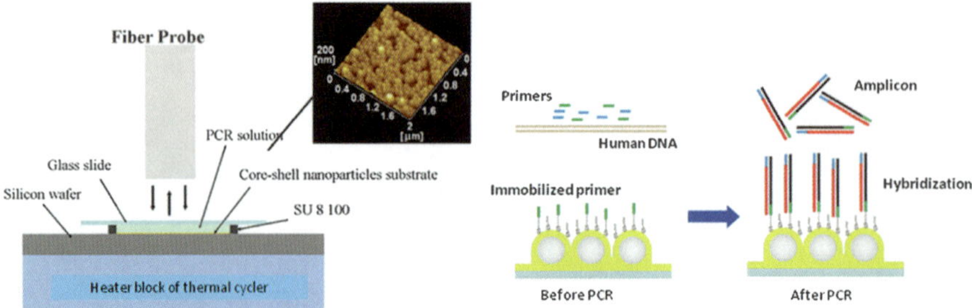

Fig. 5 Schematic of experimental setup for optical detection of PCR-based DNA amplification on a solid support modified with Au-coated NPs

[32, 56, 57]. Conventional real-time PCR technology generally uses fluorescence released from dye molecules, enabling the measurement of amplified DNA products after each cycle with a small value of standard deviation [58, 59]. The requirement for complex optical components and the inhibitory effects from some fluorescent reagents and probes make this approach difficult, in particular, if we want to consider using them at point-of-care testing (POCT), in a miniaturized portable instrument.

Ha and colleagues investigated real-time monitoring of PCR amplification using Au-coated nanostructured biochip with functionalized thiolated primers. A standard soft-lithography was used to design and fabricate a PCR chamber on a silicon wafer [60]. During its construction, a PCR chamber of 150 μm in thickness was bordered by SU-8 100 wall (a high contrast, epoxy-based photoresist material). A Au-coated NP substrate at the bottom of the PCR chamber was fabricated as described in Sect. 2.1. Figure 5 shows the experimental setup. The sensing capacities of the nanostructured substrate were characterized using refractive index of a number of solutions (from ethanol to toluene), and different lengths of alkanethiol molecules were immobilized on the nanostructured surface. After immobilization of 5′-thiolated primers on the surface, simultaneous DNA amplification and detection were performed label-less via the relative reflected intensity (RRI) of the Au-coated nanostructured substrate. When human genomic DNA at several concentrations of 0.2, 0.5, and 1 ng L^{-1} was included in the initial DNA samples, the increases in the RRI peak values were clearly observed with the increasing PCR cycle numbers. We could obtain optical signals, which were divergent from the background in our PCR biochip, after only ~3–4 cycles. This is much lower than that of the fluorescent real-time PCR analysis (around 23–25 cycles).

2.4 Interference LSPR Biosensor

In Sect. 2.1, we presented the basic design and fabrication procedure of our Au NP-based LSPR chips and demonstrated biomedical application in the ensuing sections. Although LSPR-based detection is sensitive and selective when an appropriate recognition element is immobilized on the surface, the LSPR signal from these

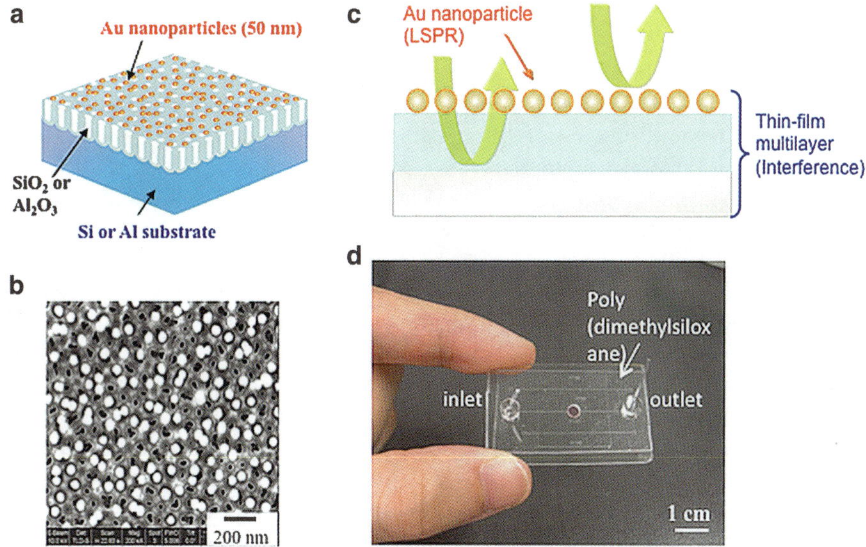

Fig. 6 Schematic (**a**) and a scanning electron micrograph (**b**) of an interference LSPR (iLSPR) biosensor. A schematic of the reflection light at normal incidence (**c**). An iLSPR biosensor integrated into a microfluidics chip (**d**)

devices is a broad spectral band with weak intensity, and thus they are useful for only for a limited number of biomolecules. In order to broaden the sensing applications of individual plasmonic devices, our group further advanced LSPR chip performance through a novel design and fabrication which we called interference LSPR (iLSPR) [61–63].

2.4.1 Development of iLSPR Biosensors

It is known that thin-film multilayers of a silicon dioxide/silicon (SiO_2/Si) substrate or an alumina/aluminum (Al_2O_3/Al) substrate show well-resolved Fabry–Perot fringes under light illumination. Since the Fabry–Perot fringe pattern undergoes wavelength shifts upon molecule binding, such structures have been used for biosensors [64]. A two-dimensional assembly of Au NPs, which could be regarded as a thin film, still retained the optical characteristics of the original NPs. From a combination of these principles and findings, we took a pioneering step to design a multilayered nanostructure with Au NP surface (Fig. 6a–c). Experimental reflection spectra of the iLSPR substrate were numerically confirmed by simulations using a combination of complex Fresnel coefficients and the Maxwell–Garnett effective medium theory [65, 66]. When the refractive index of the surrounding medium increased, an obvious spectral band coupled with LSPR appeared in the interference pattern of the reflection spectrum. What we describe here is development of a vanguard combination of plasmonic metal NPs and photonic thin-film multilayers on an interference LSPR (iLSPR) substrate for biomolecular detection [61].

Briefly, iLSPR substrates were fabricated by immobilizing Au NPs on an SiO_2/Si substrate. First, the SiO_2/Si substrate was cut into the desired size and soaked in a neutral pH detergent solution for 12 h. After cleaning the substrates in solutions of hydrogen peroxide (30%), ammonia (28%), and milliQ water for 30 min at 80°C, the substrates were immersed in 10% (v/v) solution of 3-aminopropyltrimethoxysilane for 15 min. After, they were rinsed with ethanol and dried at 120°C for 2 h. Subsequently, Au NPs were dispersed as a monolayer on the modified SiO_2/Si substrates through adsorption for 24 h, and the monolayer was heated under vacuum conditions for 2 h. Bonding of the NPs to the surface was strong enough to resist detachment from the surface because of later chemical modifications.

The developed chip was functionalized using alkane thiol molecules and the addition of the molecules resulted in increased in relative reflectance. Following this promising result, the iLSPR chip was integrated into a microfluidics chip (Fig. 6d). A standard soft-lithography technique was used to fabricate a microfluidic chip to cover the iLSPR substrate, using PDMS. This chip contained a microchannel and a chamber. The inlet and outlet of the microfluidic device were connected to fluorinated ethylene propylene (FEP) and sealed with very small amounts of PDMS to prevent liquid leaks. The microfluidic iLSPR chip was used to detect real-time label-free interactions between biotin and avidin molecules.

2.4.2 iLSPR Biosensor for Specific Biomolecular Detection

Following the successful design and fabrication of the iLSPR chip, a three-layer interference LSPR substrate constructed with plasmonic Au NPs and photonic thin-film multilayers of porous alumina/aluminum substrate was developed [62, 63]. Briefly, the iLSPR sensor was fabricated by immobilizing Au NPs on the surface of a porous aluminum oxide (Al_2O_3) layer formed on an aluminum (Al) substrate. Coupling the plasmon band with interference bands enhanced changes in the LSPR band that surrounds media with different refractive indexes, leading to increased sensor sensitivity. The sensor was verified to be highly sensitive and capable of detecting substantial change in refractive index that occurs upon the binding of analyte molecules. The thickness of the porous Al_2O_3layer could be controlled by adjusting electrochemical anodization time. This led us to conclude that the reflection spectra of the iLSPR substrates are dependent upon the thickness of the Al_2O_3layer. The results in this work showed that the porous nanostructure partially trapped two NPs, producing a small number of dimer NPs on the substrate surface. The spectral bands attributed to the longitudinal plasmon of the dimers were found at the suitable Al_2O_3thickness. Using theoretic simulation, we clarified the experimental reflection spectra of our iLSPR substrates.

We confirmed the significant reflectance change of the plasmon band attributed to dimers of Au NPs by monitoring biotin–avidin-binding events as shown in Fig. 7a–c. In addition, they also developed an immunosensor system based on the iLSPR

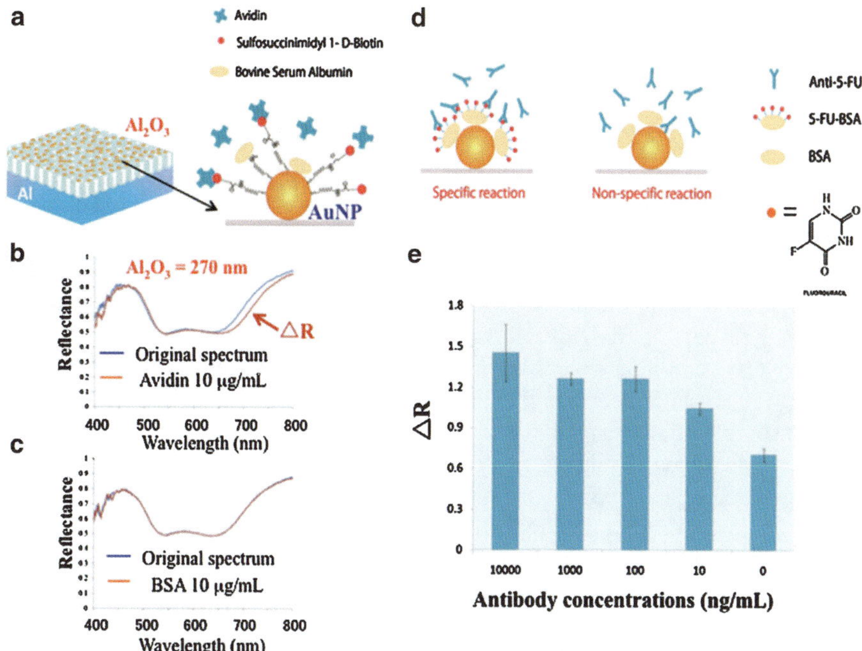

Fig. 7 Diagram of the iLSPR-based biosensor with its functionalized surface (**a**). The reflection spectra changes of biotin–avidin (**b**) and biotin–BSA (**c**) interactions. Diagram of the anti-5-fluorouracil detection (**d**). The dependence of reflectance variations (ΔR) on anti-5-fluorouracil concentrations (**e**)

substrate by examining specific interactions of 5- fluorouracil (5-FU) and anti-5-FU (Fig. 7d–e). The iLSPR sensors provided a limit of detection of 10 ng/mL, similar to the sensitivity obtained from recent ELISA results.

2.5 Detection of Melittin Binding to Planar Membrane Bilayer Using Electrochemical LSPR

We have so far described detection techniques that use physiologically relevant buffer systems in order to emulate the biological environment. These techniques are very sensitive and extremely useful for a range of applications including biomedical diagnosis. However, the physiological environment is more complicated. For detection of molecular interactions with the aim of understanding patho- and physiological mechanisms, an environment closer to the biological cell may be most useful. Model membrane systems serve as a half-way house between cell-based and buffer-based environments. Model membrane systems possess, as their main strength, the ability to enable the researcher to manipulate a "biological" micro-vesicle under a controlled environment [67]. They have been shown to effectively capture physicochemical events upon exposure to biological molecules such as amyloid beta and amylin which are implicated in protein misfolding pathologies [68, 69]. They are also used to study the effect of oxidative stress on membrane fluidity, dynamics [70]. A membrane-based electrochemical-LSPR sensor was developed in order to closely mimic the physiological environment. The nanosensor was used to detect the binding of a peptide protein, melittin, to the bilayer membrane [71].

Melittin is a pore-forming peptide toxin. Such toxins are cytotoxic, acting on a plasma membrane for the purpose of invading host cells [72]. Melittin, a noncell-selective lytic peptide from the venom of the honey bee has a direct effect on human erythrocytes lysis with the perturbation of the membrane, leading to the hemoglobin leakage [73, 74]. Thus, Tamiya's group developed a model membrane-based nanosensor in order to study the pore-forming effect of melittin on membrane with the aim of providing a feasible label-free sensitive way to detecting and studying some desirable properties in the artificial systems [71].

First, core-shell structure NPs substrate was fabricated as described in Sect. 2.1, with slight modification. The shell thickness was varied and performance of the chip determined at each thickness. This core-shell structure NPs substrate could be used simultaneously as a working gold electrode and LSPR-exciting device. All electrochemical measurements were performed by using a three-electrode system with a platinum wire as the counter electrode and an Ag/AgCl electrode as the reference electrode. The optical and electrochemical characteristics of the core-shell structure NP substrates were evaluated using a simple collinear optical system and Autolab PGSTAT 100 system. An absorbance peak at ~530 nm and a typical cyclic voltammogram of this substrate were clearly observed due to the rather regular NPs surface, confirming the good analytical performance of LSPR and electrochemistry analyses on the same surface.

Lipid vesicles were prepared by dissolving dimyristoylphosphatidylcholine (DMPC) in pure chloroform. The organic solvent was then removed with a nitrogen stream to form a thin lipid layer, and the samples were kept continuously in a vacuum desiccator for 12 h. The thin film was hydrated using 100 mM phosphate buffer saline (pH 7.5, containing 0.1 M NaCl) to a final lipid concentration of 0.5 mg/mL. The lipid vesicle solution was then sonicated for 1 h and used within 24 h. Next, a self-assembled alkanethiol layer was achieved by introducing 1 mM decanethiol solution onto the substrate surface for 1 h. The lipid vesicles were fused onto the alkanethiol-modified surface for 2 h. The result was a planar membrane bilayered Au-capped nano-shell biosensor (a hybrid bilayer membrane, HBM).

Melittin was introduced to the biosensor at various concentrations, and the binding was determined by following the change in absorbance values. Electrochemical analysis was used to probe the surface of the membrane. Following the amperometric response of a redox probe, [Fe(CN)6]3-/4-, enhancement of the signal was observed, indicating that melittin interacted with the biomimetic membrane, causing the leakage of this layer. Using impedance spectroscopy (IS), the interfacial electron-transfer properties at the core-shell structure NPs surface were altered by the adsorptions of 1-decanethiol, HBM, and melittin. It is routinely used to probe the electrode surface features because the interfacial electron-transfer at the electrode surface can be changed by modifying various biomaterial layers on the surface. The group observed that the electron-transfer resistance at the gold surface increased upon the formation of alkanethiol and HBM layers due to the blocking of the redox probe to the electrode surface by densely arranged successive layers. After interaction with HBM, melittin caused a gradual decrease in the charge-transfer resistance, as the HBM was being breached.

3 Concluding Remarks

In this chapter, we have attempted to demonstrate the use of gold nanoparticles for design and fabrication of a label-free Au-capped LSPR sensor chip. The fabricated Au nanosensor chips were immobilized with recognition ligands including antibodies, aptamers, and oligonucleotides for selective and sensitive detection of molecules of biomedical relevance. The flexibility of the fabricated chip allowed us to (1) modify the chip with at most ease to allow improved performance, (2) develop a multi-array chip to allow simultaneous detection of different biomolecules, (3) integrate microfluids, for real-time detection, and (4) design an interference LSPR biosensor with desirable characteristics that overcame the intrinsic problems with most LSPR chips. The LSPR signal from these devices is a broad spectral band with weak intensity.

We concede that we have only merely touched the surface of the range of LSPR chips and applications to biomedical diagnosis. Further, we have not at all covered the utilization of Au NPs in electrochemical biosensors in this chapter because it is another huge area and deserves a separate account.

References

1. Novoselov KS, Geim AK, Morozov SV et al (2004) Electric field in atomically thin carbon films. Science 306(5696):666–669
2. Huang X, Jain PK, El-Sayed IH, El-Sayed MA (2007) Gold nano-paricles: interesting optical properties and recent applications in cancer diagnostics and therapy. Nanomedicine 5:681–693
3. Lan J, Zhou X, Liu G, Yu J, Zhang J, Zhi L, Nie G (2011) Enhancing photocatalytic activity of one-dimensional KNbO3 nanowires by Au nanoparticles under ultraviolet and visible-light. Nanoscale 3:5161–5167. doi:10.1039/C1NR10953G
4. Kerman K, Kraatz HB (2009) Electrochemical detection of protein tyrosine kinase-catalysed phosphorylation using gold nanoparticles. Biosens Bioelectron 24:1484–1489
5. Selvaraju T, Das J, Jo K et al (2008) Nanocatalyst-based assay using DNA-conjugated Au nanoparticles for electrochemical DNA detection. Langmuir 24:9883–9888
6. Haes AJ, Chang L, Klein WL, Van Duyne RP (2005) Detection of a biomarker for Alzheimer's disease from synthetic and clinical samples using a nanoscale optical biosensor. J Am Chem Soc 127:2264–2271
7. Jensen T, Kelly L, Lazarides A, Schatz GC (1999) Electrodynamics of noble metal nanoparticles and nanoparticle clusters. J Cluster Sci 10:295–317

8. Haynes CL, Van Duyne RP (2001) Nanosphere lithography: a versatile nanofabrication tool for studies of size-dependent nanoparticle optics. J Phys Chem B 105:5599–5611
9. Endo T, Kerman K, Nagatani N, Hiep HM, Kim D-K, Yonezawa Y, Nakano K, Tamiya E (2006) Multiple label-free detection of antigen-antibody reaction using localised surface plasmon resonance-based core-shell structured layer nanochip. Anal Chem 78:6465–6475
10. Vestergaard M, Kerman K, Kim D-K, Hiep HM, Tamiya E (2008) Detection of Alzheimer's tau protein using localised surface plasmon resonance-based immunochip. Talanta 74:1038–1042
11. Haes AJ, Hal PW, Chang L, Klein WL, VanDuyne RP (2004) A localised surface plasmon resonance biosensor: first steps toward an assay for Alzheimer's disease. Nano Lett 4: 1029–1034
12. Kerman K, Chikae M, Yamamura S, Tamiya E (2007) Gold nanoparticle-based electrochemical detection of protein phosphorylation. Anal Chim Acta 588:26–33
13. Chen Z-P, Peng Z-F, Zhang P, Jin X-F, Jiang J-H, Zhang X-B, Shen G-L, Yu R-Q (2007) A sensitive immunosensor using colloidal gold as electrochemical label. Talanta 72:1800–1804
14. Ambrosi A, Castaneda MT, Killard AJ, Smyth MR, Alegret S, Merkoci A (2007) Double-codified gold nanolabels for enhanced immunoanalysis. Anal Chem 79:5232–5240
15. Pumera M, Castaneda MT, Pividor MI, Eritja R, Merkoci A, Alegret S (2005) Manetically trigged direct electrochemical detection of DNA hybridization using Au_{67} quantum dot as electrical tracer. Langmuir 21:9625–9629
16. Pumera M, Aldavert M, Mills C, Merkoci A, Alegret S (2005) Direct voltammetric determination of gold nanoparticles using graphite-epoxy composite electrode. Electrochim Acta 50: 3702–3707
17. Gonzales-Garcia MB, Costa-Garcia A (1995) Adsorptive stripping voltammetric behaviour of colloidal gold and immunogold on carbon paste electrode. Bioelectrochem Bioenerg 38: 389–395
18. Kerman K, Kobayashi M, Tamiya E (2004) Recent trends in electrochemical DNA biosensor technology. Meas Sci Technol 15:R1–R11
19. Wang J (2006) Electrochemical biosensors: towards point-of-care cancer diagnostics. Biosens Bioelectron 21:1887–1892
20. Vestergaard M, Kerman K, Tamiya E (2006) An overview of label-free electrochemical protein sensors. Sensors 6:3442–3458
21. Bini A, Minunni M, Tombelli S, Centi S, Mascini M (2007) Analytical performance of aptamer-based sensing for thrombin detection. Anal Chem 79:3016–3019
22. Hao E, Li SY, Bailey RC, Zou SL, Schatz GC, Hupp JT (2004) J Phys Chem B 108:1224–1229
23. Park TH, Mirin N, Lassiter J, Hafner J, Halas NJ, Nordlander P (2008) Plasmonic properties of nanoholes. ACS Nano 2:25–32
24. Shoute LC, Bergren AJ, Mahmoud AM, Harris KD, McCreery RL (2009) Appl Spectrosc 63:133–140
25. Su H, Li Y, Li X-Y, Wong KS (2009) Optical and electrical properties of Au nanoparticles in two-dimensional networks: an effective cluster model. Opt Express 17:22223–22234
26. Hutter E, Fendler JH (2004) Adv Mater 16:1685–1706
27. Nath N, Chilkoti A (2002) Anal Chem 74:504–509
28. Haes AJ, Van Duyne RP (2004) Anal Bioanal Chem 379:920
29. Shen XW, Hiuang CZ, Li YF (2007) Localized surface plasmon resonance sensing detection of glucose in the serum samples of diabetes sufferers based on the redox reaction of chlorauric acid. Talanta 72:1432–1437
30. Zhou W, Ma Y, Yang H, Ding Y, Luo X (2011) A label-free biosensor based on silver nanoparticles array for clinical detection of serum p53 in head and neck squamous cell carcinomat. Int J Nanomedicine 6:381–386
31. Kreuzer MP, Quidant R, Salvador JP, Marco MP, Badenes G (2008) Colloidal-based localized surface plasmon resonance (LSPR) biosensor for the quantitative determination of stanozolol. Anal Bioanal Chem 391:1813–1820

32. Endo T et al (2005) Anal Chem 77:6976–6984
33. Kato K, Barsukov LY, Derrick JP, Kim H, Tanaka R, Yoshimo A, Shiraishi M, Shimada I, Arata Y, Roberts GCK (1995) Structure 3:79
34. Blennow K (2004) J Int Med 256:224
35. Sjogren M, Vanderstichele H, Agren H, Zachrisson O, Edsbagge M, Wikkelson C, Nagga K, Andreasen N, Davidsson P, Vanmechelen E, Blennow K (2001) Clin Chem 47:776
36. Hu YY, He SS, Wang X, Duan QH, Grundke-Iqbal I, Iqbal K, Wang J (2002) Am J Pathol 160: 1269
37. Sobow T, Flirski M, Liberski PP (2004) Acta Neurobiol Exp 64:53
38. Vandermeeren M, Mercken M, Vanmechelen E, Jan S, Van de Voorde A, Martin J-J, Cras P (1993) Detection of tau proteins in normal and Alzheimer's disease cerebrospinal fluid with a sensitive sandwich enzyme-linked immunosorbent assay. J Neurochem 61:1828–1834
39. Sunderland T, Linker G, Nadeem M, Putnam KT, Friedman DL, Kimmel LH, Bergeson J, Manetti GJ, Zimmermann M, Tang B, Bartko JJ, Cohen RM (2003) J Am Med Soc 289:2094
40. Vestergaard M, Tamiya E (2007) Anal Sci 23:1443–1446
41. Tuerk C, Gold L (1990) Systematic evolution of ligands by exponential enrichment: RNA ligands to bacteriophage T4 DNA polymerase. Science 249:505–510
42. Wilson DS, Szostak JW (1999) In vitro selection of functional nucleic acids. Annu Rev Biochem 68:611–647
43. Jayasena SD (1999) Aptamers: an emerging class of molecules that rival antibodies in diagnostics. Clin Chem 45:1628–1650
44. Cho EJ, Collett JR, Szafranska AE, Ellington AD (2006) Optimization of aptamer microarray technology for multiple protein targets. Anal Chim Acta 564:82–90
45. Musheev MU, Krylov SN (2006) Selection of aptamers by systematic evolution of ligands by exponential enrichment: addressing the polymerase chain reaction issue. Anal Chim Acta 564: 91–96
46. Jenison RD, Gill SC, Pardi A, Polisky B (1994) High-resolution molecular discrimination by RNA. Science 263:1425–1429
47. Li B, Du Y, Wei H, Dong S (2007) Reusable, label-free electrochemical aptasensor for sensitive detection of small molecules. Chem Commun 3780–3783
48. Ha HM, Saito M, Nakamura Y et al (2010) RNA aptamer-based optical nanostructured sensor for highly sensitive and label-free detection of antigen–antibody reactions. Anal Bioanal Chem 396:2575–2581
49. Miyakawa S, Nomura Y, Sakamoto T, Yamaguchi Y, Kato K, Yamazaki S, Nakamura Y (2008) Structural and molecular basis for hyperspecificity of RNA aptamer to human immunoglobulin G. RNA 14:1154–1163
50. Ha HM, Nakayama T, Saito M, Yamamura S, Takamura Y, Tamiya E (2008) A microfluidic chip based on localized surface plasmon resonance for real-time monitoring of antigen–antibody reactions. Jpn J Appl Phys 47:1337–1341
51. Kerman K, Vestergaard C, Tamiya E (2009) Electrochemical DNA biosensors: protocols for intercalator-based detection of hybridization in solution and at the surface. Biosens Biodetect 2:99–113
52. Wang J (1998) DNA biosensors based on peptide nucleic acid (PNA) recognition layers. A review. Biosens Bioelectron 13:757–762
53. Ratilainen T, Holmen A, Tuite E et al (1998) Hybridization of peptide nucleic acid. Biochemistry 37:12331–12342
54. Kerman K, Vestergaard M, Nagatani N et al (2006) Electrochemical genosensor based on peptide nucleic acid-mediated PCR and asymmetric PCR techniques: electrostatic interactions with a metal cation. Anal Chem 78:2182–2189
55. Gaylord BS, Massie MR, Feinstein FC et al (2005) SNP detection using peptide nucleic acid probes and conjugated polymers: applications in neurodegenerative disease identification. Proc Natl Acad Sci U S A 102:34–39
56. Poiesi C, Albertini A, Ghielmi S et al (1993) Cytokine 5:539–545

57. Heid CA, Stevens J, Livak KJ, Williams PM (1996) Genome Res 6:986–994
58. Beer NR, Wheeler EK, Lee-Houghton L, Watkins N, Nasarabadi S, Hebert N, Leung P, Arnold DW, Bailey CG, Colston BW (2008) Anal Chem 80:1854–1858
59. Sanburn N, Cornetta K (1999) Gene Ther 6:1340–1345
60. Hiep HM, Kerman K, Endo T et al (2010) Nanostructured biochip for label-free and real-time optical detection of plolymerase chain reaction. Anal Chim Acta 661:111
61. Ha HM, Yoshikawa H, Saito M et al (2009) An interference localized surface plasmon resonance biosensor based on the photonic structure of Au nanoparticles and SiO2/Si multi-layers. ACS Nano 3:446–452
62. Ha HM, Yoshikawa H, Tamiya E (2010) Interference localized surface plasmon resonance nanosensor tailored for the detection of specific biomolecular interactions. Anal Chem 82: 1221–1227
63. Ha HM, Yoshikawa H, Tamiya E et al (2010) Immobilization of gold nanoparticles on aluminum oxide nanoporous structure for highly sensitive plasmonic sensing. Jpn J Appl Phys 49:06GM02
64. Lin VSY, Motesharei K, Dancil KPS, Sailor MJ, Ghadiri MR (1997) A porous silicon-based optical interferometric biosensor. Science 278:840–843
65. Macleod HA (1986) Thin-film optical filters. Macmillan, New York
66. Kreibig U, Vollmer M (1995) Optical properties of metal clusters. Springer, Berlin
67. Vestergaard M, Hamada T, Takagi M (2008) Use of model membranes for the study of amyloid beta: lipid interactions and neurotoxicity. Biotechnol Bioeng 99:753–763
68. Engel MF, Khemtemourian L, Kleijer CC, Meedijk HH, Jacobs J, Verkleij EJ, de Kruijff B, Killian JA, Höppener JWM (2008) Membrane damage by human islet amyloid polypeptide through fibril growth at the membrane. Proc Natl Acad Sci U S A 105:6033–6038
69. Morita M, Vestergaard M, Hamada T et al (2010) Real-time observation of model membrane dynamics induced by Alzheimer's amyloid beta. Biophys Chem 147:81
70. Vestergaard M, Yoda T, Hamada T et al (2010) The effect of oxycholesterols on thermo-induced membrane dynamics. Biochim Biophys Acta, Biomembr 1808:2245–2251
71. Ha HM, Endo T, Saito M et al (2010) Label-free detection of melittin binding to a membrane using electrochemical-localized surface plasmon resonance. Anal Chem 80:1859–1864
72. Parker MW, Feil SC (2005) Prog Biophys Mol Biol 88:91–142
73. Tosteson MT, Holmes SJ, Razin M, Tosteson DC (1985) J Membr Biol 87:35–44
74. Naito A, Nagao T, Norisada K, Mizuno T, Tuzi S, Saito H (2000) Biophys J 78:2405–2417

DNA Sensors Employing Nanomaterials for Diagnostic Applications

Manel del Valle and Alessandra Bonanni

Abstract This chapter describes DNA sensors (genosensors) that employ electrochemical impedance signal as transduction principle. With this principle, hybridization of a target gene with the complementary probe is the starting point to detect clinical diagnostic-related genes or gene variants. Electrochemical impedance spectroscopy permits, then, a labeless detection, by simple use of a redox probe. As current topic, it will focus on the use of nanocomponents to improve sensor performance, mainly carbon nanotubes integrated in the sensor platform, or nanoparticles, for signal amplification. The different formats and variants available for detecting genes in diagnostic applications will be reviewed.

Keywords Carbon nanotube, DNA biosensor, Electrochemical impedance spectroscopy, Genosensor, Gold nanoparticles, Quantum dots

Contents

M. del Valle (✉)
Sensors and Biosensors Group, Chemistry Department, Universitat Autònoma de Barcelona, 08193 Bellaterra, Barcelona, Spain
e-mail: manel.delvalle@uab.es

A. Bonanni
Division of Chemistry and Biological Chemistry, School of Physical and Mathematical Sciences, Nanyang Technological University, SPMS-CBC-04-41, 21 Nanyang Link, Singapore 637371, Singapore
e-mail: a.bonanni@ntu.edu.sg

A. Tuantranont (ed.), *Applications of Nanomaterials in Sensors and Diagnostics*, Springer Series on Chemical Sensors and Biosensors (2013) 14: 189–216
DOI 10.1007/5346_2012_38,
Published online: 8 December 2012

Abbreviations

AC	Alternating current
AuNP	Gold nanoparticle
C	Capacitance
CNT	Carbon nanotube
CPE	Constant phase element
CPE	Carbon paste electrode
DNA	Deoxyribonucleic acid
dsDNA	Double-stranded DNA
EDAC	N-(3-Dimethylaminopropil)-N-ethylcarbodiimide hydrochloride
EIS	Electrochemical impedance spectroscopy
GCE	Glassy carbon electrode
H1N1	Influenza A – H1N1 gene
HIV	Human immunodeficiency virus
hpDNA	Hairpin DNA
IgG	Immunoglobulin G
LOD	Limit of detection
MWCNT	Multi-walled carbon nanotube
NHS	N-Hydroxysuccinimide
PCR	Polymerase chain reaction
PEG	Polyethylene glycol
PNA	Peptide nucleic acid
QCM	Quartz crystal microbalance
QD	Quantum dot
R	Resistance
R_{et}	Electron transfer resistance
RNA	Ribonucleic acid
SPR	Surface plasmon resonance
ssDNA	Single-stranded DNA
strept-AuNPs	Streptavidin-coated gold nanoparticles
SWCNT	Single-walled carbon nanotube
TEM	Transmission electron microscopy
Z	Impedance
Z_i	Imaginary component of impedance
Z_r	Real component of impedance
αHL	α-Hemolysin nanopore
ϕ	Phase angle
ω	Radial frequency

1 Introduction

In this chapter we describe current variants of DNA sensors (genosensors) [1–3] that employ electrochemical impedance signal for detecting the hybridization event of a target DNA. In this way, the clinical diagnostic-related sought gene or gene variant can be detected in a very simple way, with an electrically addressable device, and, potentially, without the use of any label. The chapter will describe existing variants for the measure and different formats for the assay. To improve the performance of these devices, current nanobiotechnology utilizes nanocomponents, either employed at the transducer level or integrated in the procedure itself, to improve detection or to amplify its signal. Carbon nanotubes (CNTs) and nanowires, or even gold nanoparticles, can be used to produce or to modify the transducing electrodes, fostering their electrical characteristics or helping in the immobilization of the recognition element. Metal nanoparticles or even quantum dots may be used in some of the existing formats, if a better signal-to-noise ratio is required. The chapter ends with a summary of existing applications related to clinical diagnostic and discussion of late trends.

The determination of nucleic acid sequences from humans, animals, bacteria and viruses is the departure point to solve different problems: investigation about food and water contamination caused by microorganisms, detection of genetic disorders, tissue matching, forensic applications, etc. [2–4]. With a gene assay, either by a laboratory method or by a genosensor, one can ascertain the presence of a certain gene in a sample, which in turn may provide highly interesting information such as: (1) a specific gene is found, e.g. this individual is carrier of a genetically inherited disease; (2) a certain living species is present, e.g. food contamination, with cases as *Salmonella* in egg make out or *Listeria* in meat; another interesting examples can be cited as in fight against food fraud, biothreat protection, as the detection of Anthrax, or environmental protection, e.g. finding the source of an Avian Influenza outbreak; and (3) the identity of an individual is unravelled, like in crime suspect identification, in establishing paternity or degree of kinship, or in animal breeding.

All these interesting applications, which were very difficult to achieve in the past, or very laborious and time-delayed, for example if they needed microbiological culture, now can be approached with genosensor schemes, with goals of allowing a simpler and wider use.

The standard gene assay in this moment is the hybridization assay with the use of a fluorescent labelled string [4]. A single-strand DNA probe is placed over a surface and is used to hybridize with the sought DNA, or DNA target. The use of a labelled DNA sequence is used to show if the hybridization took place, a functionality that can be attained in different ways, for example in competition with the analyte gene. Other labelling strategy commonly used in designing genosensors, apart from the use of fluorescent markers [5, 6], is the use of redox active enzymes [7, 8], magnetic particles [9] or nanoparticles of different nature [10, 11]. An indirect labelling scheme consists of the use of redox couple which intercalates into DNA double helix, such as metal complexes [12, 13] or organic dyes [14, 15], or the use of redox indicators in solution which improves impedance performance [2].

When looking for a label-free approach, that is, when no modification on the DNA string used for capturing or detecting the sought gene is performed, several alternatives are also available. A first option is using the electrochemical properties of DNA by measuring the signal due to the direct oxidation of DNA bases [16, 17]. The other alternatives imply the use of transducing techniques which are sensitive to surface changes and able to detect the hybridization event. Some of these techniques are the quartz crystal microbalance (QCM) [18–20], surface plasmon resonance (SPR) [21, 22] or electrochemical impedance spectroscopy (EIS) [2, 3, 23]. For the latter, several examples of application to the labeless detection of specific DNA sequences in different fields have been demonstrated [3]. However, an amplification step is often necessary to achieve a defined response with very low analyte concentrations. In this case, approaches used to enhance the signal related to the use of nanocomponents will be also treated in this chapter. Besides, current research that focus on impedimetric genosensors using nanocomponents, considering the experimental principle, design of the device or use for its operation, and including nanotubes, nanoparticles, quantum dots or nanopores will be reviewed.

1.1 EIS Background

Impedance spectroscopy is a powerful method for characterizing the complex electrical resistance of a system, being capable to detect surface phenomena and also changes of bulk electrical properties [24]. Then, it is becoming an invaluable method in electrochemical research, where a constant growth of applications has been noticed during the last decade.

EIS has been intensively used, for example, for the elucidation of corrosion mechanisms [25], for studying electrode kinetics [26], the electrochemical double layer or batteries [27] or in solid-state electrochemistry, for characterizing charge transport across membranes [28]. In the field of sensors it may be used for characterization and optimization purposes. When used with biosensors, it is particularly well suited to the detection of binding events on the transducer surface. In fact, EIS is irreplaceable for characterizing surface modifications, such as those that occur during the immobilization of biomolecules on the transducer. We will present a short introduction to the basic principles of electrochemical impedance spectroscopy to help better understanding the signals generated in the biosensing event.

After applying an AC potential (E_t) to a system, its impedance Z is generally determined by relating the observed current crossing it (I_t), see Fig. 1. Experimentally, this is determined by applying an AC voltage perturbation with small amplitude (5–10 mV) and detecting the generated current intensity response, and the process is repeated for a number of frequencies.

$$E_t = E_0 \cdot \sin(\omega \cdot t) \quad (1)$$

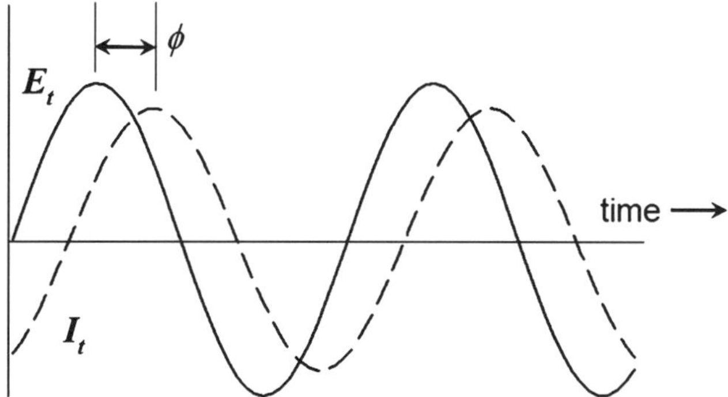

Fig. 1 Representation of the AC excitation signal, and the sinusoidal current response shown by a generic electrical circuit

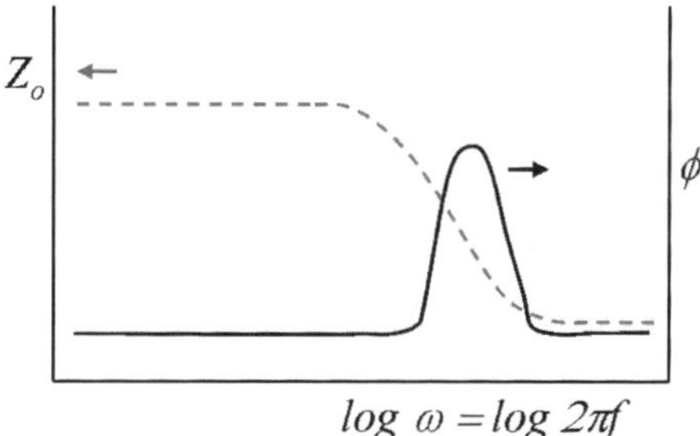

Fig. 2 Bode plot of the frequency characteristics of a given electrical circuit

$$I_t = I_0 \cdot \sin(\omega \cdot t + \phi) \tag{2}$$

From this definition, the impedance Z, also known as AC resistance, is the quotient of the voltage and current (Ohm's law for AC current):

$$Z = \frac{E_t}{I_t} = \frac{E_0 \cdot \sin(\omega \cdot t)}{I_0 \cdot \sin(\omega \cdot t + \phi)} = Z_0 \cdot \frac{\sin(\omega \cdot t)}{\sin(\omega \cdot t + \phi)} \tag{3}$$

And from this equation, it is evident that final impedance may be derived in terms of a magnitude Z_o and a phase angle ϕ. When these two magnitudes are plotted versus the scanned frequency, a characteristic representation is obtained, known as Bode plot (Fig. 2).

More informative for the sensor practitioner than the Bode plot is the Nyquist plot. This is constructed first by applying the Euler's equivalence between trigonometry and complex numbers; in this, the impedance is now written as:

$$Z = Z_r + jZ_i \tag{4}$$

being $j = \sqrt{-1}$.

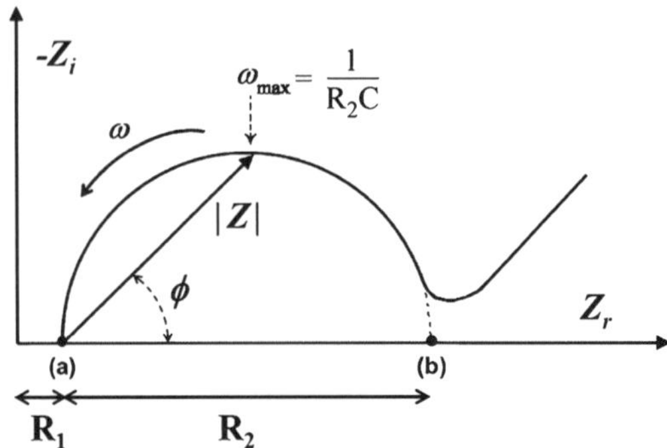

Fig. 3 Nyquist plot obtained for a typical reversible electrochemical reaction

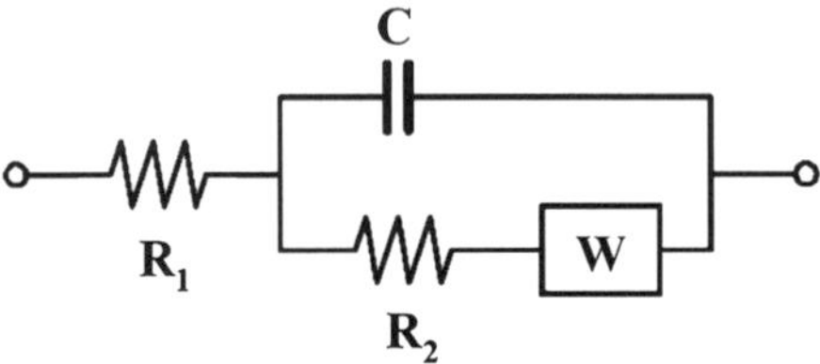

Fig. 4 Randles' equivalent circuit of a standard electrochemical reaction

And the Nyquist plot is derived when plotting for each scanned frequency, the imaginary part of the impedance ($-Z_i$) versus the real component (Z_r). Both Bode and Nyquist plots are responsible for the terminology "spectroscopy", given their use of frequency as the independent variable, thus recalling the situation with an electromagnetic spectrum. Figure 3 illustrates the typical Nyquist plot observed for a standard reversible electrochemical reaction taking place at a usual electrode. One of the valuable properties of the EIS technique is that from the shape and magnitude observed in the Nyquist plot (or alternatively in the Bode plot) one can derive which kind of electrical circuit is responsible for the profiles seen, and even calculate the electrical parameters involved. In fact, the pattern in Fig. 3 is very familiar to any electrochemist or any impedance practitioner, and it might be obtained with an electrical circuit like the one in Fig. 4. This electrical circuit, capable of providing an EIS spectrum which is equivalent to that previously seen (and so-called equivalent circuit), is well known and receives the name of Randles' equivalent circuit.

The interesting thing is that the individual elements present in it have physical meaning, illustrating the power of the EIS technique. In it, R_1 is the resistance of the solution, R_2 is the electron transfer resistance, that is, the kinetic impediment for the electrochemical reaction, C is the capacitance of the double layer, and finally, the 45° diagonal at the lower frequencies, called Warburg term, is related to the diffusion of species towards the electrode. The possibility of assigning individual elements to a circuit and finding values of their parameters is what gives to this technique the power of discriminating individual phenomena and also measuring its intrinsic characteristics. One final comment is that in many current situations, the capacitor term C is replaced by a special term, called constant phase element (CPE), originated in the lack of ideality of the electrode systems under test.

$$Z_{\mathrm{CPE}} = (j \cdot \omega)^{-\alpha}/C \quad (5)$$

where ω is the radial frequency, C the capacitance and α an empirical coefficient, related to the ideality of the system. For a CPE situation, the exponent $\alpha < 1$, since $\alpha = 1$ corresponds to the ideal capacitor. Generally the double layer between the solution and the electrode surface in an electrochemical cell is better fitted by a CPE than a capacitor.

Now turning into the genosensing application, the goal is not the electrochemical characterization of a system, but deriving the sensor signal related to hybridization of a DNA fragment. This means relating the change of one of the impedance elements, a resistance or a capacitance, depending on the specific format and design of the sensor, to the presence of the DNA gene sought and/or its concentration. Measurements can entail scanning the whole spectra range, or, probably, can be made simpler; once the system is characterized, it may be sufficient to determine the impedance at one selected frequency or within a certain frequency range.

For the typical genosensing application, a DNA probe, complementary to the one being sought, is immobilized on a working electrode, and the interaction with the target DNA (analyte) is monitored. Here the impedance of the working electrode (biosensor modified with the biological component) must be controlling the overall changes, for which auxiliary electrodes of sufficiently large area and a high concentration of saline background are preferred. Measurements with these surface-modified sensing electrodes are normally accomplished with the help of a redox-active compound, which is used as a probe. The observed phenomenon is then the electrochemical reaction of the probe, which is affected by changes of the biologically modified surface. Hybridization is therefore translated into a change of the electron transfer resistance R_{et}, the analytical signal for this impedimetric biosensor. When the redox-active compound is not used, the alternative is to monitor changes on the capacitive impedance component (since R_{et} will become extremely large). Thus, a binding event at the electrode can be detected by following the change in R_{et} in the first case, or the change in the capacitance in the second case. The first situation, in which the electrochemical reaction of the redox probe is involved, is also referred to as Faradaic impedance, while the second situation, not involving directly a redox reaction is referred to as non-Faradaic. After representation of the impedance over a sufficiently ample frequency range, and/or altering the surface area of the devices, individual events can be separated, and the corresponding region where the impedance is dominated by the impedance element under investigation can be identified.

1.2 EIS Sensing Applications

Nowadays, EIS is a reference technique for characterization and study of any electrochemical process at the electrode–electrolyte interface [29]. Although the information that it can provide is also attainable from series of experiments

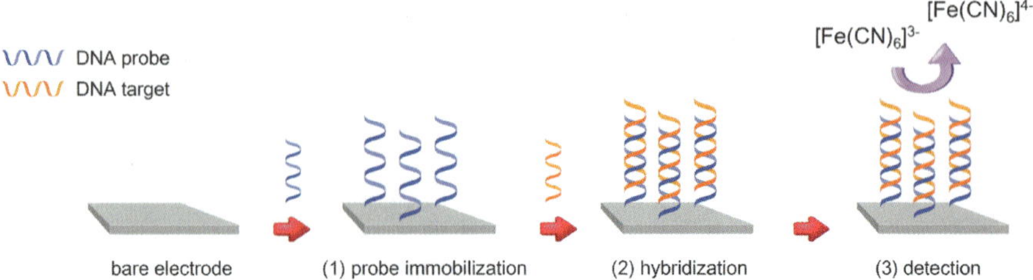

Fig. 5 Steps followed in an EIS genosensing experiment

employing the cyclic voltammetry technique at different potential scanning speeds, the powerful deductions that can be derived with the use of equivalent circuits makes EIS specially interesting to describe any electrochemical process.

Impedance spectroscopy is then mandatory for studies related to corrosion [30] semi-conducting electrodes [31], coatings [32], batteries and fuel cells [33], electrochemical kinetics and mechanism [34], biomedical and biological systems [35] and solid-state systems [36].

Due to its ability of directly probing the interfacial properties of a modified electrode, the technique is rapidly developing as a tool for studying biorecognition events at the electrode surface [23, 29, 37, 38]. In particular, EIS is becoming an attractive electrochemical tool for numerous applications either in immune-sensing [39, 40] or in genosensing field [2, 3, 41], especially in the last decade.

Generally speaking, the analysis of DNA using biosensors consists normally in a capture format, and can be described by these essential steps: (1) DNA probe immobilization onto the electrode surface, (2) Hybridization with a complementary target sequence and (3) Detection. These are schematized in Fig. 5. Normal sizes of the oligomers employed are ca. 25-mer for the probe, 20–50-mer for the target, and ca. 25 for additional fragments. The lengths specified are those typically used for PCR primers or for genetic assays, as the associated permutations assure a sufficiently high specificity. In some cases, additional steps are required in the protocol, such as sample preparation (i.e. PCR amplification), the use of other specific stages for signal amplification, or the use of systems for data treatment (i.e. Artificial Neural Network). Each genosensing step is then open to its monitoring by EIS, allowing for verification of its completeness. Figure 6 shows a typical evolution of the R_{et} observed for the ferrocyanide/ferricyanide redox probe during the steps of a genosensing experiment. In this case the redox species is considered a marker, not a label, since it is merely an accessory used for obtaining the signal and is only indirectly related to the sensing event. Each step in the experiment, associated with changes in the surface of the electrode, is responsible for altering the kinetics of this electrochemical reaction. This is due mainly to two chief effects: (1) the steric hindrance offered by the DNA probe in first instance, and the hybrid with the target, once formed; and (2) the electrical repulsion between the anionic backbone of the DNA double string and the anionic redox probe. This is the reason for the choice of the redox probe, among other options with neutral or cationic markers.

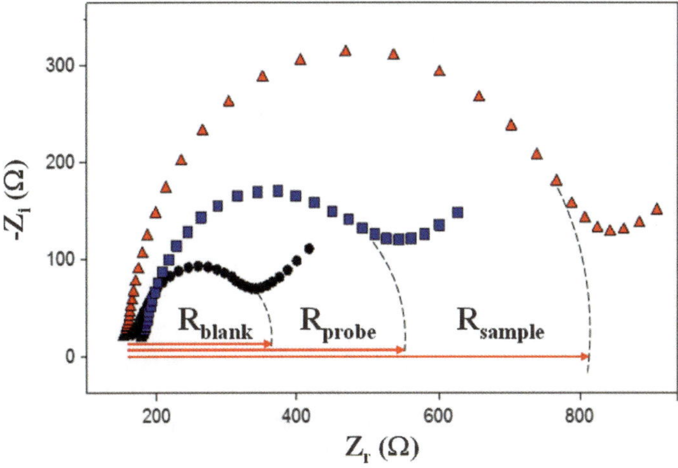

Fig. 6 Signals recorded in a typical EIS genosensing experiment

In some protocols, in order to enhance the difference in the signal obtained between the probe immobilization and the hybridization with a complementary sequence, instead of using DNA, a peptide nucleic acid (PNA) probe may be employed [42]. PNA is an artificially synthesized polymer which hybridizes equivalently to DNA, but in which the backbone is composed of repeating N-(2-aminoethyl)-glycine units linked by peptide bonds, instead of deoxyribose sugar backbone. With this change PNA results uncharged due to absence of hydrolysable phosphate groups, and all R_{et} variation observed in the biosensing process may be mainly attributable to hybrid formation [43].

In fact, the detailed observation of the evolution of the EIS signal during the genosensing process gives solution to one important problem with these sensors. The problem is that different electrodes may show slightly different impedance values, if their reproducibility of construction is not good enough. Moreover, this problem becomes worse in many occasions, as different measurements are performed with different electrode units or with the same unit after renewal of the sensing surface. This issue, which is in fact originated in the high sensitivity of the technique, poses difficulties in the representation and/or comparison of results between replicated experiments, the decision of positive or negative test, or also the quantitative estimation of target DNA.

The found solution to this problematic resides in normalizing the readings to the blank measurement, which is the one that may vary with differently produced electrodes. Hence, a solution is to express the parameter of interest (i.e. charge transfer resistance or capacitance) relative to the value given by the bare electrode [44]. Results are represented then as the relative R_{et} variation between net values obtained after DNA immobilization and hybridization. This relative variation is represented as a ratio of delta increments versus the bare electrode, as sketched on Eq. (6):

$$\Delta_{\text{ratio}} = \frac{\Delta_{\text{s}}}{\Delta_{\text{p}}} \tag{6}$$

being $\Delta_s = R_{et}(\text{sample}) - R_{et}$ (blank) and $\Delta_p = R_{et}$ (probe) $- R_{et}$ (blank). This elaboration was required for the comparison of data coming either from different electrode units or from the same unit after regeneration of surface. Briefly, when hybridization occurred Δ_s/Δ_p value should be >1 for the hybridization experiments and close to 1 for negative controls with non-complementary targets (that means $\Delta_s = \Delta_p$, i.e. no variation of R_{et} value because no hybridization occurred).

Apart of the generic sensing scheme, which in fact is the simplest concept, two additional variants should be commented. The first is the use of labelled targets, of interest when there is the need to increase the detection ability (i.e. decrease the detection limits); the second variant entails the design of more complex formats, i.e. a sandwich format with three or more DNA fragments.

For the use of labelled targets, e.g. the use of biotin to which many other functional groups may be linked is one possibility to amplify, or visualize the hybridization event by complementary techniques, i.e. fluorescence, amperometry or electron microscopy. For example, Ma and Madou [45] developed an enzymatic amplification scheme, employing a biotinylated oligonucleotide bound to a streptavidin-modified enzyme, in order to increase the sensitivity of the DNA sensor. Their approach took profit of the enzymatic precipitation of an insoluble compound on the sensing interface after hybridization, which caused an important impedance change. In a related protocol, Patolsky and Willner [46] also exploited the biocatalysed precipitation of an insoluble product on the transducer, to provide a mean to confirm and amplify the detection of a single-base mutation. The sensitivity of the method enabled the quantitative analysis of the mutant of Tay–Sachs genetic disorder without the need of PCR amplification. The same authors employed tagged, negatively charged, liposomes to amplify DNA sensing performance for hybridization and base mismatches detection [47]. One objection that may be stated here is that detection of a biotin-labelled (or any other label) DNA is unpractical, as the DNA analyte in a sample will not be biotinylated. But one should not forget that direct detection is not the unique possibility here, and in fact, an indirect, competitive assay may be used. In it, a fixed amount of biotinylated probe may be employed together with the sample, and any presence of the sought gene in the latter will produce a decrease in the finally observed signal.

An amplification that may be accomplished from a different strategy is to take profit of the ability of the double-strand DNA, or of some of its specific base-pairing points to interact with different species. For example, Bonanni et al. improved the sensitivity obtained for the detection of SNP correlated with kidney disease by performing the detection in presence of Ca^{2+} [48]. In this case, the specific binding

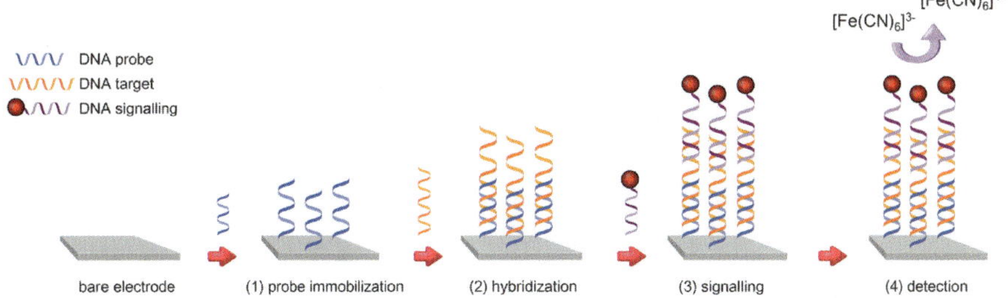

Fig. 7 Steps followed in a sandwich format EIS genosensing experiment, in this case using a labelled signalling probe

of the metal ions in the presence of A–C nucleotide mismatch induced a further impedance change, thus improving the discrimination between the mutated and healthy gene, as the signal amplification was achieved only for the former.

The second strategy that deserves comment is the use of multi-stage sandwich protocols, which simultaneously look for avoiding the use of labelled targets, to employ competitive schemes and to increase sensitivity. In essence, they employ a capture probe that hybridizes first with the sought gene in a sample, but in this case being larger than in previous examples, and using only a first half of its sequence for its capture. The second half is then free for fixing a third DNA string, named in this case signalling DNA, which may be directly detected by EIS in a labeless approach, or incorporate further labels to improve the detection. This general sandwich scheme is schematized in Fig. 7.

2 Use of Nanomaterials for Genosensing

The use of nanostructured materials for sensors and biosensor design and operation [49] is nowadays a very active field of research, where a wide variety of nanoscale or nanostructured materials of different sizes, shapes and compositions are now available [50]. The huge interest in nanomaterials is driven by their many desirable properties. In particular, the ability to tailor the size and structure and hence the properties of nanomaterials offers excellent prospects for designing novel sensing systems [51, 52] and for enhancing the performance of bioanalytical assays [53–55]. The similarity of dimensions between the involved molecules and the nanocomponents employed in these nanobiosensors is in part responsible for the increased efficiency and improved signal-to-noise ratios observed [56]. The use of these nanomaterials suggests their operation as effective mediators to facilitate the electron transfer between the active sites of probe DNA and surface of the electrodes. Moreover, the decrease in dimensions involved may show important advantages for integration of addressable arrays on a massive scale, which sets them apart from other sensors technologies available today.

The most widely utilized nanomaterials in impedance sensors are gold (Au) nanoparticles and CNTs [3]. Au nanoparticles have been employed in impedance sensors to form electrodes from nanoparticle ensembles and to amplify impedance signals by forming nanoparticle–biomolecule conjugates in the solution phase [57, 58]. CNTs have been employed for impedance sensors within composite electrodes and as nanoelectrode arrays [59, 60]. The advantages of nanomaterials in impedance sensors include increased sensor surface area, electrical conductivity and connectivity, chemical accessibility and electrocatalytic effect.

2.1 Use of AuNPs

One of the major trails of advance in nowadays nanotechnology is the use of nanoparticles. The unique chemical and physical properties of nanoparticles make them extremely suitable for designing new and improved sensing devices, especially electrochemical sensors and biosensors. Many kinds of nanoparticles, such as metal, oxide and semiconductor nanoparticles have been used for constructing electrochemical sensors and biosensors [55]. Owing to their small size (normally in the range of 1–100 nm), nanoparticles exhibit unique chemical, physical and electronic properties that are different from those of bulk materials and can be used to improve performance of sensing devices. Some important functions provided by nanoparticles include the immobilization of biomolecules, the catalysis of electrochemical reactions, the enhancement of electron transfer between electrode surfaces and proteins, the labelling of biomolecules and even their actuation as reagents. Of the different choices, one of the very relevant roles is the labelling of biomolecules, as they can retain their bioactivity and interact with their counterparts, and nanoparticles may be used for supplying the measurable signal.

Metal nanoparticle labels can be used in both immunosensors and DNA sensors. The most frequently used nanoparticles are those made of gold (AuNPs), given the extraordinary properties they present. Main use of AuNPs in genosensing is related to hybridization tagging, with added advantages of sensitivity enlargement [55]. Different from the amperometric detection [61], multiple tagging with nanoparticles of different nature is not useful for multiplexed detection of different genes in the same sample, given their nature may not be discriminated by EIS.

In a typical example of application of this type of nanobiosensor, Moreno-Hagelsieb et al. used a gold nanoparticle labelled oligonucleotide DNA target in order to amplify the capacitance signal between interdigitated aluminium electrodes imprinted over an oxidized silicon wafer [62]. As already commented, one does not expect to find gold-tagged nanoparticles in a generic sample, but their use can allow, for example, a competitive assay. In addition, a silver enhancement treatment, also useful for the electron microscopy detection, was performed offering a further signal amplification strategy. In a similar work Bonanni et al. used streptavidin-coated gold nanoparticles (strept-AuNPs) to amplify the impedimetric signal generated in a biosensor for the detection of DNA hybridization [58]. In this approach, a biotinylated target sequence was employed for the first capture by

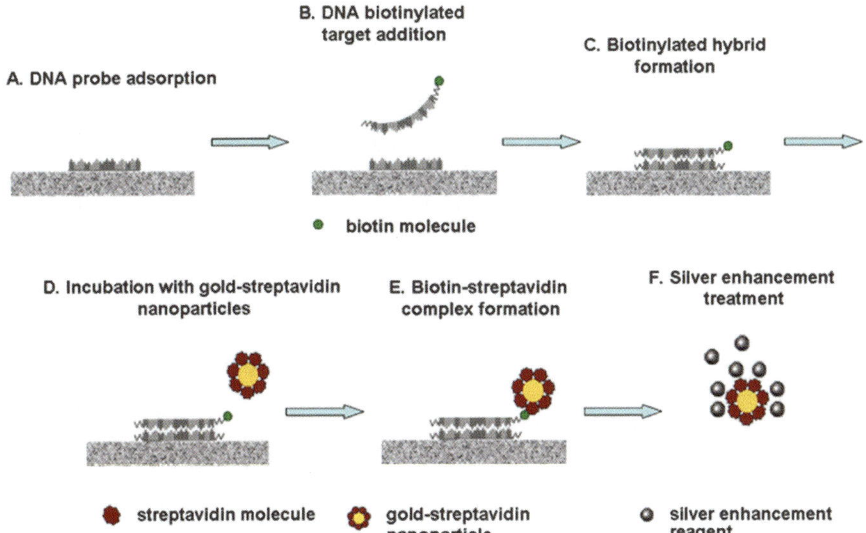

Fig. 8 Steps followed in a direct hybridization assay with impedimetric genosensor and amplification using strept-AuNPs and silver enhancement

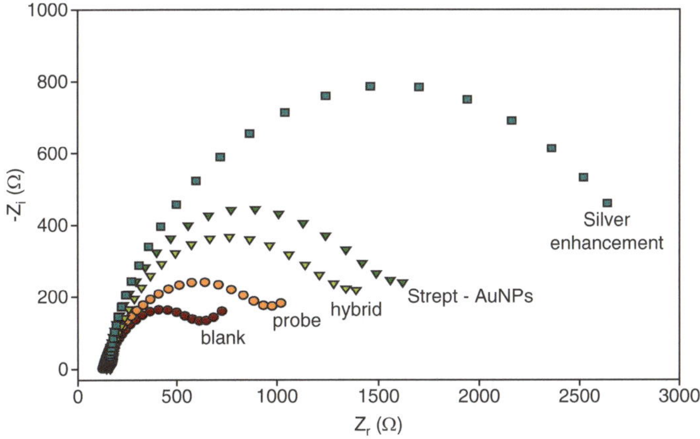

Fig. 9 Evolution of impedimetric signal after the different stages employed in the amplified genosening scheme employing strept-AuNPs

hybridization, followed by the conjugation with strept-AuNPs. The obtained impedimetric signal resulted 90% amplified in the presence of strept-AuNPs. Figure 8 schematizes the steps involved in the use of AuNPs for amplifying the impedimetric signal; the electrode used here was a simple epoxy graphite composite electrode [63, 64], and the immobilization used a simple adsorption.

Figure 9 is an illustration of the gain in impedimetric signal if the described protocol using amplification with gold nanoparticles is followed. After immobilization of the DNA probe, with little increase in R_{et}, the hybridization with the target

DNA, the primary positive signal, represents an increment similar to the fixation of the probe. The conjugation with the strept-AuNPs, still bringing a large amount of steric hindrance to the detection, is not translated in a large change in signal. It is only the amplification of the latter, with the catalytic reduction of silver onto the gold nanoparticles, which brings the largest gain in $R_{et.}$ Although not shown in here, it is obvious that even the use of non-biotinylated target or non-complementary biotinylated target produced very little signal, even lower in comparison once amplified.

In addition to labelling, a very interesting alternative of using AuNPs is to use them to construct three-dimensional networks with the nanoparticles dispersed throughout the sensing interface, in a nanostructured or molecular imprint approach, and that can be used to enhance impedance detection for biosensors. This may be accomplished through repeated use of a bifunctional gold coupling reagent, such as cysteamine or 4-aminothiophenol, where the amino group can bind to a biomolecule and the thiol group can bind to Au nanoparticles, for layer-by-layer formation of the Au nanoparticle network. Impedance detection of human immunoglobulin (IgG) using such a three-dimensional Au-nanoparticle network was recently reported using 6 nm diameter Au nanoparticles and cysteamine as the bifunctional reagent [65]. Some of the added advantages are the increased surface area for sensing, the improved electrical connectivity through the AuNPs network, the chemical accessibility to the analyte through these networks, and also the electrocatalysis.

2.2 *Use of QDs*

Quantum dots are nanometric scale semiconductor crystals (mainly sulphides, selenides or tellurides of heavy metals Cd, In, Zn or Tl) with unique properties originated in the quantum confinement effect that are advantageous for the development of novel bioassays, chemical sensors and biosensors [66]. Although mainly applied as fluorescent tags for biomolecules, where they bring out their exceptional properties, they have been also exploited in electrochemical sensors, normally with amperometric transduction, in which the heavy metal content after their dissolution can be detected by Anodic Stripping Voltammetry [67].

In our scope of interest, Xu et al. described a novel, sensitive DNA hybridization detection protocol, based on DNA-quantum dots nanoconjugates coupled with EIS detection. For this purpose, suitable DNA probes were covalently immobilized onto a self-assembled mercaptoacetic acid monolayer modified gold electrode; then, after hybridization with the target ssDNA-CdS nanoconjugate, they observed a remarkably increase in R_{et} value only when complementary DNA sequence was used in comparison with a three-base mismatched or non-completely matched sequences. The results showed that CdS nanoparticle labels on target DNA improved the sensitivity by two orders of magnitude when compared with nonlabelled DNA sequences [68].

For the case of the impedimetric technique, a very interesting work was reported by Travas-Sedjic's laboratory [57]. In this work, hybridization with a complementary DNA sequence is assayed employing a CdS nanoparticle label, showing a significant improvement in sensor sensitivity. In this variant, DNA probe was immobilized through entrapment during electropolymerization of conducting polymer (polypyrrole). Authors stated a limit of detection of the DNA probe of 1 nM. One important feature of their sensor is that it could be regenerated by removing hybridized DNA with NaOH, suggesting the possibility of sensor reuse.

In a similar work, Kjallman et al. employed a CdTe nanoparticle for the modification of a hairpin DNA probe. The stem–loop structured probes and the blocking poly(ethylene glycol) (PEG) molecules were self-assembled on the gold electrode through S–Au bonding, to form a mixed monolayer employed as the sensing platform. EIS was next used for characterization of the interfacial electrochemical characteristics of the modified gold electrode before and after hybridization with the target DNA [69]; this permitted to detect the target DNA with detection limit of 4.7 fm and even discrimination of non-complementary oligomers. Depending on the probe DNA to PEG ratio, the genosensor showed completely opposite response trends with regard to the change in charge transfer resistance and in the impedance at the electrode interface.

2.3 Use of CNTs

CNTs can be considered one of the most commonly used building blocks of nanotechnology [70]. CNTs are allotropes of carbon from the fullerene structural family, and can be conceived as sp^2 carbon atoms arranged in graphene sheets that have been rolled up into hollow tubes. Thanks to their extraordinary properties, like tensile strength, thermal and electrical conductivity or anisotropic conductivity behaviour, they are attracting much interest among all applied sciences and technologies. Analytical chemistry is one of the fields taking benefit of several advantages that CNTs bring for applications like chromatography, sensors, biosensors, and nanoprobes. There can be distinguished two main types of CNTs. The multi-walled CNTs (MWCNTs) behave as conductors and show electrical conductivities greater than metals. These interesting properties suggest that their incorporation into any electrical transduction scheme may be beneficial. Also, there is a second type of CNTs, the single-walled CNTs (SWCNTs), that depending on the tube diameter and chirality may behave electronically as either metals or semiconductors, complicating their use in sensing schemes. CNT synthesis methods create a mixture that includes amorphous carbon, graphite particles and CNTs, so synthesis is typically followed by a difficult and critical separation process. For electrochemical applications, CNTs are typically activated in strong acids, which opens the CNT ends and forms oxygenated species, making the ends hydrophilic and increasing the aqueous solubility of CNTs [52]. The electrochemical behaviour of CNTs varies considerably with the methods used for preparation and purification, including oxidation treatment. For analytical applications, and in part due to difficulties in their handling, CNTs are most often used to modify other electrode materials, or as part of a composite electrode.

As a first typical application, Xu et al. [71] incorporated multi-walled carbon nanotubes (MWCNTs) into composite electrodes used for impedance detection of DNA hybridization with a redox marker. In these studies, MWCNTs were co-polymerized with polypyrrole atop a glassy carbon electrode and then ssDNA was covalently immobilized. The complementary oligonucleotide was detected with the impedance technique by the accompanying change in R_{et}.

In the work of Caliskan et al. graphite electrodes were surface-modified with carboxylic acid functionalized SWCNTs; next, amino terminated DNA probes were covalently linked with the carbodiimide (EDAC)-N-hydroxysuccinimide (NHS) reaction to form an amide bond with the terminal acid groups. Finally, DNA target hybridization was monitored employing EIS and/or voltammetry [72]. The sequence chosen as study case was a specific gene for hepatitis B virus.

In the similar work in our laboratory, we employed a screen-printed, carboxyl functionalized MWCNT electrode, in which the detected gene was the sequence identifying the genetically modified organism Bt maize, given the high demand for analysis of transgenic food products [73]. For this purpose, the capture probe for the transgenic insect resistant Bt maize was covalently immobilized using the above carbodiimide chemistry; hybridization with DNA sample was followed, and impedance measurement performed in a solution containing the redox marker ferrocyanide/ferricyanide. A signal amplification protocol could also be performed, using a biotinylated complementary target to capture streptavidin-modified gold nanoparticles, thus increasing the final impedimetric signal (LOD improved from 72 to 22 fmol, maintaining a good reproducibility (RSD $<$ 12.8% in all examined cases).

An equivalent procedure was followed for an impedimetric genosensor devised for screening the Influenza A virus outbreak on spring 2009, which created a great social alarm [74]. Although the pathogenic H1N1 virus is a RNA virus, the diagnostic tools are normally prepared for its reverse transcripted DNA, given the higher availability of custom DNA synthesis. Figure 10 shows the preparation and detection scheme, in this case a sandwich capture format. First, the aminated DNA probe was immobilized using the carbodiimide chemistry (EDAC/NHS) to the carboxylated SWCNT-modified electrode. Then, hybridization with a previously formed duplex between the virus DNA and a biotinylated DNA probe was followed, to which further amplification employing strept-AuNPs was possible.

Figure 11 illustrates evolution of impedimetric signal along the process, in a very similar sequence as in Sect. 2.1: small increases for probe immobilization, noticeable increase for hybridization with the duplex, and possibility of amplification employing Strept-AuNPs. With these, a different strategy than before was used, which was a catalytic gold reduction onto the AuNPs instead of the classical silver reduction, in this case just to show a second amplification alternative.

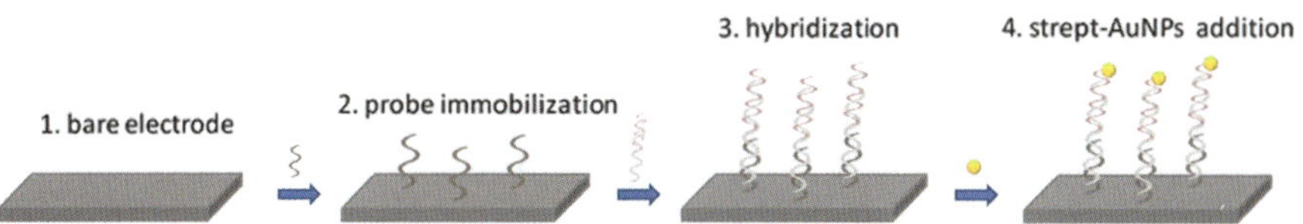

Fig. 10 Steps followed in a sandwich format EIS genosensing experiment, in this case using a labelled signalling probe

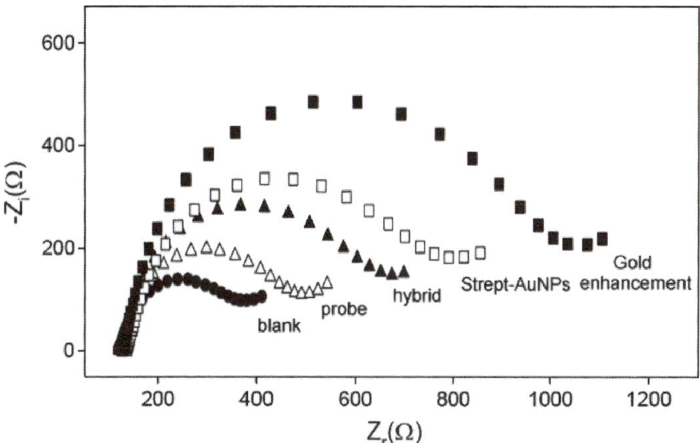

Fig. 11 Evolution of impedimetric signal after the different stages employed in the amplified genosening scheme employing strept-AuNPs

In a similar work [75], a genosensor for the impedimetric detection of the triple base deletion in a cystic fibrosis (CF)-related DNA synthetic sequence was shown. Screen-printed carbon electrodes containing carboxyl functionalized MWCNTs were used for the immobilization of an amino-modified oligonucleotide probe, complementary to the cystic fibrosis mutant gene. The complementary target (the mutant sequence) was then added and its hybridization allowed, later monitored by EIS. Results were contrasted against a non-complementary DNA sequence and a three-mismatch sequence corresponding to the wild DNA gene, present in healthy people. A further step employing a signalling biotinylated probe was performed for signal amplification using strept-AuNPs. With the developed protocol, a very sensitive detection of the triple base deletion in a label-free CF-related DNA sequence was possible, achieving an LOD around 100 pM.

A timely material very recently used to design biosensors and very much related to CNTs is graphene. Graphene is a two-dimensional lattice of carbon atoms arranged following an honeycomb pattern, and has become a star material sparked with the 2010 Physics Nobel prize, awarded to Novoselov and Geim (Manchester University) [76]. Graphene is an exceptional material in many regards, from huge charge mobility to strength and flexibility, offering a spectrum of applications ranging from flexible electronics to supercapacitors, composite materials and also biosensors, with amperometric or impedimetric transduction.

For example, the work of Muti et al. [77] used graphene oxide integrated on a graphite electrode for the enhanced monitoring of nucleic acids and for the sensitive and selective detection of the label-free DNA hybridization related to hepatitis B virus (HBV) sequences. The electrochemical behaviour of a graphene oxide-modified graphite electrode was firstly investigated using EIS and differential pulse voltammetry (DPV). The sequence selective DNA hybridization was determined voltammetrically in the case of hybridization between amino linked probe and its complementary (target), being capable of differentiating the noncomplementary target or a target/mismatch mixture (1:1).

In the work of Wang et al. [78] a reduced graphene oxide-modified glassy carbon electrode is used to detect the methicillin-resistant *Staphylococcus aureus* DNA, in this case using EIS detection. DNA probe is successfully anchored on the graphene-modified surface by simply adsorption. Hybridization with target DNA increased largely the measured impedance, with a detected amount of 100 fM.

A recent work from Bonanni and Pumera [79] investigated the suitability of different graphene surfaces for hairpin impedimetric genosensing. Electrodes modified by graphene nanoribbons were used. The hairpin DNA (hpDNA) probes were immobilized on the graphene-modified electrode surface by physical adsorption. The π-stacking interactions between the ring of nucleobases and the hexagonal cells of graphene made the platform a stable substrate for genosensing. Sensing mechanism was based on the partial release of the hpDNA probes from the graphene surface which occurs as a consequence of hybridization with complementary target, and translated in a significant decrease in R_{et}. Different DNA sequences correlated with Alzheimer's disease were used in this work, for example the mutated Apolipoprotein E gene. When hybridization was less effective, as in the case of the mutant target, a lower amount of the hpDNA probes are expected to be released, thus resulting in a less significant R_{et} decrease.

2.4 Use of Nanopores

As the last nanotechnology element to comment, the use of nanopores or nanochannels for detecting flux of ions biomolecules has to be mentioned [80]. Molecular-scale pore structures, called nanopores, can be assembled by protein ion channels through genetic engineering or be artificially fabricated on solid substrates using current nanofabrication technologies. When target molecules interact with the functionalized lumen of a nanopore, they characteristically block the ion pathway. The resulting conductance changes allow for identification of single molecules and quantification of target species. Detection can be accomplished through many different transduction mechanisms, mainly electrochemical.

A model example is the glass nanopore-terminated probe for single-molecule DNA detection designed by Takmakov et al. [81]. An array of nanopores was first prepared by anodization of aluminum, generating pores of ca. 10 nm. The inner pores were modified with biotin molecules via covalent attachment using aminosilane/succinimide chemistry. A first model detection was done employing the biotin-streptavidin pair, detected via impedance spectroscopy of the redox probe with a gold electrode formed at the bottom of the pore network. The same principle of pore blockage was also used to detect DNA hybridization onto the DNA probes immobilized inside the pores of the device.

Nanopores are key elements in the emerging technique of 4th generation DNA sequencers [82]. In these, a voltage is used to drive molecules through nanopores separating two solutions. When nucleotide bases, ssDNA or dsDNA, are threaded through these nanopores, a specific current (ionic current or other signal) can be monitored, which can be specific for the mononucleotide interacting with the nanopore. This is in essence the technology behind the sequencers being developed by companies like Oxford Nanopore Technologies in the UK.

Compared with conventional sequencing technologies, the nanopore single-molecule approach is simpler and more cost-efficient. It does not need fluorescent labelling or amplification of the sample DNA, obviating the use of restriction enzymes or redundancy. A huge potentiality can be foreseen, as it represents a direct sequencing, just like reading a teletype that further can be parallelized. The most usable technology at this moment [83] (nothing commercially available up to now) is the use of exonuclease enzyme to fragment the ssDNA and α-Hemolysin (αHL), a protein natively used in bacteria wall pores as the nanopore. αHL defines a 2 nm wide channel, with inner peptide fragments able to interact with passing species; when immobilized in a nanopore of the proper dimension (e.g. a nano-fabricated silicon structure) and forcing the unidirectional movement of bases through potential biasing, this protein interacts and permits identification of the four A,C,T,G bases, in principle through measurement of characteristic pico-currents [84]. Reasonable mononucleotide base throughputs with acceptable signal-to-noise ratio are ca. 25 s^{-1}, a translocation velocity not easy to accelerate because of worsening of sensitivity [85]. Other nanopore protein structures, such as the porin A from *Mycobacterium smegmatis* have also been demonstrated to produce DNA translocation for the sequencing purpose [86].

More stable nanopore systems can be potentially devised on graphene [87, 88]. With the same aim, completely nanofabricated systems have been proposed by IBM researchers, in this case with FET structures built along the nanopore and using capacitance detection [89]. Also, a coupled nanopore-hybridization strategy has been described, in which a library of ca. 10-mer probes align with ssDNA fragments and pass electrophoretically driven, a technology that has been named hybridization-assisted nanopore sequencing [90].

3 Application of Impedimetric Genosensors for Medical Diagnostics

Table 1 displays a summary of employed nanomaterials and applications of above-mentioned impedimetric genosensors, as summary of the use of these types of genosensors for diagnostic and other important applications. Among the topics

Table 1 Selection of examples of impedimetric genosensors employing nanomaterials and their applications from recent literatures

Working electrode	Nanocomponent used	Application	LOD	Reference
Al/Al_2O_3	AuNPs	Cytochrome P450 2p2 gene	2 pM	[91]
Graphite epoxy composite	AuNPs	Arbitrary sequence	120 nM	[58]
Graphite	SWCNTs	Hepatitis B virus	50 nM	[72]
MWCNTs	AuNPs	Transgenic maize	2 nM	[73]
Gold	AuNPs	Arbitrary sequence	5 nM	[92]
Glassy carbon	AuNPs/polyaniline nanotubes	PAT gene (transgenic crops)	300 fM	[93]
Glassy carbon	AuNPs	PAT gene	24 pM	[94]
Gold	CdS nanoparticles	Arbitrary sequence	1 nM	[57]
Al/Al_2O_3	AuNPs	HIV gene	200 pM	[62]
Glassy carbon	MWCNTs	Arbitrary sequence	50 pM	[71]
Glassy carbon	MWCNTs	Arbitrary sequence	5 pM	[95]
Carbon paste	SWCNTs	PAT and NOS genes	300 fM	[96]
MWCNTs	AuNPs	Influenza A virus – H1N1 gene	500 nM	[74]
Carbon paste	Polyaniline nanofibers, AuNPs, CNTs	Genetically modified beans	500 fM	[97]
Glassy carbon	Nano-MnO_2/chitosan	HIV gene	1 pM	[98]
Glassy carbon	CeO_2 nanoparticles, SWCNTs	(PEPCase) gene	200 fM	[99]
Carbon paste	AuNPs/TiO_2	Cauliflower mosaic virus gene	200 fM	[100]
MWCNTs	AuNPs	Cystic fibrosis gene related sequence	100 pM	[75]
Gold	CdTe nanoparticles	Arbitrary sequence	5 fM	[69]
Gold	CdS nanoparticles	Arbitrary sequence	5 pM	[68]
Graphite	Graphene	Hepatitis B virus	160 nM	[77]
Graphite	Graphene	Alzheimer's disease-related Apo-E gene	3 pM	[79]
Graphite	Graphene	Methicillin-resistant *Staphylococcus aureus*	100 fM	[78]

covered, several applications are devoted to the detection of transgenic plants and genetically modified organisms. One of the genes detected in varied applications is the PAT gene [93, 94], specific for transgenic crops. Some other works are centered on the simultaneous determination of PAT and NOS genes [96]. The determination of genetically engineered maize, transgenic Bt corn, was shown to be possible [73]. The equivalent detection of genetically modified beans employing a specifically devised DNA sensor is also present in the literature [97].

Other important applications, regarding the medical field, include the identification of certain gene or nucleotide polymorphism correlated with specific diseases. One of the first works described from the laboratory of Itamar Willner in Jerusalem was the sandwich determination of a gene related to the Tay–Sachs mutation that would be utilizable as a biosensor device to diagnose this genetically carried disease [101]. Similar works capable of detecting a gene cause of a inherited disease were those to detect cystic fibrosis [75]. Very recent efforts have also attempted to correlate certain gene with Alzheimer's disease and propose its detection [79]. The detection of its genetic material can be also the base for the confirmation of certain virus infections, and in this sense, impedimetric genosensors for detection of human immunodeficiency virus (HIV) to evaluate people suffering from AIDS [62, 98]. Nanobiosensors capable of detecting the hepatitis B virus [72, 77] have also been elaborated. After the pandemic Influenza A declaration in 2009, genetic assays were quickly prepared to diagnose and control the expansion of the disease; with these information, a nanobiosensor to detect its H1N1 virus genetic material was also developed [74].

The identification of microbiological species is also the other clear trend when classifying the nanobiosensors reported in the literature. As already commented in the diagnosis of illnesses of viral origin, HIV [62, 98], hepatitis B [77], Avian Influenza [102] or H1N1 Influenza [74] viruses are some of the available DNA biosensors. Also diseases to other organisms different to humans can be incorporated in this list, for example the cauliflower mosaic virus [100].

But there are not only viruses that can be detected by examining their genetic material; the presence of bacteria, or its specific variant may be evaluated also by examining their genetic material. For example, the work of Wang et al. that discriminated the strain of *S. aureus* resistant to antibiotic methicillin [78]. Also interesting is the work in the literature describing the identification of *Salmonella* spp employing capacitive detection [101], after its immunocapture with monoclonal antibodies grafted to AuNPs, these entrapped in electropolymerized ethylenediamine.

4 Outlook and Perspectives

The impedimetric genosensing topic is nowadays an active research area, where many formats and designs are reported in order to improve performance of existing biosensors. Research is still to be done in order to obtain devices with better reproducibility and stability, although any objection here can be balanced with the low detection limits achieved. Moreover, researchers should still increase efforts to get better electrode assemblies for their use in real samples, overcoming all problems associated with the complexity of matrices in various natural or commercial samples. Progress on these analytical features will accelerate their routine use, and even enable the massive production of devices using some of the principles stated in this chapter. Electrochemical impedance sensors are particularly promising for portable, on-site or point-of-care applications, in combination with simplified discrete-frequency instruments. However, there are certain impediments for solving these future applications and for the successful commercialization of useful devices, as minimizing effects of non-specific adsorption or automating all operation steps. And precisely these areas are the ones that can take more benefits from the incorporation of nanocomponents into genosensors.

A first challenge is the fabrication of useful electrochemically addressed genosensor arrays. The electrochemical impedance technique is fully compatible with multiplexed detections in electrically addressable DNA chips, which is one of the clear demands in genosensing for the next years [103]. Array sizes on the order of 10 have been described, but to be clinically useful, arrays of ca. 50 sequences are necessary. For example, a genetic disease like cystic fibrosis involves detection of around 25 different mutations plus the positive and negative controls. Microfabricated platforms can be of great help here, although issues like the mechanical reliability of the electrical contact, or reproducibility of construction and operation are still to be improved.

A second problem is related to sensitivity. DNA analysis is nowadays closely connected with PCR amplification, which is the step providing the major gain in it. Thus, two separate stages are needed, PCR, and afterwards, biosensing. Platforms are needed to integrate the two, allowing for really fast, intervention-less gene analysis. Microfluidic systems, of the lab-on-a-chip type, can be the solution here [104]. With such a platform, the goal of detecting a few viable pathogen microorganisms in a clinical sample in less than one hour might be a reality [4, 105].

The conversion of all the information which is generated with the unravelling and understanding of the functionalities yielded by the human genome is showing new achievements every day. Many of the properties found can be translated into genosensing applications to help in clinical practice and diagnostic, with small, cheap and decentralized analytical devices. But the challenge is even greater with the proteome. We are just in the beginning of its deciphering, for which high-throughput screening methods are in constant demand. Perhaps the principles used by electrochemical genosensors may be of help in the immense workload still to be done, to catalogue the human proteome in its biologically active form and to relate it to disease and cell state. Aptamer sensors, as already described, may be one starting point here. Most of the formats and strategies that have been described in this chapter are also extensible to specific detection of proteins, when the aptamer–protein interaction is exploited [106]. And impedance transduction is one of the simplest, more directly achievable transduction schemes available for their operation.

5 Conclusions

This chapter has presented current technology typically employed with genosensors which employ the EIS as the detection technique. Its operational principles and the essential protocols employed for impedimetric genosensing have been introduced. Although impedance is commonly used to investigate a variety of electrochemical systems, including fundamental redox studies, corrosion, electrodeposition, batteries and fuel cells, only recently it has been applied in the field of biosensors. Given its ability to monitor R_{et} and the double layer C, it is possible to derive applications for different types of sensing schemes, with numerous recognition agents, by direct signal acquisition, or with the use of simple and cheap redox markers. One chief advantage of impedimetric genosensing is that it can provide potentially label-free assays, as hybridization with the DNA probe immobilized on a surface can be directly monitored. In general, impedimetric genosensors are extremely simple in operation, and capable of achieving low detection limits even when used without any amplification. If combined with additional signal amplification strategies, their absolute detection limits may be comparable to other genosensing strategies. The contribution of nanostructured materials in the development of genosensors is an active research area of activity, and the use of nanoparticles, nanotubes, graphene or other nanostructured materials has been pointed out as some of the significant research with impedimetric nanosensors.

Acknowledgements Financial support for this work was provided by *Spanish Ministry of Science and Innovation*, MCINN (Madrid) through projects Consolider-Ingenio CSD2006-00012 and CTQ2010-17099 and by program ICREA Academia.

References

1. Cosnier S, Mailley P (2008) Recent advances in DNA sensors. Analyst 133(8):984–991
2. Lisdat F, Schäfer D (2008) The use of electrochemical impedance spectroscopy for biosensing. Anal Bioanal Chem 391(5):1555–1567
3. Bonanni A, del Valle M (2010) Use of nanomaterials for impedimetric DNA sensors: a review. Anal Chim Acta 678(1):7–17
4. Teles FRR, Fonseca LP (2008) Trends in DNA biosensors. Talanta 77:606–623
5. Piunno PAE, Krull UJ, Hudson RHE, Damha MJ, Cohen H (1995) Fiber-optic DNA sensor for fluorometric nucleic acid determination. Anal Chem 67(15):2635–2643
6. Fang L, Lü Z, Wei H, Wang E (2008) A electrochemiluminescence aptasensor for detection of thrombin incorporating the capture aptamer labeled with gold nanoparticles immobilized onto the thio-silanized ITO electrode. Anal Chim Acta 628(1):80–86
7. Lermo A, Campoy S, Barbé J, Hernández S, Alegret S, Pividori MI (2007) In situ DNA amplification with magnetic primers for the electrochemical detection of food pathogens. Biosens Bioelectron 22(9–10):2010–2017
8. Díaz-González M, de la Escosura-Muñiz A, González-García MB, Costa-García A (2008) DNA hybridization biosensors using polylysine modified SPCEs. Biosens Bioelectron 23(9):1340–1346
9. Ferreira HA, Cardoso FA, Ferreira R, Cardoso S, Freitas PP (2006) J Appl Phys 99:1–3
10. Wang J, Xu D, Kawde AN, Polsky R (2001) Metal nanoparticle-based electrochemical stripping potentiometric detection of DNA hybridization. Anal Chem 73(22):5576–5581
11. Wang J, Liu G, Polsky R, Merkoçi A (2002) Electrochemical stripping detection of DNA hybridization based on cadmium sulfide nanoparticle tags. Electrochem Commun 4(9):722–726
12. Millan KM, Saraullo A, Mikkelsen SR (1994) Voltammetric DNA biosensor for cystic fibrosis based on a modified carbon paste electrode. Anal Chem 66(18):2943–2948
13. Maruyama K, Motonaka J, Mishima Y, Matsuzaki Y, Nakabayashi I, Nakabayashi Y (2001) Detection of target DNA by electrochemical method. Sens Actuators B Chem 76(1–3):215–219
14. Yan F, Erdem A, Meric B, Kerman K, Ozsoz M, Sadik OA (2001) Electrochemical DNA biosensor for the detection of specific gene related to *Microcystis* species. Electrochem Commun 3(5):224–228

15. Hashimoto K, Ito K, Ishimori Y (1994) Sequence-specific gene detection with a gold electrode modified with DNA probes and electrochemically active dye. Anal Chem 66 (21):3830–3833
16. Wang J, Kawde AN (2002) Amplified label-free electrical detection of DNA hybridization. Analyst 127(3):383–386
17. Thorp HH (1998) Cutting out the middleman: DNA biosensors based on electrochemical oxidation. Trends Biotechnol 16(3):117–121
18. O'Sullivan CK, Guilbault GG (1999) Commercial quartz crystal microbalances – theory and applications. Biosens Bioelectron 14(8–9):663–670
19. Wang L, Wei Q, Wu C, Ji J, Wang P (2007) A QCM biosensor based on gold nanoparticles amplification for real-time bacteria DNA detection. In: Proceedings of the 2007 International Conference on Information Acquisition, ICIA, 2007, pp 46–51
20. Lazerges M, Perrot H, Zeghib N, Antoine E, Compere C (2006) In situ QCM DNA-biosensor probe modification. Sens Actuators B Chem 120(1):329–337
21. Schuck P (1997) Use of surface plasmon resonance to probe the equilibrium and dynamic aspects of interactions between biological macromolecules. Annu Rev Biophys Biomol Struct 26:541–566
22. Homola J, Yee SS, Gauglitz G (1999) Surface plasmon resonance sensors: review. Sens Actuators B Chem 54(1):3–15
23. Katz E, Willner I (2003) Probing biomolecular interactions at conductive and semiconductive surfaces by impedance spectroscopy: routes to impedimetric immunosensors, DNA-sensors, and enzyme biosensors. Electroanalysis 15(11):913–947
24. Macdonald JR (1987) Impedance spectroscopy. Wiley, New York
25. Mansfeld F, Han LT, Lee CC, Chen C, Zhang G, Xiao H (1997) Analysis of electrochemical impedance and noise data for polymer coated metals. Corros Sci 39(2):255–279
26. Armstrong RD, Bell MF, Metcalfe AA (1978) The A.C. impedance of complex electrochemical reactions. Electrochemistry 6:98–127
27. Roy SK, Orazem ME (2008) Analysis of flooding as a stochastic process in polymer electrolyte membrane (PEM) fuel cells by impedance techniques. J Power Sources 184 (1):212–219
28. Orazem M, Tribollet B (2008) Electrochemical impedance spectroscopy. Wiley, New York
29. Park SM, Yoo JS (2003) Electrochemical impedance spectroscopy for better electrochemical measurements. Anal Chem 75(21):455A–461A
30. Srisuwan N, Ochoa N, Pébère N, Tribollet B (2008) Variation of carbon steel corrosion rate with flow conditions in the presence of an inhibitive formulation. Corros Sci 50 (5):1245–1250
31. Tzvetkov B, Bojinov M, Girginov A, Pébère N (2007) An electrochemical and surface analytical study of the formation of nanoporous oxides on niobium. Electrochim Acta 52 (27 SPEC. ISS.):7724–7731
32. Nogueira A, Nóvoa XR, Pérez C (2007) On the possibility of using embedded electrodes for the measurement of dielectric properties in organic coatings. Prog Org Coat 59(3):186–191
33. Wagner N (2005) Electrochemical power sources-fuel cells. In: Barsoukov E, Macdonald JR (eds) Impedance spectroscopy: theory, experiment, and applications, 2nd ed, pp 497–537, Wiley, New York
34. Seland F, Tunold R, Harrington DA (2006) Impedance study of methanol oxidation on platinum electrodes. Electrochim Acta 51(18):3827–3840
35. Kharitonov AB, Alfonta L, Katz E, Willner I (2000) Probing of bioaffinity interactions at interfaces using impedance spectroscopy and chronopotentiometry. J Electroanal Chem 487 (2):133–141
36. Vladikova D, Raikova G, Stoynov Z, Takenouti H, Kilner J, Skinner S (2005) Differential impedance analysis of solid oxide materials. Solid State Ionics 176(25–28):2005–2009
37. Kell DB, Davey CL (1990) Conductimetric and Impedimetric devices. In: Cass AEG (Ed) Biosensors A practical approach, pp 125–154, IRL Press, Oxford

38. Berggren C, Bjarnason B, Johansson G (2001) Capacitive biosensors. Electroanalysis 13(3):173–180
39. Prodomidis MI (2010) Impedimetric immunosensors – a review. Electrochim Acta 55:4227–4233
40. Willner I, Willner B (1999) Electronic transduction of photostimulated binding interactions at photoisomerizable monolayer electrodes: novel approaches for optobioelectronic systems and reversible immunosensor devices. Biotechnol Prog 15(6):991–1002
41. Park JY, Park SM (2009) DNA hybridization sensors based on electrochemical impedance spectroscopy as a detection tool. Sensors 9(12):9513–9532
42. Degefa TH, Kwak J (2008) Electrochemical impedance sensing of DNA at PNA self assembled monolayer. J Electroanal Chem 612(1):37–41
43. Liu J, Tian S, Neilsen PE, Knoll W (2005) In situ hybridization of PNA/DNA studied label-free by electrochemical impedance spectroscopy. Chem Commun (23):2969–2971
44. Bonanni A, Esplandiu MJ, Pividori MI, Alegret S, del Valle M (2006) Impedimetric genosensors for the detection of DNA hybridization. Anal Bioanal Chem 385(7):1195–1201
45. Ma KS, Zhou H, Zoval J, Madou M (2006) DNA hybridization detection by label free versus impedance amplifying label with impedance spectroscopy. Sens Actuators B Chem 114(1):58–64
46. Patolsky F, Lichtenstein A, Willner I (2001) Detection of single-base DNA mutations by enzyme-amplified electronic transduction. Nat Biotechnol 19(3):253–257
47. Patolsky F, Lichtenstein A, Willner I (2001) Electronic transduction of DNA sensing processes on surfaces: amplification of DNA detection and analysis of single-base mismatches by tagged liposomes. J Am Chem Soc 123(22):5194–5205
48. Bonanni A, Pumera M, Miyahara Y (2010) Rapid, sensitive, and label-free impedimetric detection of a single-nucleotide polymorphism correlated to kidney disease. Anal Chem 82(9):3772–3779
49. Huang X-J, Choi Y-K (2007) Chemical sensors based on nanostructured materials. Sens Actuators B Chem 122(2):659–671
50. Poole CP, Owens FJ (2003) Introduction to nanotechnology. Wiley, New York
51. Wang J (2005) Nanomaterial-based electrochemical biosensors. Analyst 130(4):421–426
52. Merkoçi A, Pumera M, Llopis X, Pérez B, del Valle M, Alegret S (2005) New materials for electrochemical sensing VI: carbon nanotubes. TrAC, Trends Anal Chem 24(9):826–838
53. Katz E, Willner I, Wang J (2004) Electroanalytical and bioelectroanalytical systems based on metal and semiconductor nanoparticles. Electroanalysis 16(1–2):19–44
54. Sadik OA, Mwilu SK, Aluoch A (2010) Smart electrochemical biosensors: from advanced materials to ultrasensitive devices. Electrochim Acta 55(14):4287–4295
55. Luo X, Morrin A, Killard AJ, Smyth MR (2006) Application of nanoparticles in electrochemical sensors and biosensors. Electroanalysis 18(4):319–326
56. Suni II (2008) Impedance methods for electrochemical sensors using nanomaterials. TrAC, Trends Anal Chem 27(7):604–611
57. Peng H, Soeller C, Cannell MB, Bowmaker GA, Cooney RP, Travas-Sejdic J (2006) Electrochemical detection of DNA hybridization amplified by nanoparticles. Biosens Bioelectron 21(9):1727–1736
58. Bonanni A, Esplandiu MJ, del Valle M (2008) Signal amplification for impedimetric genosensing using gold-streptavidin nanoparticles. Electrochim Acta 53(11):4022–4029
59. Agüí L, Yáñez-Sedeño P, Pingarrón JM (2008) Role of carbon nanotubes in electroanalytical chemistry: a review. Anal Chim Acta 622(1–2):11–47
60. Wang J (2005) Carbon-nanotube based electrochemical biosensors: a review. Electroanalysis 17(1):7–14
61. Guo S, Wang E (2007) Synthesis and electrochemical applications of gold nanoparticles. Anal Chim Acta 598(2):181–192
62. Moreno-Hagelsieb L, Lobert PE, Pampin R, Bourgeois D, Remacle J, Flandre D (2004) Sensitive DNA electrical detection based on interdigitated Al/Al_2O_3 microelectrodes. Sens Actuators B Chem 98(2–3):269–274

63. Erdem A, Pividori MI, del Valle M, Alegret S (2004) Rigid carbon composites: a new transducing material for label-free electrochemical genosensing. J Electroanal Chem 567 (1):29–37
64. Pacios M, del Valle M, Bartroli J, Esplandiu MJ (2008) Electrochemical behavior of rigid carbon nanotube composite electrodes. J Electroanal Chem 619–620:117–124
65. Wang M, Wang L, Yuan H, Ji X, Sun C, Ma L, Bai Y, Li T, Li J (2004) Immunosensors based on layer-by-layer self-assembled Au colloidal electrode for the electrochemical detection of antigen. Electroanalysis 16(9):757–764
66. Frasco M, Chaniotakis N (2009) Semiconductor quantum dots in chemical sensors and biosensors. Sensors 9(9):7266–7286
67. Huang H, Li J, Tan Y, Zhou J, Zhu J-J (2010) Quantum dot-based DNA hybridization by electrochemiluminescence and anodic stripping voltammetry. Analyst 135(7):1773–1778
68. Xu Y, Cai H, He PG, Fang YZ (2004) Probing DNA hybridization by impedance measurement based on CdS-oligonucleotide nanoconjugates. Electroanalysis 16(1–2):150–155
69. Kjällman THM, Peng H, Soeller C, Travas-Sejdic J (2010) A CdTe nanoparticle-modified hairpin probe for direct and sensitive electrochemical detection of DNA. Analyst 135 (3):488–494
70. Merkoçi A (2006) Carbon nanotubes in analytical sciences. Microchim Acta 152(3–4 SPEC. ISS.):157–174
71. Xu Y, Jiang Y, Cai H, He PG, Fang YZ (2004) Electrochemical impedance detection of DNA hybridization based on the formation of M-DNA on polypyrrole/carbon nanotube modified electrode. Anal Chim Acta 516(1–2):19–27
72. Caliskan A, Erdem A, Karadeniz H (2009) Direct DNA hybridization on the single-walled carbon nanotubes modified sensors detected by voltammetry and electrochemical impedance spectroscopy. Electroanalysis 21(19):2116–2124
73. Bonanni A, Esplandiu MJ, del Valle M (2009) Impedimetric genosensors employing COOH-modified carbon nanotube screen-printed electrodes. Biosens Bioelectron 24(9):2885–2891
74. Bonanni A, Pividori MI, del Valle M (2010) Impedimetric detection of influenza A (H1N1) DNA sequence using carbon nanotubes platform and gold nanoparticles amplification. Analyst 135(7):1765–1772
75. Bonanni A, Esplandiu MJ, del Valle M (2010) Impedimetric genosensing of DNA polymorphism correlated to cystic fibrosis: a comparison among different protocols and electrode surfaces. Biosens Bioelectron 26(4):1245–1251
76. Geim AK, Novoselov KS (2007) The rise of graphene. Nat Mater 6(3):183–191
77. Muti M, Sharma S, Erdem A, Papakonstantinou P (2011) Electrochemical monitoring of nucleic acid hybridization by single-use graphene oxide-based sensor. Electroanalysis 23(1):272–279
78. Wang Z, Zhang J, Chen P, Zhou X, Yang Y, Wu S, Niu L, Han Y, Wang L, Chen P, Boey F, Zhang Q, Liedberg B, Zhang H (2011) Label-free, electrochemical detection of methicillin-resistant *Staphylococcus aureus* DNA with reduced graphene oxide-modified electrodes. Biosens Bioelectron 26(9):3881–3886
79. Bonanni A, Pumera M (2011) Graphene platform for hairpin-DNA based impedimetric genosensing. ACS Nano 5(43):2356–2361
80. Gu L-Q, Shim JW (2010) Single molecule sensing by nanopores and nanopore devices. Analyst 135:441–451
81. Takmakov P, Vlassiouk I, Smirnov S, Brevnov D, Atanassov P (2005) Impedimetric detection of biomolecules via ionic conductance through alumina nanoporous arrays. Meeting Abstracts MA 2005–02:2524
82. Rhee M, Burns MA (2006) Nanopore sequencing technology: research trends and applications. Trends Biotechnol 24(12):580–586
83. Venkatesan BM, Bashir R (2011) Nanopore sensors for nucleic acid analysis. Nat Nanotechnol 6(10):615–624

84. Clarke J, Wu H-C, Jayasinghe L, Patel A, Reid S, Bayley H (2009) Continuous base identification for single-molecule nanopore DNA sequencing. Nat Nanotechnol 4 (4):265–270
85. Astier Y, Braha O, Bayley H (2006) Toward single molecule DNA sequencing: direct identification of ribonucleoside and deoxyribonucleoside 5'-monophosphates by using an engineered protein nanopore equipped with a molecular adapter. J Am Chem Soc 128 (5):1705–1710
86. Niedringhaus TP, Milanova D, Kerby MB, Snyder MP, Barron AE (2011) Landscape of next-generation sequencing technologies. Anal Chem 83(12):4327–4341
87. Garaj S, Hubbard W, Reina A, Kong J, Branton D, Golovchenko JA (2010) Graphene as a subnanometre trans-electrode membrane. Nature 467(7312):190–193
88. Merchant CA, Healy K, Wanunu M, Ray V, Peterman N, Bartel J, Fischbein MD, Venta K, Luo Z, Johnson ATC, Drndić M (2010) DNA translocation through graphene nanopores. Nano Lett 10(8):2915–2921
89. Luan B, Peng H, Polonsky S, Rossnagel S, Stolovitzky G, Martyna G (2010) Base-by-base ratcheting of single stranded DNA through a solid-state nanopore. Phys Rev Lett 104 (23):238103
90. Ling XS (2011) Solid-state nanopores: methods of fabrication and integration, and feasibility issues in DNA sequencing. In: Iqbal SM (ed) Nanopores: sensing and fundamental biological interactions. Springer, New York, pp 177–201
91. Moreno-Hagelsieb L, Foultier B, Laurent G, Pampin R, Remacle J, Raskin JP, Flandre D (2007) Electrical detection of DNA hybridization: three extraction techniques based on interdigitated Al/Al2O3 capacitors. Biosens Bioelectron 22(9–10):2199–2207
92. Fu Y, Yuan R, Xu L, Chai Y, Zhong X, Tang D (2005) Indicator free DNA hybridization detection via EIS based on self-assembled gold nanoparticles and bilayer two-dimensional 3-mercaptopropyltrimethoxysilane onto a gold substrate. Biochem Eng J 23(1):37–44
93. Feng Y, Yang T, Zhang W, Jiang C, Jiao K (2008) Enhanced sensitivity for deoxyribonucleic acid electrochemical impedance sensor: gold nanoparticle/polyaniline nanotube membranes. Anal Chim Acta 616(2):144–151
94. Yang J, Yang T, Feng Y, Jiao K (2007) A DNA electrochemical sensor based on nanogold-modified poly-2,6-pyridinedicarboxylic acid film and detection of PAT gene fragment. Anal Biochem 365(1):24–30
95. Xu Y, Ye X, Yang L, He P, Fang Y (2006) Impedance DNA biosensor using electropolymerized polypyrrole/multiwalled carbon nanotubes modified electrode. Electroanalysis 18 (15):1471–1478
96. Jiang C, Yang T, Jiao K, Gao H (2008) A DNA electrochemical sensor with poly-l-lysine/single-walled carbon nanotubes films and its application for the highly sensitive EIS detection of PAT gene fragment and PCR amplification of NOS gene. Electrochim Acta 53(6):2917–2924
97. Zhou N, Yang T, Jiang C, Du M, Jiao K (2009) Highly sensitive electrochemical impedance spectroscopic detection of DNA hybridization based on Au nano-CNT/PAN nano films. Talanta 77(3):1021–1026
98. Liu ZM, Li ZJ, Shen GL, Yu RQ (2009) Label-free detection of DNA hybridization based on MnO_2 nanoparticles. Anal Lett 42(18):3046–3057
99. Zhang W, Yang T, Zhuang X, Guo Z, Jiao K (2009) An ionic liquid supported CeO2 nanoshuttles-carbon nanotubes composite as a platform for impedance DNA hybridization sensing. Biosens Bioelectron 24(8):2417–2422
100. Zhang Y, Yang T, Zhou N, Zhang W, Jiao K (2008) Nano Au/TiO2 hollow microsphere membranes for the improved sensitivity of detecting specific DNA sequences related to transgenes in transgenic plants. Sci China B Chem 51(11):1066–1073
101. Yang G-J, Huang J-L, Meng W-J, Shen M, Jiao X-A (2009) A reusable capacitive immunosensor for detection of Salmonella spp. based on grafted ethylene diamine and self-assembled gold nanoparticle monolayers. Anal Chim Acta 647:159–166

102. Fan H, Ju P, Ai S (2010) Controllable synthesis of CdSe nanostructures with tunable morphology and their application in DNA biosensor of Avian Influenza Virus. Sens Actuators B Chem 149(1):98–104
103. Levine PM, Gong P, Levicky R, Shepard KL (2009) Real-time, multiplexed electrochemical DNA detection using an active complementary metal-oxide-semiconductor biosensor array with integrated sensor electronics. Biosens Bioelectron 24(7):1995–2001
104. Mastichiadis C, Niotis AE, Petrou PS, Kakabakos SE, Misiakos K (2008) Capillary-based immunoassays, immunosensors and DNA sensors – steps towards integration and multi-analysis. TrAC, Trends Anal Chem 27(9):771–784
105. Sun H, Zhang Y, Fung Y (2006) Flow analysis coupled with PQC/DNA biosensor for assay of *E. coli* based on detecting DNA products from PCR amplification. Biosens Bioelectron 22(4):506–512
106. Radi AE, Sánchez JLA, Baldrich E, O'Sullivan CK (2005) Reusable impedimetric aptasensor. Anal Chem 77(19):6320–6323

Nanoprobes for In Vivo Cell Tracking

Juyeon Jung and Bong Hyun Chung

Abstract Cell-based immunotherapy has emerged as a promising therapy for the treatment of cancer. Measuring the changes in tumor volume and tumor markers post treatment has been the most common means of evaluating the therapeutic efficacy. In order to assess the consequences of a given therapy in real time, the development of efficient molecular probes and imaging modalities are urgently needed. Efficient molecular probes and imaging modalities will provide qualitative and quantitative real-time images with long-term stability in physiological conditions as well as low toxicity and high sensitivity for in vivo monitoring of the transplanted cells. Therapeutic cells can be intrinsically or extrinsically modified with proper molecular probes, amplified in vitro, and transferred back into the host. In this chapter, we will discuss the relative strengths and weaknesses of multiple molecular imaging modalities as well as recent advances in molecular imaging probes. We will also address their application in relation to in vivo tracking of dendritic cells (DCs), natural killer (NK) cells, and T cells. Noninvasive molecular imaging techniques have great potential in the diagnostic and prognostic assessments of patients.

Keywords Immune cell tracking, Molecular imaging modalities, Nanoprobes

Contents

J. Jung and B.H. Chung (✉)
BioNanotechnology Research Center, Korea Research Institute of Bioscience and Biotechnology (KRIBB), Daejeon 305-806, Republic of Korea
e-mail: chungbh@kribb.re.kr

A. Tuantranont (ed.), *Applications of Nanomaterials in Sensors and Diagnostics*, Springer Series on Chemical Sensors and Biosensors (2013) 14: 217–236
DOI 10.1007/5346_2012_48,
Published online: 15 February 2013

1 Introduction

In the past decade, major advances in molecular imaging have driven the development of noninvasive imaging technologies for biomedical research that allows in vivo investigation of specific cellular and molecular cascades in living subjects. Molecular imaging attempts to visualize and characterize complex biological processes of disease present within the context of physiologically authentic environments. Such noninvasive in vivo real-time monitoring by molecular imaging techniques allows for continuous tracking of specific cell targeting in the action site, drug delivery inside the target cells, and the evaluation of treatment in a rapid, reproducible, and quantitative manner. Therefore, molecular imaging has a great potential to enhance our understanding of drug delivery and localization of therapeutic cells in preclinical and clinical settings. The convergence of imaging techniques, cell biology, molecular biology, chemistry, medicine, pharmacology, medical physics, biomathematics, and bioinformatics contributes to the development of molecular imaging.

Currently available noninvasive image modalities can be broadly divided into morphological imaging and molecular imaging modalities. Morphological imaging modalities include magnetic resonance imaging (MRI), computed tomography (CT), and ultrasounds which are characterized by high spatial resolution. However, lower sensitivity than radioactive and optical methods and the requirement of expensive equipments limit applications of these techniques. Molecular imaging modalities including optical imaging (i.e., fluorescence and bioluminescence), positron emission tomography (PET), and single photon emission computed tomography (SPECT) are highly sensitive and can detect tumors at early stages. Optical imaging techniques have more potential for real-time imaging as compared to the morphological imaging modalities but will have limited spatial resolutions (Table 1).

Development and implementation of noninvasive imaging technologies for biomedical research have successfully advanced. These imaging techniques have had a significant influence on the way we detect and characterize diseases in both basic research and clinical applications. In spite of these successes, no single molecular imaging modality meets and/or provides complete information effectively in either preclinical studies or clinical applications. Therefore a combination of different imaging modalities, using multimodal molecular probes, can provide

Table 1 Comparisons of molecular imaging modalities

	Optical imaging	MRI	PET/SPECT	Ultrasound	CT
Spatial resolution	1–5 mm	25–100 μm	1–2 mm	50–500 μm	50–200 μm
Tissue penetration	<5 mm	No limit	No limit	mm–cm	No limit
Temporal resolution	Second–minute	Minute–hour	Second–minute	Second–minute	Minute
Sensitivity (mol/l)	10–15 to 10–17	10–3 to 10–5	10–10 to 10–12	ND	ND
Imaging probes	Fluorochromes, photoproteins	Iron oxide particles, gadolinium,	18F, 11C, 15O, 99mTc, 111In	Microbubbles	NA
Amount of probe used	μg–mg	μg–mg	ng	μg–mg	NA
Clinical application	Under development	Yes	Yes	Yes	Yes

both the spatial resolution and the adequate sensitivity in living subjects to enable detection of biochemical events and even minimal clinical changes over time (e.g., PET-CT and PET-MRI).

Over the past two decades, nanoparticle technology has attracted a great deal of attention in bioimaging and biosensing research because nanoparticles have unique sizes in the same range of dimensions as antibodies, membrane receptors, nucleic acids, and proteins among other biomolecules. Their biomimetic features, broad optical properties, and high surface to volume ratio make nanoparticles powerful tools for imaging, diagnosis, and therapy [1].

The ideal nanoparticles as nanoprobes should have the ability to produce an image and be targetable depending on their particle compositions, sizes, surface charges, surface functionalities, biocompatibilities, contrast sensitivities, and stabilities [2]. Several nanoparticles in preclinical studies and clinical trials have been developed to have controllable sizes and morphology, long-term stability in physiological conditions, low toxicity in cells, high sensitivity in real time, and target specificity. Individual nanoparticles participate in multiple applications such as monitoring with an imaging modality, targeting, and drug delivery.

Considerable attention has been paid to the development of immunotherapy which uses autologous, allogeneic, or xenogeneic cells to replace or renew damaged tissue and treat diseases. Products for cell therapy involve ex vivo propagation, expansion, selection, activation, or genetic alteration. Introducing naturally occurring biological molecules (e.g., immune cells) to boost the immune system seems to be the most promising therapeutic approach since it is natural and patient friendly. However, there is very little success in immunotherapy compared to the over 17,000 clinical trials of cell therapy due to the lack of technologies monitoring transplanted therapeutic immune cells with qualitative and quantitative images in real time. Development of novel molecular imaging probes with controllable sizes, low toxicity, compatible immunogenicity, target specificity, physiological stability, and longer circulation time in vivo for tracking transplanted immune cells to determine their location, cell numbers, and functional lifespan in patients has been urgently needed.

2 Molecular Imaging Modalities

2.1 Optical (Fluorescence and Bioluminescence) Imaging

Optical imaging modalities for in vivo molecular imaging include fluorescence and bioluminescence which are highly sensitive in the range of 10^{-15} to 10^{-17} mol/l but have low spatial resolutions of 1–5 mm. Optical imaging modalities are suited for drug development and target validation processes due to the ease of the operation, short acquisition times (10–60 s), simultaneous measurement, and relatively affordable cost. However, the limited depth penetration of these imaging techniques halts applications of optical agents in human.

Fluorescence imaging uses an external light source to excite fluorochromes such as green fluorescent protein, red fluorescence protein, and near-infrared (NIR) fluorescent probes. Light is then reemitted by the fluorochrome at a longer wavelength of lower energy. Fluorescence imaging is not easily applicable in biological tissue because the light intensity decays with the depth of the tissue, resulting in low signal to background ratios and autofluorescence [3].

Bioluminescence imaging uses photons emitted by a metabolic reaction between an enzyme (e.g., luciferase from firefly *Photinus pyralis*) and its specific substrate (e.g., D-luciferin). No background light emission is found due to lack of excitation light source required for the detection of the photons released, leading to high sensitivity detecting low levels of gene expression. However it is hard to ensure that the substrate is transported to all possible bioluminescent cells [4].

Despite the many advantages of optical imaging there is no current optical method which sufficiently provides the imaging of early stage tumor growth or metastasis due to the strong absorbance and scattering of the illuminating light by tissue near the target [5, 6].

2.2 Positron Emission Tomography

In PET imaging, a compound (natural biological molecule/drug) is labeled with positron-emitting radioisotopes such as ^{11}C, ^{13}N, ^{15}O, ^{64}Cu, ^{124}I, and ^{18}F and administered to the subject in trace amounts. The emitted positrons collide with an electron of a neighboring atom and the interaction of both particles is converted into energy in the form of two γ-rays, emitting in opposite directions. An array of detectors detects these γ-rays and converts them into visible light and finally into electrons. This results in the localization and quantification of the radiolabeled compound in the living subject.

PET can visualize functional physiological change and biochemical change with high sensitivity in the range of 10^{-11} to 10^{-12} mol/l. It is independent of the depth of the compound. The spatial resolution of PET, depending on the size of the single detector component, is ~4–8 mm^3 in clinical and ~2 mm^3 in small animals [7].

Quantitative kinetic data can be acquired repetitively and it allows for the drawing of conclusions in many cancer cases solely by means of PET images [8]. However, it is an expensive approach to analyzing mouse models in preclinical studies because a dedicated mouse PET scanner, a cyclotron, and radiochemistry facility are required.

2.3 Single Photon Emission Computed Tomography

In SPECT imaging, a compound is labeled with one or more γ-rays emitting isotopes such as ^{99}mTc, ^{111}In, ^{123}I, ^{131}I, ^{67}GA, and ^{201}TI. In contrast to PET imaging, this allows two or more compounds with different labeling radioisotopes (for example, ^{99}mTc and ^{111}In) to be distinguished within the same study.

The advantages of SPECT over PET include lower costs, simultaneous imaging by different isotopes, and spatial resolution with most mouse imaging systems in the order of 1–2 mm and the availability of many molecular probes already in clinical use. However, SPECT is one to two orders of magnitudes less sensitive than PET [9].

2.4 Magnetic Resonance Imaging

The physical principles of MRI are that certain nuclei, such as hydrogen or phosphorous, have magnetic dipoles (unpaired nuclear spins) and align themselves when these nuclei are exposed to a high static magnetic field. The MRI method converts relaxation time differences to image contrast differences. The two orientations have slightly different energy levels with respect to the applied field and one of the two orientations allows and produces a net magnetization (Fig. 1). Approximately 1 out of 10^5 of the dipole moments in the field is aligned while others are in random orientation, resulting in the low signal intensities of MRI.

Since water molecules are abundant in a biological system, MRI derives signals from the hydrogen nuclei in general. The versatility and uniqueness of MRI is derived from the fact that the ^{1}H are sensitive to its local physicochemical micro-environment thus the MRI experiment can be tailored to exploit these properties in the interrogation pathophysiology.

In general, MRI applications have the advantages of noninvasive, clinically transferable, high spatial resolution, and inherent soft tissue contrast. On the downside, MRI is several orders of magnitudes less sensitive than imaging modalities such as radionuclides and optical molecular probes.

The use of nanoparticles can be engineered to have magnetic characteristics that can be detected by MRI at low concentrations. For example, superparamagnetic iron oxide nanoparticles (SPIONS) have been widely researched for use as MRI contrast agents and have proven effective in increasing contrast in MRI [11].

2.5 Ultrasound Imaging

Ultrasound is the most widely used method both in clinics and for animal studies because of its low cost, accuracy, real-time administration of drugs, and no use of ionizing radiation. Ultrasound uses high-frequency (>20 kHz) sounds, which are emitted from a transducer. The sound waves are propagated through tissues of different densities and the returning echoes are used to build up images of the plane in the body scanned.

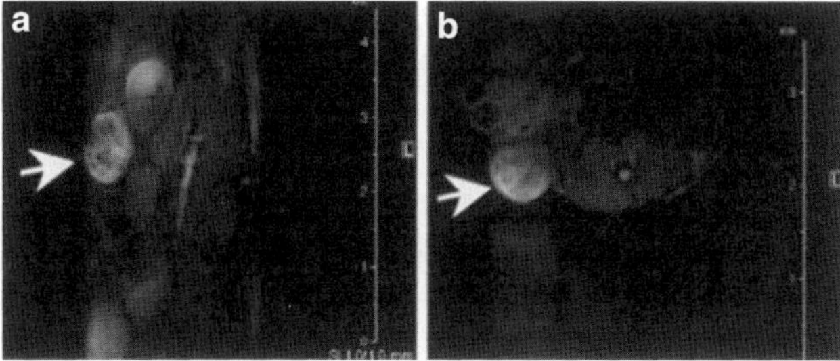

Fig. 1 MR images of a mouse containing a PC3 tumor in transplanted human bone tissue [10]. Reproduced with permission from Nemeth et al. [10]

Targeted labels have been developed for this technique and contrast agents (e.g., gas-containing microbubbles) are used for targeted molecular imaging of specific cell-surface receptors, especially within the vascular compartment to increases their echogenicity [12, 13].

The real-time properties of ultrasound makes it very effective for the drug development process and studying disease development, particularly in cardiac, obstetric, and vascular diseases. However, the drawbacks of this technique are that the performance of ultrasound imaging, ultrasound-guided diagnostics, or therapeutic interventions mostly depend on the experience and skills of the operator. In addition, targeted imaging is limited to vascular compartments. Innovations providing automation for better objectivity and three-dimensional imaging approaches are currently under development and will make the ultrasound a very powerful molecular imaging technique in the future.

2.6 Computed Tomography

Images in CT are obtained when an object differentially absorbs X-rays as they pass through the body [14]. Based on the X-ray images, high-resolution tomographic anatomical 2D and 3D cross-sectional images are reconstructed with a spatial resolution of up to ~6 μm in small animals. They reveal the characteristics of the internal structure of an object, dimensions, shape, internal defects as well as density, which result in generating high-resolution volumetric images. However, CT lacks sensitivity in detecting soft tissue. The combination of CT with PET provides images that identify the anatomic context of irregular metabolic activity within the body and thus, provides more accurate diagnosis. PET-CT imaging has rapidly emerged as a reference standard for functional and molecular imaging at the clinical level (Fig. 2).

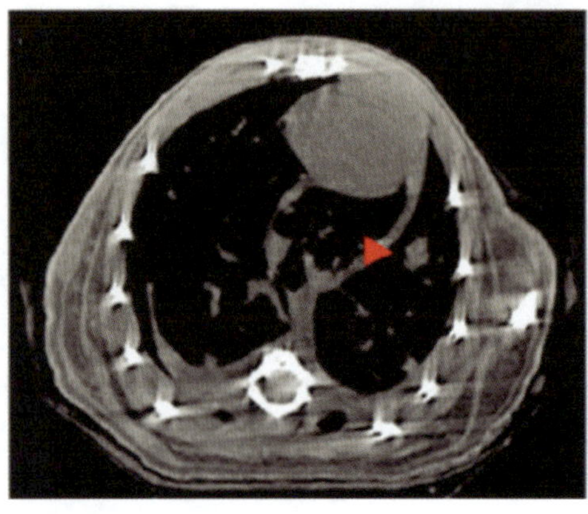

Fig. 2 High-resolution computed tomography (CT) of a mouse showing (*vertical line*) metastatic lung tumor (*arrowhead*) [15]. Reproduced with permission from Kennel et al. [15]

3 Molecular Imaging in Immunotherapy

3.1 Immune Cells in Immunotherapy

Cancer immunotherapy aims to establish immune-mediated control of tumor growth using effecter cells such as dendritic cells (DCs), natural killer cells (NK cells), and T cells. Dendritic cells (DCs) reside in most of the body and recognize foreign antigens in tissues which participate in proteolytic intracellular cleavage and increasing antigen expressing molecules [16]. DC activation, trafficking to lymph nodes, and stimulating T cells result in the generation of antigen-specific cytotoxic T lymphocyte (CTL), migration of CTLs to the tumor site, and the destruction of the cancer [17–19]. Cytotoxic T cells are involved in the cell-mediated immune response and destroy virus-infected cells, whereas NK cells comprising of 5–20% of peripheral lymphocytes in an innate immune system play a major role in defending the host against malignant cells [20, 21]. In many studies, cancer patients containing DCs, T cells, and NKs demonstrated much improved overall survival rates as compared to those without DCs, T cells, and NKs [22, 23].

In general, immunotherapy uses immune cells from an animal or a patient that are modified and amplified in vitro. The amplified cell populations are then transferred back into the host to eliminate the tumor. They, then, should be carefully monitored noninvasively, instantly, and in real time to determine the success or failure of the treatment. In addition, to enhance antigen presenting in T cells located within lymph nodes, DCs need to traffic through the vascular and lymphatic system. The best administration route (intradermally, subcutaneously, intravenously, or directly into the lymph node) for DCs-based therapy is not known [24]. In fact, evaluation of immunotherapy has mostly relied on a reduction of tumor volume and that of tumor markers after treatment [25–27]. In the context of these concerns, different molecular probes and modalities should be considered to track the transplanted cells and induced immune responses for qualitative and quantitative images. Proper molecular probes should have long-term stability in physiological conditions, low toxicity in cells, high sensitivity in real time, and universal application to different cells.

3.2 *Nanoparticles for Molecular Imaging*

Nanoparticles are promising vehicles for directing the molecular imaging probes because of their unique physicochemical properties and low toxicity. The high surface area to volume ratio of nanoparticles provides for both the ease and extent of modification as molecular probes.

Paramagnetic gadolinium, superparamagnetic iron oxides (SPIOs, 50–200 nm diameters), ultrasmall superparamagnetic iron oxides (USPIOs, ~35 nm diameter), and cross-linked iron oxide (CLIO) have been investigated as probes for indirect methods to detect cells such as in MRI. These nanoparticle-based MRI cell tracking methods can efficiently address the location of the injected cells and the morphological and physiological changes of the host tissues due to their strong contrast effect. Iron oxide particles have been coated with dextran, carboxydextran, polystyrene, or other substrates to ensure stability and solubility in biological media and also to minimize effects on cell function upon cellular uptake.

To date, FDA-approved (U)SPIO, a liver agent (Feridex, Bayer HealthCare), is the only pharmaceutical-grade MRI contrast agent used in lymph nodes in late-phase clinical trials (USPIO; Combidex-USA; Sinerem-Europe) [28]. Many studies have been done in the search for efficient SPIO labeling of cells and their application to detect cells after transplantation [29–32] or systemic injection [33–37]. Monocrystalline iron oxide nanoparticles and USPIO with a longer blood half-life were also developed for imaging in lymph node.

Success in immunotherapy requires an accurate injection of cells into the target tissue as well as MRI-guided real-time cell injection. Nanoparticle-labeled cells can offer more precise injection of cells.

This chapter discusses some of the recent advances in the use of nanoprobes for cell tracking in vivo after transplantation, mainly focusing on three types of cells in immunotherapy (i.e., DCs, T cells, and NK cells).

3.3 *Molecular Imaging Probes for Tracking Immune Cell Fate In Vivo*

3.3.1 Dendritic Cells

Autologous dendritic cells in humans as cancer vaccines were labeled with SPIO and ^{111}In-labeled oxine by De Vroes et al. [38]. In vitro-produced DCs were then administrated into lymph nodes of stage-III melanoma patients under the observance of an ultrasound and the biodistribution of SPIO-labeled dendritic cells were monitored in vivo by MR imaging along with scintigraphy and (immune) histopathology. MRI tracking of DCs labeling with iron oxide is clinically safe and excellent in soft tissue contrast, making this method ideal for monitoring cell therapies in humans.

DCs were also labeled with SPIO and protamine sulfate for tracking in vivo migration of DCs into the lymph node by MRI. Significant signal reduction was observed in the lymph node where SPIO-DC had been injected into footpads of mice, indicating the migration of T cells into the draining lymph nodes. This study again showed application of SPIO in noninvasive molecular imaging by MRI in vivo at high resolution [39].

Labeling cells with ^{19}F offers a high signal sensitivity of 89% in comparison to ^{1}H and high specificity for labeled cells because of its lack of an endogenous background signal in vivo. ^{19}F-labeled cells can be directly detected by ^{19}F-MRI. However, labeling cells with ^{19}F requires the need for high concentrations of ^{19}F to achieve a minimal detection limit, long scanning times, and separate ^{1}H images for anatomical localization of ^{19}F detected cells. DCs were labeled with cationic perfluoropolyether (PFPE) agents and the migration of the cells was tracked through to the regional lymph nodes following their injection into the footpad of mice by 19F MRI [40].

Fluorescence quantum dots (QDs) and their application as molecular imaging probes have been extensively investigated because of their nanometer dimensions and attractive optical characteristics, which includes high resistance to photobleaching. Due to the relatively reduced absorbance and scattering in biological tissue in the NIR region of the spectrum (700–1,000 nm), NIR-emitting QDs are currently attracting considerable attention [9].

Two major limitations in using fluorescence for in vivo imaging are autofluorescence and low depth penetration. Fluorescence-based imaging of cells labeled with near-infrared fluorophores (NIRFs) could be an alternative way to label the cells as NIR imaging allows for the detection of fluorophore-labeled cells at several centimeters tissue depth without disturbing cellular function.

Conjugation of antibodies, small peptides, and polymers with NIR excitable fluorescent contrast agents for the detection of tumors have been demonstrated in vivo [41–44]. An epidermal growth factor (EGF) and NIR fluorophore conjugate, Cy5.5, was injected into mice bearing epidermal growth factor receptor (EGFR) positive tumors. EGF-Cy5.5 was monitored in vivo by NIR images demonstrating that NIR imaging can be used to identify EGFRs in tumors and to monitor EGFR-directed therapies (Fig. 3).

NIR-emitting fluorescent QDs have been developed and used as fluorescent contrast agents to monitor DCs migration to lymph nodes, resulting in the increase of fluorescence intensity [46].

Bimodal imaging probes, with both the MR and optical imaging modes containing disulfonated indocyanine green (ICG), FDA-approved cyanine NIR dye as the optical imaging modality, iron oxide nanoparticles coated with biocompatible

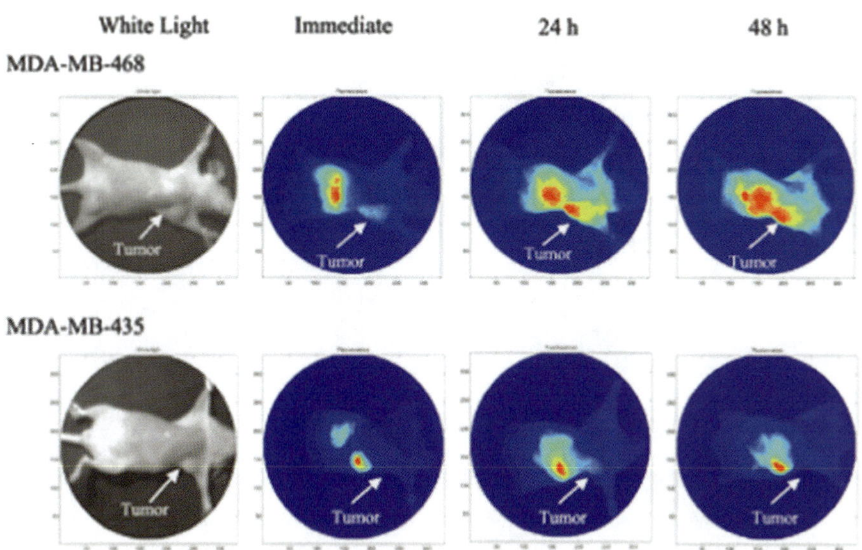

Fig. 3 NIR fluorescence images of mice bearing an EGFR positive tumor (MDA-MB-468) and EGFR negative tumor (MDA-MB-435) after being injected with EGF-Cy5.5 [45]. Reproduced with permission from Ke et al. [45]

polymer and poly(lactide-*co*-glycolide) (PLGA) have been developed to overcome the low sensitivity of an MRI modality as compared to those of PET and SPECT. PLGA can be broken down into nontoxic materials and metabolized in the human. In addition, degradation of PLGA can be controlled by the ratio of PLA (poly[lactic acid]) to PGA (poly[glycolic acid]) which makes it a good carrier of sold drug nanoparticles for controlling their release. The migration of DCs labeled with dual-functional PLGA nanoparticles through lymphatic drainage was detected by real-time NIR fluorescence imaging and the homing of DCs into lymph nodes was tracked through noninvasive MRI techniques (from footpad to the lymph node). Both migrated and non-migrated DCs were detected by NIR fluorescence imaging 2 days after the cell transplantation (1×10^6 cells) [47].

Christian et al. developed near-infrared emissive and fully biocompatible polymersomes (NIR-OBs) which contain poly(ethylene oxide (1,300)-*b*-butadiene (2,500)) (OB) and porphyrin-based NIRF (oligomeric (porphinato) zinc(II) chromophore). In order to optimize the intracellular delivery of NIRFs and HIV-derived Tat, a highly cationic peptide shown to enhance cellular delivery of nanoparticles were combined to the polymersomes [48–50]. Labeling DCs with Tat-NIR-emissive polymersomes allows for tracking both intravenously and subcutaneously injected DCs in the popliteal lymph node (Fig. 4).

DCs labeled with a bimodal imaging nanoprobe containing IRDye800-coated perfluoroctyl bromide (PFOB) and nanoemulsions were used both in vitro and in vivo. When the DCs were injected subcutaneously into the mice, ^{1}H-based

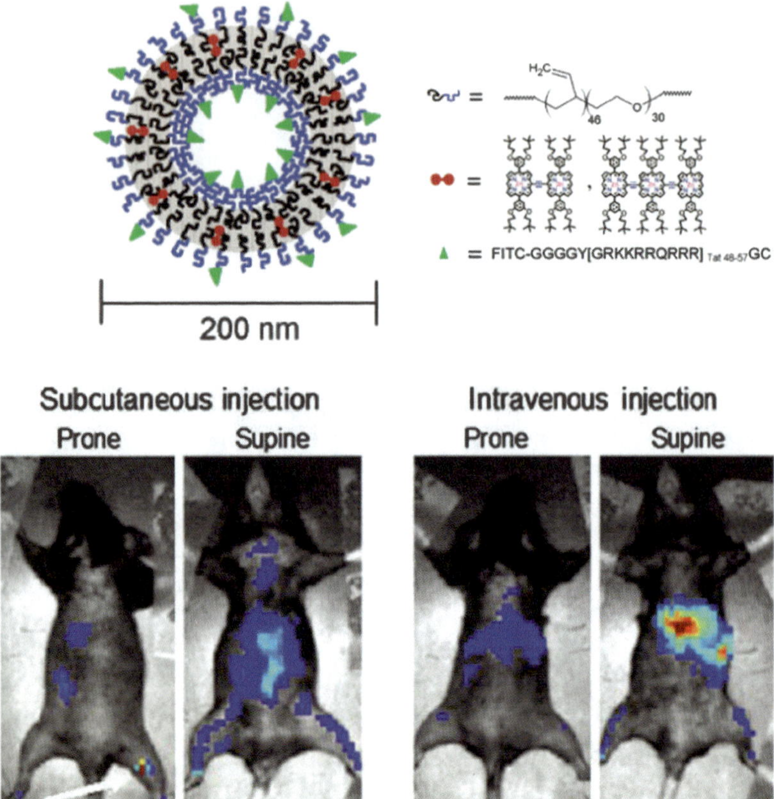

Fig. 4 Tracking of migrating Tat-NIR-emissive polymersomes-labeled DCs in vivo after administration [51]. Reproduced with permission from Christian et al. [51]

MRI provided a whole-body image whereas the ^{19}F-based MR imaging technique showed only signals generated from the injected DCs labeled with IRDye800-coated PFOB nanoemulsion (Fig. 5).

3.3.2 T Cells Tracking

A HSV1-sr39TK reporter gene was transduced into T cells by retroviral injection for monitoring T-cell antitumor responses and the labeled cells were administrated into mice bearing an antigen-positive tumor. The migration of labeled T cells was visualized with ^{18}F-FHBG under a PET camera [53].

HIV-derived Tat peptides have been attached to highly derivatized CLIO particles in order to achieve efficient intracellular labeling of CTLs [33, 54–56]. The feasibility of using MRI to monitor T-cell homing in vivo after administration

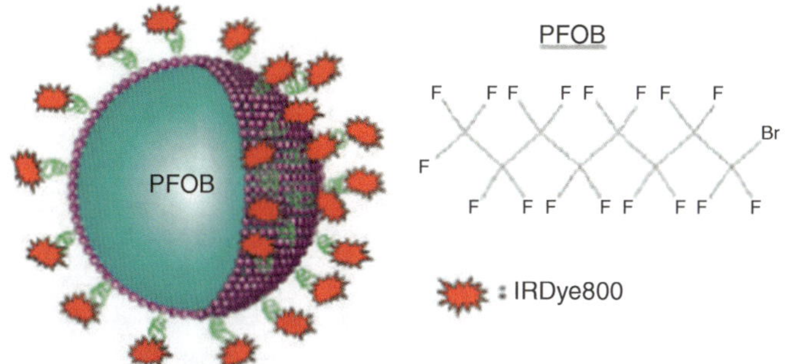

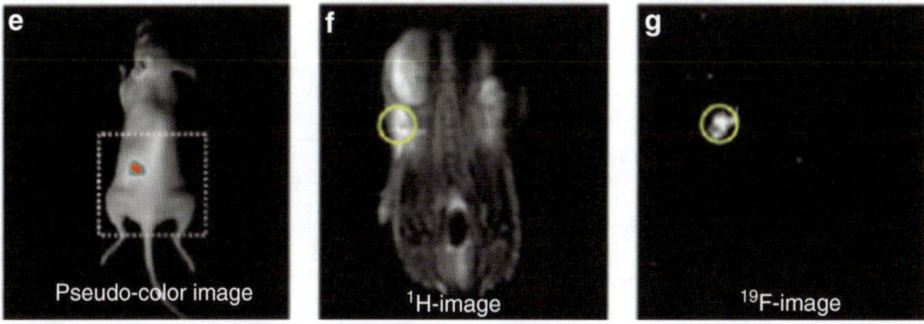

Fig. 5 Bimodal imaging of a mouse injected with DCs labeled with IRDye800-coated PFOB nanoemulsions and monitored by an NIR image and ^{19}F-based MR image [52]. Reproduced with permission from Lim et al. [52]

was studied by injecting FITC-CLIO-Tat-labeled T cells intravenously. Homing of T cells into the spleen and the biodistribution of FITC-CLIO-Tat-labeled T cells in vivo were monitored by MRI [57].

A physiologically inert nanoparticle (highly derivatized CLIO nanoparticle; CLIO-HD) has been developed for labeling of a variety of cell types that now allows for in vivo MRI tracking of systemically injected cells at near single-cell resolution. Kircher et al. have displayed the utility of MRI cell tracking in immunotherapy with CD8+ CTLs labeled with CLIO-HD by high-resolution imaging of T-cell recruitment to intact tumors in vivo for the first time. B16-OVA melanoma and CLIO-HD-labeled OVA-specific CD8+ T cells were used and it was revealed that CLIO-HD is uniquely suited for quantitative and repetitive MRI of T cells following adoptive transfer in vivo (Fig. 6).

T cells are also labeled ex vivo with a PFPE nanoparticle tracer agent, transferred and detected in vivo using ^{19}F MRI that selectively visualizes only the labeled cells with no background signals whereas ^{1}H MR images taken simultaneously provide anatomical information. In vivo ^{19}F MRI data showed transferred T cells homing to the pancreas and a computational algorithm provided T cell counts in the pancreas. Both unambiguous detection of labeled cells and quantification have become directly

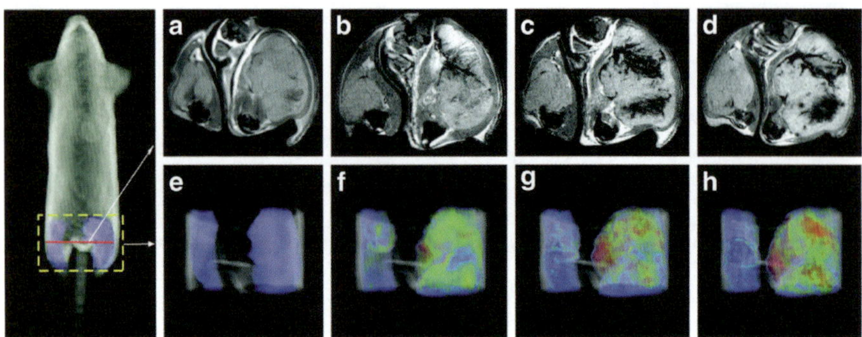

Fig. 6 Monitoring time course of CLIO-HD OT-I CD8+ T cell homing to B16-OVA tumor by MRI [33]. Reproduced with permission from Kircher et al. [33]

possible from the in vivo images produced by this technique. The labeling procedure did not affect the ability of T cell migration in vivo and, therefore, it can potentially be used for monitoring the trafficking of cellular therapeutics [58].

3.3.3 NK Cell Tracking

NK cells also hold promise for cell-based therapies as they exhibit high cytotoxic activity on multiple malignancy types, while sparing normal cells. In fact, NK cell-based therapy by adoptive transfer of ex vivo expanded and activated autologous or donor-derived cytotoxic NK cells are currently being clinically trialed (e.g., NCI study NCT00376805) and, therefore, advancement in molecular imaging has gained tremendous attention.

The initial attempts of NK cell tracking in tumors came from PET studies of the localization of IL-2A activated NK cells labeled with [^{11}C] methyl iodide in Fall sarcoma tumor-bearing mice. These showed an accumulation of 4–30% of the activated NK cells post injection [59].

GFP-luc transfected NK cells in A-20 lymphoma xenografts were monitored by bioluminescence imaging. The luminescent signal from a genetic modification remains stable for a long time after administration. After the injection of GFP positive NK cells (5×10^6) in mice, a decrease in tumor mass was observed at day 12 [60].

NK cells labeled with iron oxide-based contrast agents (Ferumoxides and Ferucarbotran) were used to monitor the in vivo accumulation of HER2/neu targeted NK-92 cells to HER2/neu-positive breast sarcomas using MR imaging at clinically applicable fields of strength. A novel molecular imaging study demonstrated SPIO linked a genetically modified NK-92 cell line, NK-92-scFv(FRP5)-zeta, targeted

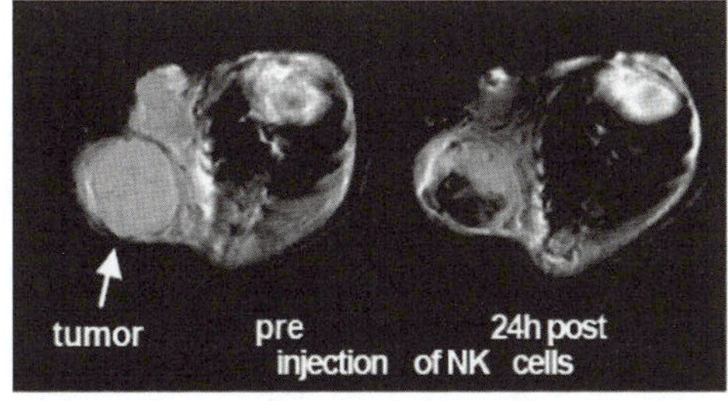

Fig. 7 MR imaging of ErbB2 directed SPIO-labeled NK cells [61]. Reproduced with permission from Daldrup-Link et al. [61]

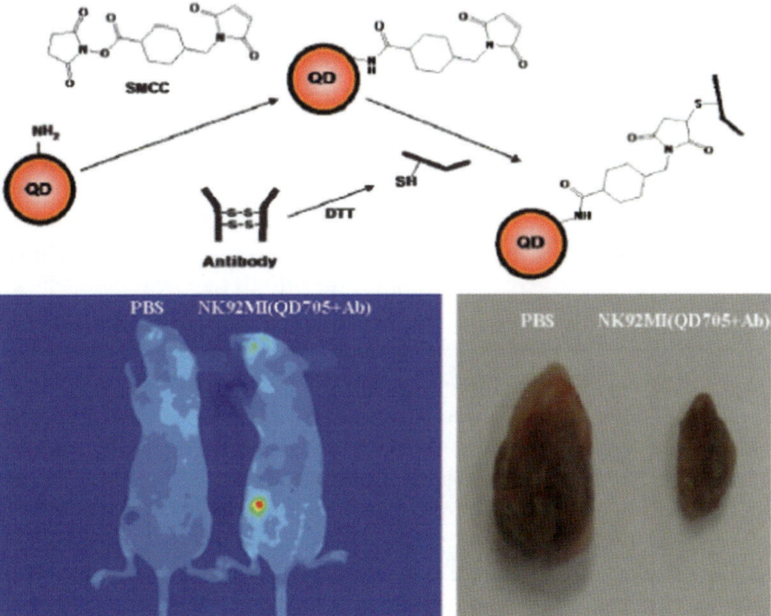

Fig. 8 NIR fluorescence imaging for tracking NK cells labeled with antibody-coated QDs in mice [62]. Reproduced with permission from Lim et al. [62]

malignant cells over-expressing HER2/neu. MR imaging of SPIO-labeled NK-92-scFv (FRP5)-zeta cells offers distinct and long-lasting recognition of HER2/neu-positive tumors in vivo (Fig. 7). Despite its high cost, MR imaging would be immediately applicable in clinics and thus deserves special attention.

Lim and colleagues labeled NK92MI cells with anti-CD56 antibody conjugated quantum dots (QD705) and tracked the labeled cells for 12 days by optical imaging. The authors observed stable cell labeling of QD705 over a period of months as well as decreased size of tumors that were treated with NK92MI cells compared with controls (Fig. 8). Again the characteristics of QDs such as high quantum yield, high sensitivity, stability, and a very narrow excitation and emission spectrum makes them ideal for the function of molecular imaging probes.

4 Concluding Remarks

The field of molecular imaging includes multiple disciplines such as cell biology, biomedical engineering, chemistry, mathematics, medicine, pharmacology, and genetics. It has been rapidly growing due to tremendous advances in imaging instruments, imaging probes, and quantification methods which have helped elucidate molecular mechanisms in biology and medicine over the last decade. Application of molecular imaging in immunotherapy has emerged as an active field of investigation and shown great promise in preclinical models and greater efficacy in clinical trials.

Rapid developments in imaging contrast agents, NIR fluorophores, isotopes, novel disease-related biomarkers, and imaging hardware allow for efficient labeling of therapeutic cells, specifically in vivo. This can provide visual and quantitative information regarding location, activation status, biodistribution, and therapeutic efficacy of the transplanted cells by visualizing noninvasively in the living subjects at the cellular and subcellular level.

Advanced technology for novel, multimodal, biocompatible, and biodegradable molecular imaging probes in combination with different imaging modalities may overcome the limitation issues concerning sensitivities and resolutions and thus could provide powerful tools for diagnostics, prognostics, and the treatment of disease in patients over multiple time points. This would also mean a greater potential to expand scientific discoveries from laboratory settings into therapeutic tools in diverse clinical settings. As it is, much of the technology is still in the proof-of-concept stages. The field of molecular imaging now faces the challenge of overcoming the safety issues in regards to adjunct compounds, such as transfection agents and immunogenicity in patients. These safety issues will need to be resolved in order to move from animal studies towards human studies. Nevertheless, further advances in molecular imaging technology will play a major role and shed new light on the treatment of human diseases for many years to come.

References

1. Yezhelyev MV, Gao X, Xing Y, Al-Haii A, Nie S, O'Reqan RM (2006) Emerging use of nanoparticles in diagnosis and treatment of breast cancer. Lancet Oncol 7:657–667
2. Coroiu I (1999) Relaxivities of different superparamagnetic particles for application in NMR tomography. J Magn Magn Mater 201:449–452
3. Weissleder R, Ntziachristos V (2003) Shedding light onto live molecular targets. Nat Med 9:123–128
4. Massod TF, Gambhir SS (2003) Molecular imaging in living subjects: seeing fundamental biological processes in a new light. Genes Dev 17:545–580
5. Tearney GJ, Brezinski ME, Bouma BE, Boppart SA, Pitris C, Southern JF, Fujimoto JG (1997) In vivo endoscopic optical biopsy with optical coherence tomography. Science 276:2037–2039
6. Taubes G (1997) Play of light opens a new window into the body. Science 276:1991–1993

7. Kim JS, Lee JS, Im KC, Kim SJ, Lee DS, Moon DH (2007) Performance measurement of the microPET focus 120 scanner. J Nucl Med 48:1527–1535
8. Price P (2000) Positron emission tomography (PET) in diagnostic oncology: is it a necessary tool today? Eur J Cancer 36:691–693
9. Weissleder R (2001) A clearer vision for in vivo imaging. Nat Biotechnol 19:316–317
10. Nemeth JA, Harb JF, Barroso U Jr, He Z, Grignon DJ, Cher ML (1999) Severe combined immunodeficient-hu model of human prostate cancer metastasis to human bone. Cancer Res 59:1987–1993
11. McAteer MA, Sibson NR, von Zur MC, Schneider JE, Lowe AS, Warrick N, Channon KM, Anthony DC, Choudhury RP (2007) In vivo magnetic resonance imaging of acute brain inflammation using microparticles of iron oxide. Nat Med 13:1253–1258
12. Modo M (2008) Noninvasive imaging of transplanted cells. Curr Opin Organ Transplant 13:654–658
13. Talu E, Hettiarachchi K, Zhao S, Powell RL, Lee AP, Longo ML, Dayton PA (2007) Tailoring the size distribution of ultrasound contrast agents: possible method for improving sensitivity in molecular imaging. Mol Imaging 6:384–392
14. Khandelwal P, Kumar K, Singh B, Singh R (2012) A review on medical image modalities. International Journal of Computer Science and Management Research 1:844–853
15. Kennel SJ, Davis IA, Branning J, Pan H, Kabalka GW, Paulus MJ (2000) High resolution computed tomography and MRI for monitoring lung tumor growth in mice undergoing radioimmunotherapy: correlation with histology. Med Phys 27:1101–1107
16. Vakkila J, Lotze MT (2004) Inflammation and necrosis promote tumour growth. Nat Rev Immunol 4:641–648
17. Steinman RM (2003) Some interfaces of dendritic cell biology. APMIS 111:675–697
18. Sogn GA (1998) Tumour immunology: the glass is half full. Immunity 9:757–763
19. Pardoll DM (1998) Cancer vaccines. Nat Med 4:525–531
20. Biron CA, Nguyen KB, Pien GC, Cousens LP, Salazar-Mather TP (1999) Natural killer cells in antiviral defense: function and regulation by innate cytokines. Annu Rev Immunol 17:189–220
21. Wu J, Lanier LL (2003) Natural killer cells and cancer. Adv Cancer Res 90:127–156
22. Zhang L, Conejo-Garcia JR, Katsaros D, Gimotty PA, Massobrio M, Regnani G, Makrigiannakis A, Gray H, Schlienger K, Liebman MN, Rubin SC, Coukos G (2003) Intratumoral T cells, recurrence, and survival in epithelial ovarian cancer. N Engl J Med 348:203–213
23. Ishigami S, Natsugoe S, Tokuda K, Nakajo A, Xiangming C, Iwashige H, Aridome K, Hokita S, Aikou T (2000) Clinical impact of intratumoral natural killer cell and dendritic cell infiltration in gastric cancer. Cancer Lett 159:103–108
24. Adema GJ, de Vries IJ, Punt CJ, Figdor CG (2005) Migration of dendritic cell based cancer vaccines: in vivo veritas? Curr Opin Immunol 17:170–174
25. Miller JS (2009) Should natural killer cells be expanded in vivo or ex vivo to maximize their therapeutic potential? Cytotherapy 11:259–260
26. Brand JM, Meller B, Von Hof K, Luhm J, Bähre M, Kirchner H, Frohn C (2004) Kinetics and organ distribution of allogeneic natural killer lymphocytes transfused into patients suffering from renal cell carcinoma. Stem Cells Dev 13:307–314
27. Matera L, Galetto A, Bello M, Baiocco C, Chiappino I, Castellano G, Stacchini A, Satolli MA, Mele M, Sandrucci S, Mussa A, Bisi G, Whiteside TL (2006) In vivo migration of labeled autologous natural killer cells to liver metastases in patients with colon carcinoma. J Transl Med 4:49
28. Harisinghani MG, Barentsz J, Hahn PF, Deserno WM, Tabatabaei S, van de Kaa CH, de la Rosette J, Weissleder R (2003) Noninvasive detection of clinically occult lymph-node metastases in prostate cancer. N Engl J Med 348:2491–2499
29. Bulte JW, Zhang S, van Gelderen P, Herynek V, Jordan EK, Duncan ID, Frank JA (1999) Neurotransplantation of magnetically labeled oligodendrocyte progenitors: magnetic resonance tracking of cell migration and myelination. Proc Natl Acad Sci USA 96:15256–15261

30. Bulte JW, Kraitchman DL (2004) Iron oxide MR contrast agents for molecular and cellular imaging. NMR Biomed 17:484–499
31. Kraitchman DL, Heldman AW, Atalar E, Amado LC, Martin BJ, Pittenger MF, Hare JM, Bulte JW (2003) In vivo magnetic resonance imaging of mesenchymal stem cells in myocardial infarction. Circulation 107:2290–2293
32. Hoehn M, Küstermann E, Blunk J, Wiedermann D, Trapp T, Wecker S, Föcking M, Arnold H, Hescheler J, Fleischmann BK, Schwindt W, Bührle C (2002) Monitoring of implanted stem cell migration in vivo: a highly resolved in vivo magnetic resonance imaging investigation of experimental stroke in rat. Proc Natl Acad Sci USA 99:16267–16272
33. Kircher MF, Allport JR, Graves EE, Love V, Josephson L, Lichtman AH, Weissleder R (2003) In vivo high resolution three-dimensional imaging of antigen-specific cytotoxic T-lymphocyte trafficking to tumors. Cancer Res 63:6838–6846
34. Anderson SA, Shukaliak-Quandt J, Jordan EK, Arbab AS, Martin R, McFarland H, Frank JA (2004) Magnetic resonance imaging of labeled T-cells in a mouse model of multiple sclerosis. Ann Neurol 55:654–659
35. Anderson SA, Glod J, Arbab AS, Noel M, Ashari P, Fine HA, Frank JA (2005) Noninvasive MR imaging of magnetically labeled stem cells to directly identify neovasculature in a glioma model. Blood 105:420–425
36. Yeh TC, Zhang W, Ildstad ST, Ho C (1995) In vivo dynamic MRI tracking of rat T-cells labeled with superparamagnetic iron-oxide particles. Magn Reson Med 33:200–208
37. Ahrens ET, Feili-Hariri M, Xu H, Genove G, Morel PA (2003) Receptor-mediated endocytosis of iron-oxide particles provides efficient labeling of dendritic cells for in vivo MR imaging. Magn Reson Med 49:1006–1013
38. de Vries IJ, Lesterhuis WJ, Barentsz JO, Verdijk P, van Krieken JH, Boerman OC, Oyen WJ, Bonenkamp JJ, Boezeman JB, Adema GJ, Bulte JW, Scheenen TW, Punt CJ, Heerschap A, Figdor CG (2005) Magnetic resonance tracking of dendritic cells in melanoma patients for monitoring of cellular therapy. Nat Biotechnol 23:1407–1413
39. Baumjohann D, Hess A, Budinsky L, Brune K, Schuler G, Lutz MB (2006) In vivo magnetic resonance imaging of dendritic cell migration into the draining lymph nodes of mice. Eur J Immunol 36:2544–2555
40. Ahrens ET, Flores R, Xu H, Morel PA (2005) In vivo imaging platform for tracking immunotherapeutic cells. Nat Biotechnol 23:983–987
41. Folli S, Westermann P, Braichotte D, Pelegrin A, Wagnieres G, van den Bergh H, Mach JP (1994) Antibody-indocyanin conjugates for immunophotodetection of human squamous cell carcinoma in nude mice. Cancer Res 54:2643–2649
42. Ballou B, Fisher GW, Deng JS, Hakala TR, Srivastava M, Farkas DL (1998) Cyanine fluorochrome-labeled antibodies in vivo: assessment of tumor imaging using Cy3, Cy5, Cy5.5, and Cy7. Cancer Detect Prev 22:251–257
43. Becker A, Hessenius C, Licha K, Ebert B, Sukowski U, Semmler W, Wiedenmann B, Grotzinger C (2001) Receptor-targeted optical imaging of tumors with near-infrared fluorescent ligands. Nat Biotechnol 19:327–331
44. Bremer C, Tung CH, Weissleder R (2001) In vivo molecular target assessment of matrix metalloproteinase inhibition. Nat Med 6:743–748
45. Ke S, Wen X, Gurfinkel M, Charnsangavej C, Wallace S, Sevick-Muraca EM, Li C (2003) Near-infrared optical imaging of epidermal growth factor receptor in breast cancer xenografts. Cancer Res 63:7870–7875
46. Noh YW, Lim YT, Chung BH (2008) Noninvasive imaging of dendritic cell migration into lymph nodes using near-infrared fluorescent semiconductor nanocrystals. FASEB J 22:3908–3918
47. Lim YT, Noh YW, Han JH, Cai QY, Yoon KH, Chung BH (2008) Biocompatible polymer-nanoparticle-based bimodal imaging contrast agents for the labeling and tracking of dendritic cells. Small 4:1640–1645

48. Gupta B, Levchenko TS, Torchilin VP (2005) Intracellular delivery of large molecules and small particles by cell-penetrating proteins and peptides. Adv Drug Deliv Rev 57:637–651
49. Torchilin VP (2002) TAT peptide-modified liposomes for intracellular delivery of drugs and DNA. Cell Mol Biol Lett 7:265–267
50. Vives E (2003) Cellular uptake of the Tat peptide: an endocytosis mechanism following ionic interactions. J Mol Recognit 16:265–271
51. Christian NA, Benencia F, Milone MC, Li G, Frail PR, Therien MJ, Coukos G, Hammer DA (2009) In vivo dendritic cell tracking using fluorescence lifetime imaging and near-infrared-emissive polymersomes. Mol Imaging Biol 11:167–177
52. Lim YT, Noh YW, Kwon JN, Chung BH (2009) Multifunctional perfluorocarbon nanoemulsions for (19)F-based magnetic resonance and near-infrared optical imaging of dendritic cells. Chem Commun (Camb) 7:6952–6954
53. Dubey P, Su H, Adonai N, Du S, Rosato A, Braun J, Gambhir SS, Witte ON (2003) Quantitative imaging of the T cell antitumor response by positron-emission tomography. Proc Natl Acad Sci USA 100:1232–1237
54. Josephson L, Tung CH, Moore A, Weissleder R (1999) High-efficiency intracellular magnetic labeling with novel superparamagnetic-Tat peptide conjugates. Bioconjug Chem 10:186–191
55. Lewin M, Carlesso N, Tung CH, Tang XW, Cory D, Scadden DT, Weissleder R (2000) Tat peptide-derivatized magnetic nanoparticles allow in vivo tracking and recovery of progenitor cells. Nat Biotechnol 18:410–414
56. Vianello F, Papeta N, Chen T, Kraft P, White N, Hart WK, Kircher MF, Swart E, Rhee S, Palù G, Irimia D, Toner M, Weissleder R, Poznansky MC (2006) Murine B16 melanomas expressing high levels of the chemokine stromal-derived factor-1/CXCL12 induce tumor-specific T cell chemorepulsion and escape from immune control. J Immunol 176:2902–2914
57. Dodd CH, Hsu HC, Chu WJ, Yang P, Zhang HG, Mountz JD Jr, Zinn K, Forder J, Josephson L, Weissleder R, Mountz JM, Mountz JD (2001) Normal T-cell response and in vivo magnetic resonance imaging of T cells loaded with HIV transactivator-peptide-derived superparamagnetic nanoparticles. J Immunol Methods 256:89–105
58. Srinivas M, Morel PA, Ernst LA, Laidlaw DH, Ahrens ET (2007) Fluorine-19 MRI for visualization and quantification of cell migration in a diabetes model. Magn Reson Med 58:725–734
59. Melder RJ, Brownell AL, Shoup TM, Brownell GL, Jain RK (1993) Imaging of activated natural killer cells in mice by positron emission tomography: preferential uptake in tumors. Cancer Res 53:5867–5871
60. Edinger M, Cao YA, Verneris MR, Bachmann MH, Contag CH, Negrin RS (2003) Revealing lymphoma growth and the efficacy of immune cell therapies using in vivo bioluminescence imaging. Blood 101:640–648
61. Daldrup-Link HE, Meier R, Rudelius M, Piontek G, Piert M, Metz S, Settles M, Uherek C, Wels W, Schlegel J, Rummeny EJ (2005) In vivo tracking of genetically engineered, anti-HER2/neu directed natural killer cells to HER2/neu positive mammary tumors with magnetic resonance imaging. Eur Radiol 15:4–13
62. Lim YT, Cho MY, Noh YW, Chung JW, Chung BH (2009) Near-infrared emitting fluorescent nanocrystals-labeled natural killer cells as a platform technology for the optical imaging of immunotherapeutic cells-based cancer therapy. Nanotechnology 20:475102

Optical Chemical Sensor and Electronic Nose Based on Porphyrin and Phthalocyanine

Teerakiat Kerdcharoen and Sumana Kladsomboon

Abstract Recently, electronic nose (e-nose) has emerged as a viable technology to detect and analyze various kinds of gases based on chemical gas sensor array and pattern recognition. Researchers worldwide make their efforts to improve the sensitivity and stability of the chemical gas sensors. Among the sensing and transduction technologies, such as metal oxide, piezoelectric, organic semiconductor, nano-composite and optical sensing, optical gas sensors present several advantages, i.e., low energy consumption and high signal-to-error ratio, etc. Specifically, the optical gas sensors based on optically active organic materials, e.g., metallo-porphyrin (MP) and metallo-phthalocyanine (MPc) molecules, have recently become very attractive and practically alternative because MP and MPc present versatile and tunable optical spectra by changing the central metal atoms such as Zn and Mg. For this type of gas sensors, an ordinary UV–Vis spectrophotometer can be easily modified to be the transducing unit for optical e-nose measurement. The gas sensing films were prepared by spin-coating and working by measuring the absorption spectral change under ambient conditions. A simple pattern recognition method such as principal component analysis (PCA) was demonstrated to be very effective for this e-nose system. The results from the PCA method indicate that both MP and MPc materials were cost-effective choices for classifying various odors. Based on the density functional theory (DFT) calculations, sensing mechanism of this type of chemical sensors can be described in terms of the ion–dipole interactions between the central metals of the sensing molecules and VOCs molecule.

Keywords Chemical gas sensors, Electronic nose, Optical gas sensors, Porphyrin and phthalocyanine

---

T. Kerdcharoen (✉) and S. Kladsomboon
Faculty of Science, Department of Physics, Mahidol University, Bangkok, Thailand
e-mail: sctkc@mucc.mahidol.ac.th

A. Tuantranont (ed.), *Applications of Nanomaterials in Sensors and Diagnostics*, Springer Series on Chemical Sensors and Biosensors (2013) 14: 237–256
DOI 10.1007/5346_2012_49,
Published online: 18 March 2013

Contents

1 Introduction

Chemical gas sensor is a device that converts chemical properties of the odorants into electronic signal. There are many types of chemical gas sensors, for instance, conductometric, gravimetric (piezoelectricity), chemoresistor, and optical sensor. Among these transduction techniques, optical measurement has been one of the most popular choices due to its variety in the measurement of optical modes, for example, absorbance, fluorescence, polarization, refractive index, and reflectance [1, 2]. Metallo-porphyrin (MP) and metallo-phthalocyanine (MPc), which are the organic compounds, have been very attractive sensing materials because their absorption spectral changes upon exposure to volatiles are active in the visible region [3]. Consequently, measuring the color changes within the absorption spectra of these sensing materials can be done conveniently using standard instrument such as UV–Vis spectrophotometer. Such equipment can be modified to incorporate gas flow chamber, data acquisition, and processing, which finally leads to an optical electronic nose system that can effectively discriminate organic compounds (VOCs) in the aroma of food and beverages [4]. The optical e-nose system developed from a general UV–Vis spectrophotometer has thus become a promising new technology for odor classification for food and agricultural industries.

This chapter is divided into six sections as follows: (1) introduction, (2) chemical gas sensor, (3) a background on porphyrin and phthalocyanine, (4) e-nose system based on UV–Vis spectrophotometer, (5) development of optical gas sensor, and (6) odor classification based on principal component analysis (PCA).

2 Chemical Gas Sensor

e-Nose has become a new technology for odor classification by mimicking the functions of the olfactory system in human. In the olfactory system, there are millions of odorant receptors in the nasal working for odor detection. Signals from the odorant receptors are accumulated and transmitted to the brain by the olfactory bulb. Then, the brain classifies the odor types using pattern recognition based on the past memory. In turn, an e-nose system consists of a gas sensor array, signal processing, and pattern recognition, reflecting the biological odorant receptors, olfactory bulb, and brain, respectively [4]. The most important area in the development of e-nose is to improve the sensitivity and stability of the chemical gas sensors.

In contrast to biological nose that involves very complex biological and neural process, odor sensing in e-nose is based on much simpler transduction techniques such as electrochemical, optical, and gravimetric approaches. Electrochemical transduction, as often implemented in the metal-oxide semiconductor gas sensor, is the most popular technology due to its established advantages such as low cost, compactness, and easy integration with integrated circuit technology. The most used sensing material in the metal-oxide gas sensors is SnO_2 thin film, as reported to detect several industrial gases such as methane, H_2, CO, H_2S, and methanol [5, 6]. Metal-oxide gas sensors work at high temperature using additional electrical power to supply the heater. Another electrochemical technology based on polymer/carbon nanotube nanocomposites was introduced to allow measurement at room temperature [7]. Gravimetric transduction employs piezoelectric technique, usually known as the quartz crystal microbalance (QCM), for sensitive mass measurement by observing the change of oscillating frequency upon adsorption of the odorants on the crystal surface. For example, Ding et al. reported a QCM-based device coated with copper(II) tetra-*tert*-butyl-5,10,15,20-tetraazaporphyrin that is sensitive to hydrocarbon gases, i.e., hexane, benzene, and toluene [8].

Organic semiconductor materials such as MP and MPc can work as gas sensor based on the change in electrical or optical properties via charge transfer on their surface. The possible sites for gas adsorption were found at the central metal atom and the conjugated π-electron system. It was found that MP and MPc are very versatile sensing materials for detection of a wide range of gases. For example, four metallo-octaethyl porphyrins (with the metal atoms of Mn, Fe, Co, and Ru) thin films were tested with four vapor samples such as 2-propanol, ethanol, acetone, and cyclohexane by recording the reflected light intensity before and after being exposed toward volatiles samples [9]. Blending of MP and MPc also leads to enhancement of the sensor, as reported by Spadavecchia and co-workers [10]. In their work, Zn(II) tetra-4-(2,4-di-*tert*-amylphenoxy)-phthalocyanine and Cu(II) tetrakis (*p*-*tert*-butylphenyl) porphyrin blended thin film was used as the sensing materials for detection of methanol, ethanol, and isopropanol vapors. Spadavecchia et al. also proposed a very novel technique based on the optical absorption using MP and MPc as sensing materials, in which it employs only one sensor film to detect and classify several gases depending on their selectivity on specific wavelength regions. Many research works have reported that MP and MPc are very effective sensing materials for optical sensing due to their versatile and tunable spectra based on variation of central metal and substituent of the molecules [11, 12].

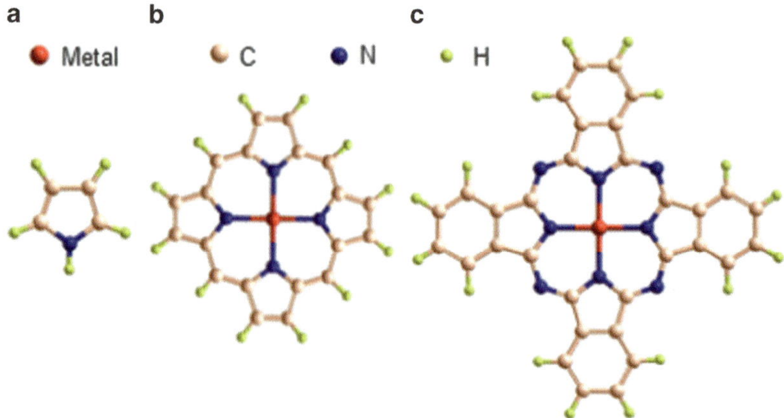

Fig. 1 The molecular structure of (**a**) pyrrole, (**b**) metallo-porphyrin, and (**c**) metallo-phthalocyanine

3 Background of Porphyrin and Phthalocyanine

Porphyrin and phthalocyanine are organic compounds that present the strong absorption spectra in the UV–Vis regions. Both porphyrin and phthalocyanine are therefore used as common coloring compounds in the dye industry.

The structures of porphyrin and phthalocyanine basically consist of four pyrrole units that produce a macrocycle as shown in Fig. 1a. The electronic structures of both organic compounds can be varied by substitution of some peripheral positions with side chains or replacement to the central atom with other metallic atoms (see in Fig. 1b, c). MP and MPc present different electronic properties because of a little but significant variation in the core molecular structures. The skeleton of MP has an extended conjugation system with 24-π electrons leading to a wide range of wavelengths for light absorption. MP presents the strong absorption spectra in a region around 400 nm so-called Soret band or B band. The strong absorption band in UV–Vis region had generally been interpreted in terms of π–π* transition between bonding and anti-bonding molecular orbital [13, 14]. MPc presents the absorption in two regions, e.g., around 300 and 700 nm. The first absorption band (around 300 nm) is described as the transition of π-electron from the highest occupied energy level to the lowest unoccupied level. The maximum absorption band (around 700 nm) is produced from the resonating of π-electrons, which is the free electron gas between two equivalent limiting structures [15].

Computational method based on density functional theory (DFT) at the B3LYP/6-31G* level of theory was used to describe the interaction energies between magnesium-tetra-phenyl-porphyrin (MgTPP) and gas molecules such as methanol, ethanol, isopropanol, acetone, and acetic acid molecules.

Figure 2 plots the interaction energies between gas molecules and MgTPP versus varying distance between the oxygen atom in gas molecule and the Mg atom in

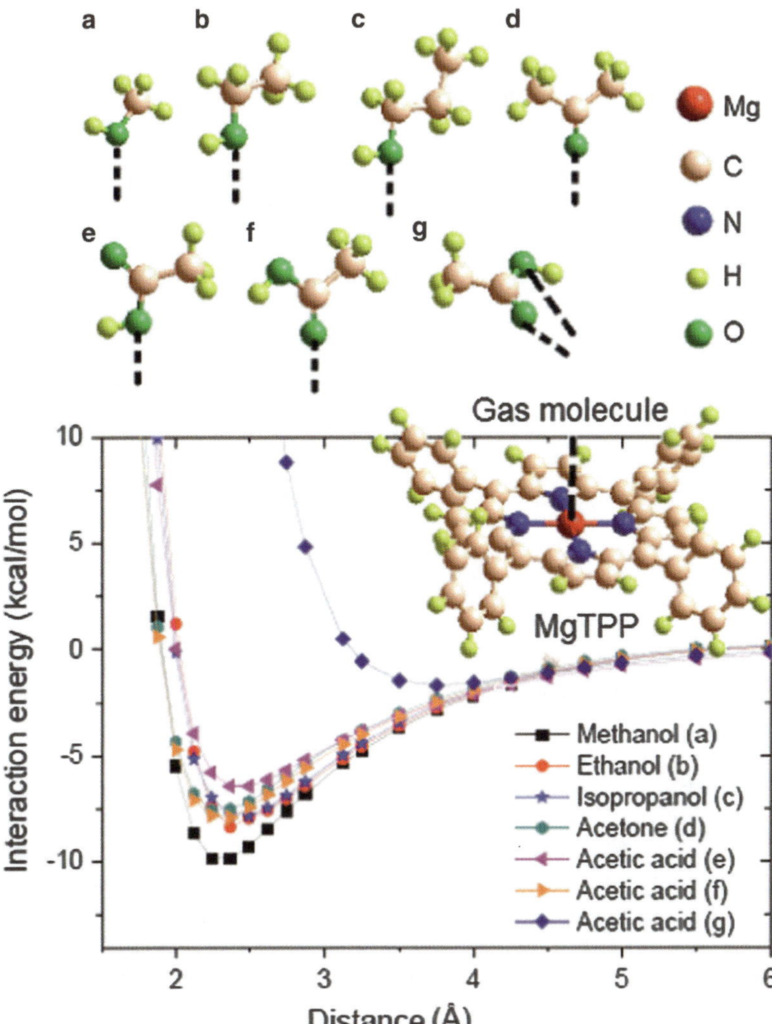

Fig. 2 Plot of interaction energies versus distance between oxygen atom in gas molecule and Mg atom in MgTPP molecule

porphyrin molecule. The interaction energies at the optimized distance indicate that MgTPP presents the stable structure with gas molecules. The results reveal that MgTPP has stronger interaction with methanol molecule at the optimized distance of 2.375 Å than other molecules.

Effect from the substitution on the central metal atom in porphyrin was tested by DFT calculation. The interaction energies (E_{Int}), changes in energy gap (ΔE_g), and changes in NBO charges of MTPP (M = Mg, Zn) were investigated as presented in Table 1. The interactions between porphyrin and VOC molecules are determined by the metal atom site via the interactions of the π-electrons and the free electrons of

Table 1 The interaction energies; E_{Int} (kcal/mol) at the optimized distance; D (Å), change in energy gap (MgTPP = 2.82 eV and ZnTPP = 2.91 eV); ΔE_g (eV), and the change in NBO charge: for MgTPP at magnesium (1.70153 a.u.), nitrogen (−0.73441 a.u.); for ZnTPP at zinc (1.61282 a.u.), nitrogen (−0.71875 a.u.); for VOCs at oxygen atomic site (methanol; O −0.741 a.u., ethanol; O −0.759 a.u. and isopropanol; O −0.753 a.u., acetone; O −0.540 a.u. and acetic acid; O −0.591 a.u.) atomic site [16]

				Delta NBO charge of		
Interaction of	D (Å)	E_{Int} (kcal/mol)	ΔE_g (eV)	Metal	O	N
MgTPP with						
Methanol	2.375	−9.92	0.105	+0.001	−0.045	+0.024
Ethanol	2.50	−8.37	0.098	+0.002	−0.041	+0.022
Isopropanol	2.25	−7.83	0.102	+0.003	−0.040	+0.023
Acetone	2.375	−7.61	0.117	+0.001	−0.065	+0.028
Acetic acid (Fig. 2f)	2.375	−7.91	0.099	+0.004	−0.060	+0.026
ZnTPP with						
Methanol	2.375	−5.79	0.121	+0.036	−0.046	+0.010
Ethanol	2.50	−5.66	0.104	+0.031	−0.035	+0.007
Isopropanol	2.50	−5.97	0.103	+0.031	−0.039	+0.010
Acetone	2.50	−4.08	0.109	+0.034	−0.047	+0.010
Acetic acid (Fig. 2f)	2.50	−3.89	0.114	+0.034	−0.048	+0.011

the metal atom in porphyrin with the electrons of the VOC molecules. Hence, electron transfer from the metal atom in porphyrin to an oxygen atom of the VOCs molecule occurs when MgTPP/ZnTPP are in contact with the VOC molecule at the optimized distance [16].

To use MP and MPc as optical gas sensor, the materials have been fabricated as the thin solid film on glass substrate. Measuring changes in the absorption or emission spectra of the thin film is a convenient technique for the optical gas sensor. Various derivatives of MP and MPc compounds have been fabricated to use as the gas sensors. For instance, zinc-5,10,15,20-tetra-phenyl-21H,23H-porphyrin (ZnTPP), manganese(III)-5,10,15,20-tetraphenyl-21H,23H-porphyrin chloride (MnTPPCl), and zinc-2,9,16,23-tetra-*tert*-butyl-29H,31H-phthalocyanine (ZnTTBPc) were widely used to detect NO_2 [17, 18], VOCs, [19] and foods [20]. Figure 3 illustrates the molecular structure of (a) ZnTPP, (b) MnTPPCl, and (c) ZnTTBPc.

4 e-Nose System Based on UV–Vis Spectrophotometer

An optical e-nose system consists of three basic components: a light source, sensing materials or gas sensor, and a detector. In this system, a UV–Vis spectrophotometer was used as the light source and light detector. All measurements were performed at room temperature and at the normal incidence of the light beam. The gas sensing of the organic thin films has been measured under the dynamic gas flow through a home-built stainless steel chamber, equipped with quartz windows for optical measurements as shown in Fig. 4. The carrier gas (in this case 99.9% nitrogen)

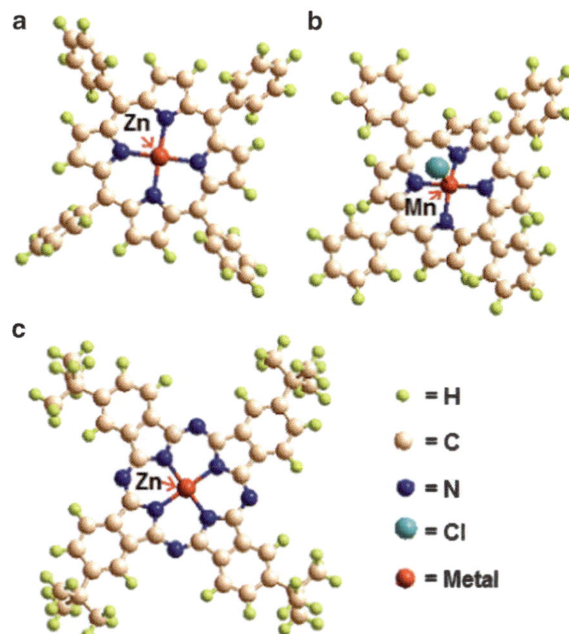

Fig. 3 Molecular structures of (**a**) ZnTPP, (**b**) MnTPPCl, and (**c**) ZnTTBPc

was supplied into the sample bottle that was immersed in a heat bath. The sample vapor was evaporated from the surface of the liquid sample and carried by the nitrogen gas. The gas flow was controlled by solenoid valve to switch between the reference gas and the sample gas every 10 min. The rate of gas flow was controlled by the flow meter. The absorption spectra of the organic thin film were collected by the UV–Vis spectrometer for every 1 s based on the function of the DAQ card.

Figure 5 shows the architecture of a data processing system of this optical e-nose system, starting with a flow of different VOC vapor into the gas sensor that fabricated from MP and MPc compounds. Then the absorption spectra from the UV–Vis spectrophotometer were collected for different sample vapors, comparison with the spectra of the unexposed films. The features extracted from the sensing signal were prepared for further pattern recognition process (in this case, PCA). Then the new data set was presented in the new orthogonal axes or principal components (PCs) such as the first PC (PC1) and the second PC (PC2), which are linear combinations of the original axes. The first PC carries most of the data variance, hence the most important information about the data [12].

5 Development of Optical Gas Sensor

To fabricate the optical gas sensor, MP and MPc must be deposited in the form of solid thin film. There are many fabrication techniques for preparing the thin film, for example, solvent casting, Langmuir–Blodgett, self-assembled monolayer,

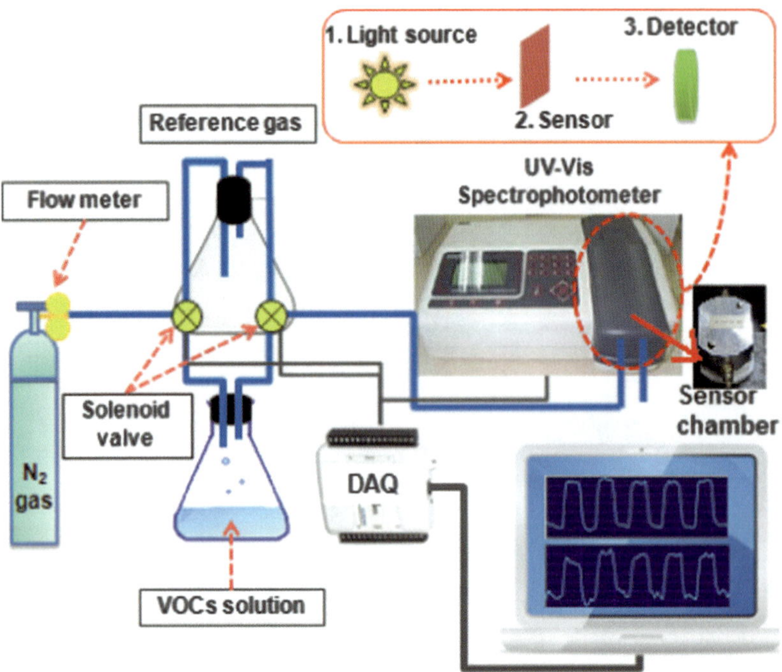

Fig. 4 Schematic diagram of the UV–Vis spectrophotometer setup

electropolymerization, and spin-coating techniques [21–23]. Among these fabrication techniques, spin-coating has been the cost-effective technique to produce regular sensing layers with moderately controllable thickness [24]. Therefore, spin-coating has been widely applied in the field of sensor technology.

In this work, the optical gas sensor uses standard UV–Vis spectrophotometer to measure absorption change when the sensing thin film interacts with the sample vapor. The absorption spectra were divided into several regions based on the response to VOCs. Thus, each sensing material yields different response at different spectral regions [3]. For example, the absorption spectra of MgTPP can be divided into six regions (R) such as R1: 300–370, R2: 370–410, R3: 410–490, R4: 490–555, R5: 555–610, and R6: 610–800 nm. For ZnTPP, the spectra can be divided into R1: 300–360, R2: 360–425, R3: 425–460, R4: 460–550, R5: 550–615, and R6: 615–800 nm) [16]. The gas sensing responses (S) can be calculated from Eq. (1):

$$S = \Delta A / A_{\mathrm{Base}} \tag{1}$$

where ΔA is the difference of the integrated area within a specific range of absorption spectra between the sample (A) and reference (A_{Base}).

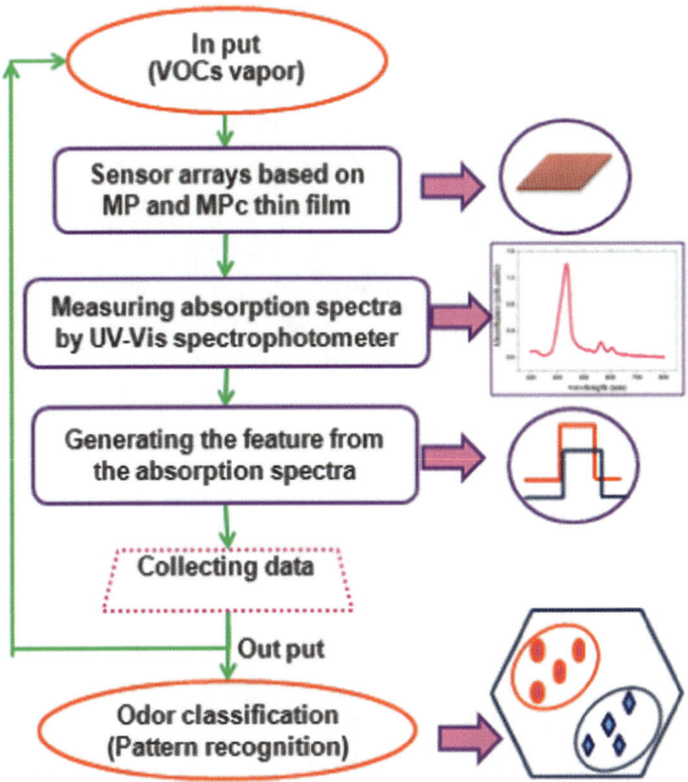

Fig. 5 The architecture of a data processing system for the optical electronic nose

5.1 The Effect of Thickness in Gas Sensor Thin Film

To prepare an efficient chemical gas sensor, the thickness of the thin film must be controlled and optimized. The thickness of the film depends on the concentration of the solution and the spin speed of spin-coating. Hence, the effect of spin speed to the MgTPP spin-coated thin film was tested by using methanol and ethanol vapors as the analyte gases [25]. In this experiment, MgTPP was prepared by spin-coating at the spin speeds of 500, 1,000, and 1,500 rpm and subjected to thermal annealing process. MgTPP thin films spun at lower speed have higher thickness, as revealed by atomic force microscopy. For instance, a typical thickness of the film spun at 500 rpm is 600 nm which decreases to 400 and 200 nm when the rotation speed increases to 1,000 and 1,500 rpm, respectively. Specifically, the spinning speed of 1,000 and 1,500 rpm produces stable thin films that yield almost equivalent sensing properties as shown in Fig. 6.

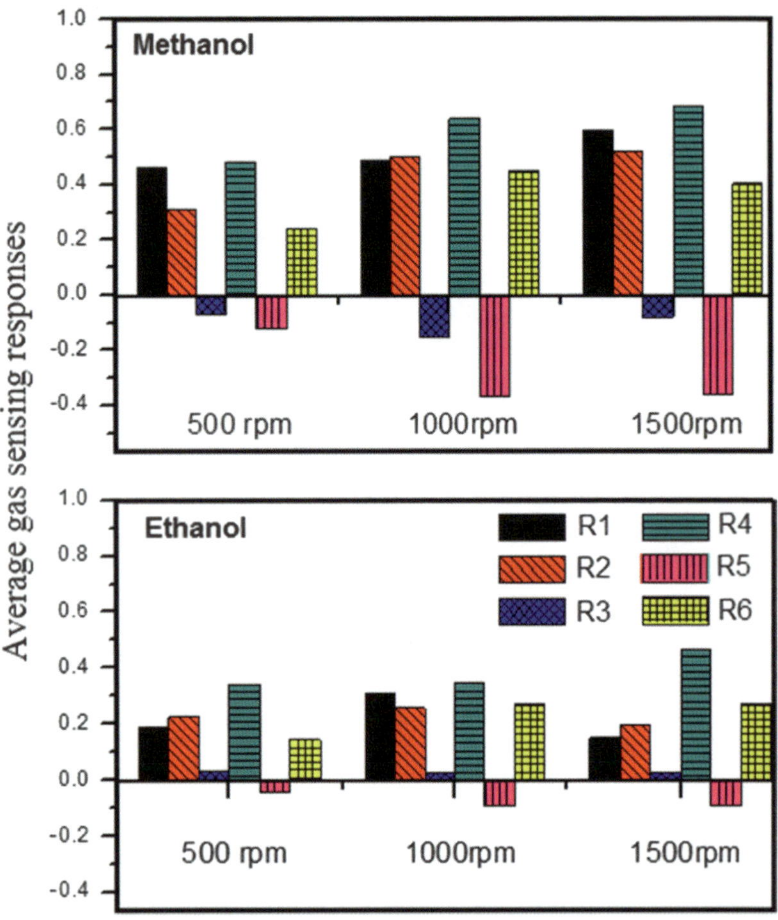

Fig. 6 The average gas sensing responses of MgTPP thin films with methanol and ethanol vapors by varying the speed of spin-coating

5.2 The Effect of Treatment Process in Gas Sensor Thin Film

There have been many research that report the processes to increase the sensitivity of gas sensors such as thermal treatments of cobalt–porphyrins–SnO_2 thin film [26] and solvent-vapor treatments of aluminum phthalocyanine chloride and lead phthalocyanine thin films [27, 28]. Thus, MgTPP thin films undergoing different treatments, i.e., thermal and solvent-vapor treatments were compared in the sensing properties to methanol and ethanol vapors [29]. In Fig. 7, the AFM results indicate larger grain sizes for both treated films, which result from molecular crystallization. The average roughness of the spin-coated film was about 2.83 nm. After thermal and solvent-vapor treatments, the grain size is 41.7 and 41.4 nm, respectively.

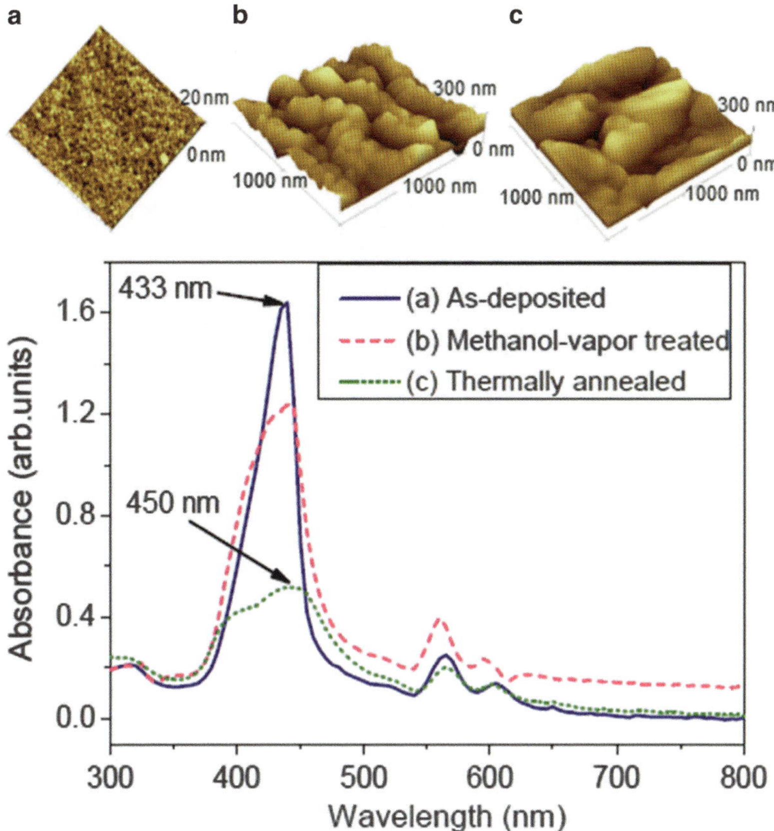

Fig. 7 UV–Vis absorption spectra and AFM images of the spin-coated thin film (spun at 1,000 rpm and concentration of 10 mg/mL) of (**a**) as-deposited, (**b**) methanol-vapor treated, and (**c**) thermally annealed MgTPP thin film

The UV–Vis absorption spectra of these thin films are shown in Fig. 7. The intensity of the main peak of the as-deposited MgTPP thin film located at 433 nm reduced after the film treatment.

In Fig. 8, it can be seen that both types of treated MgTPP films present a good sensing response with methanol for all six ranges, whereas the thermally treated film is more efficient to detect all types of VOCs, namely, methanol and ethanol.

5.3 The Effect of Central Metal Atom in Porphyrin

The effect of central metal atom in porphyrin has been interested by many research, e.g., molecular modeling of Mg- and Zn-porphyrins. The calculation results confirm that the optimized structures of these porphyrins are dependent on the central metal atomic size [30].

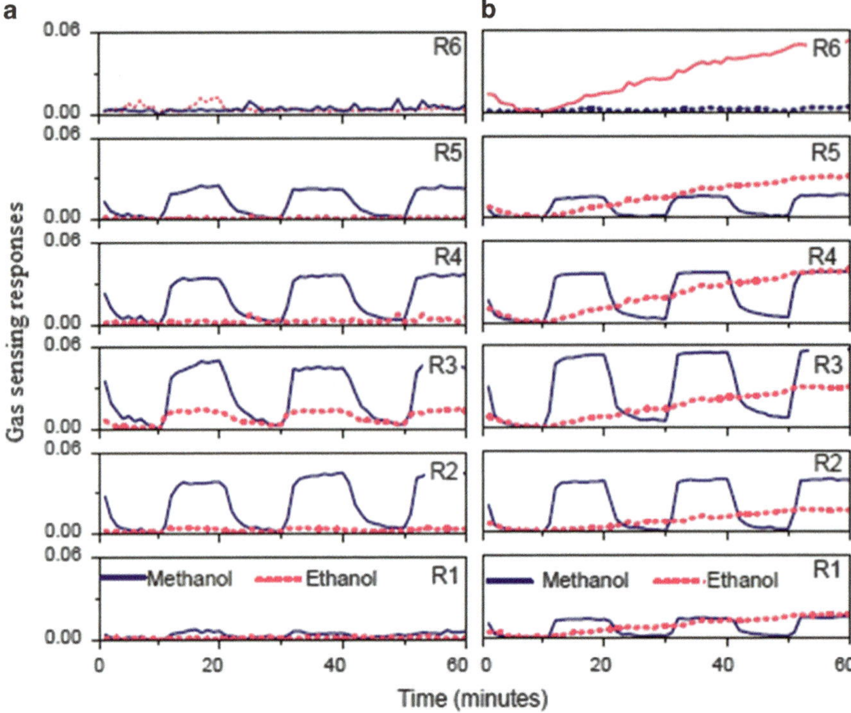

Fig. 8 The gas sensing responses (*S*) of MgTPP films under switching sample/reference gases, (**a**) thermally treated film, and (**b**) methanol-vapor treated film (R1: 300–360 nm, R2: 360–425 nm, R3: 425–510 nm, R4: 510–600 nm, R5: 600–650 nm, R6: 650–800 nm)

The effect of the central metal atom in porphyrin gas sensor was investigated experimentally as shown in Fig. 9. In this experiment, it was shown that both MgTPP and ZnTPP films express more sensing response to methanol than other alcohols. It can be seen that MgTPP is more selective to methanol comparing to ZnTPP. Thus, ZnTPP expresses higher response to ethanol and isopropyl alcohol than MgTPP [16].

5.4 The Hybrid Gas Sensor by Mixed Layer MP and MPc

The spin-coated thin films of MP and MPc are widely used as opto-chemical sensing materials because of their absorbance in the UV–Vis regions and their absorption spectral changes upon exposure to various vapor such as alcohols [24], amines, hexylamine, and octylamine [31]. The strongest response for porphyrin gas sensors corresponds to the Soret band or main peak of absorption spectra, indicating the π–π^* transition between bonding and anti-bonding molecular orbital in the porphyrin compounds [32]. Thus, the hybrid gas sensor between MP and MPc has

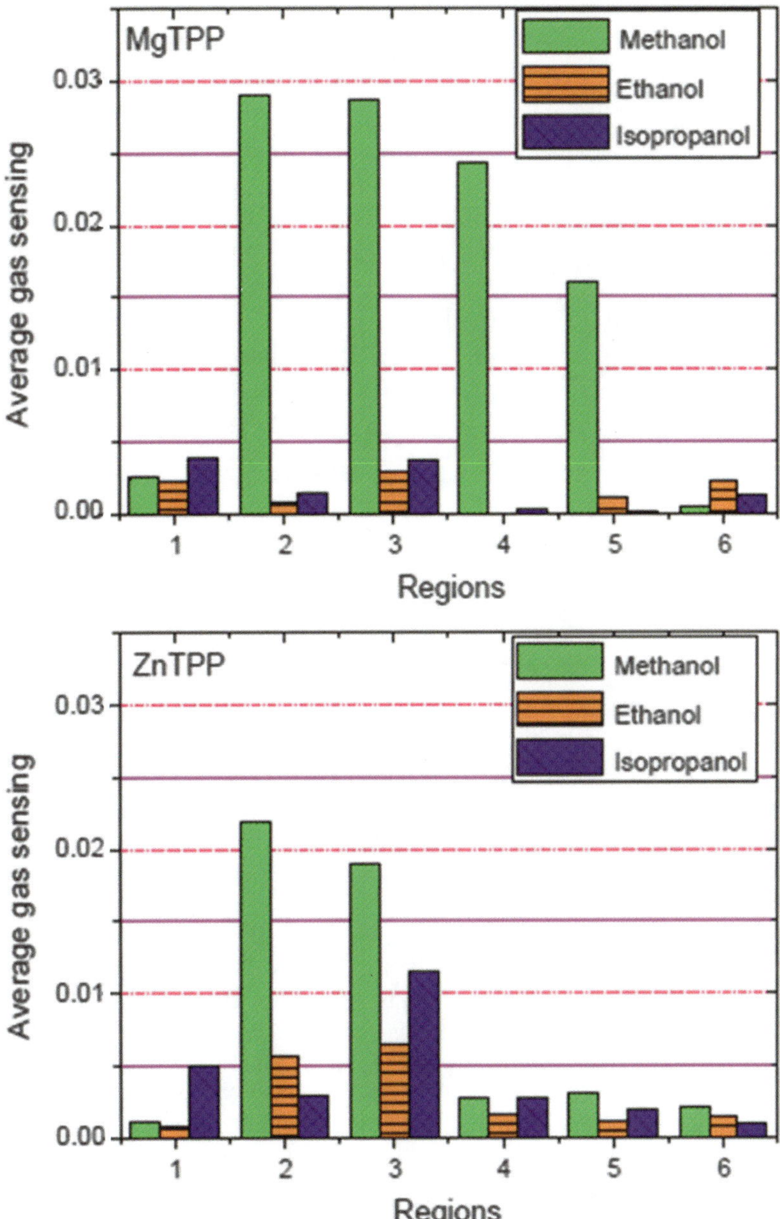

Fig. 9 The average gas sensing of MgTPP annealed film and ZnTPP spin-coated film

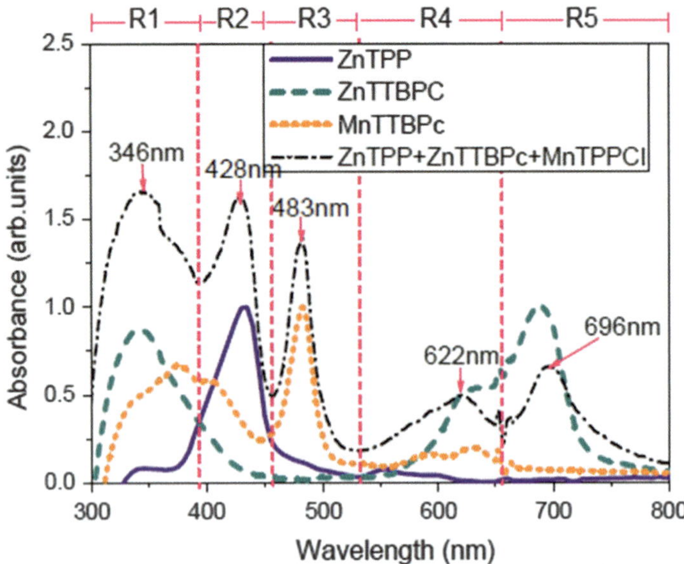

Fig. 10 Optical absorption spectra of (**a**) the ZnTTBPc, ZnTPP, and MnTPPCl spin-coated film and (**b**) the mixed layer of ZnTTBPc/ZnTPP/MnTPPCl spin-coated film

been investigated because MP and MPc present the main peak of absorption spectra in the different wavelength regions. To prepare the hybrid gas sensor thin film, ZnTPP, MnTPPCl, and ZnTTBPc were obtained from Sigma-Aldrich. The mixed solution of ZnTPP/MnTPPCl/ZnTTBPc was obtained from a 3 mL of chloroform solution that was composed of 5 mg of ZnTPP, 10 or 15 mg of MnTPPCl, and 12 mg of ZnTTBPc. Then the blended ZnTPP/MnTPPCl/ZnTTBPc spin-coated thin films were obtained by spinning at 1,000 rpm for 30 s. The optical absorption spectra of thin films were recorded by the e-nose setup (Fig. 4).

The absorption spectra of the ZnTTBPc/ZnTPP/MnTPPCl spin-coated thin films are shown in Fig. 10. The hybrid thin films show main peaks at 346, 428, 483, 622, and 696 nm, in accordance with the main peaks of the MP and the MPc. Then we divided the absorption spectra into five regions around each of the main peaks: 300–400 (R1), 400–460 (R2), 460–530 (R3), 530–655 (R4), and 655–800 nm (R5).

Figure 11a shows the average gas sensing of ZnTTBPc, ZnTPP, and MnTPPCl spin-coated thin films to 10 mol% of alcohols vapor in the nitrogen gas. The concentration of alcohols vapor was controlled by the heat bath and calculated from the weight loss of the alcohols in the sample bottles. Methanol shows the highest sensing response amongst the alcohols. ZnTPP and MnTPPCl exhibit no response to ethanol and isopropanol alcohol in the region R3. The hybrid layer of ZnTTBPc/ZnTPP/MnTPPCl displays responses to all types of alcohols as shown in Fig. 11b, c. Overall the results indicate that the hybrid spin-coated thin films fabricated from 12 mg of ZnTTBPc, 5 mg of ZnTPP, and 15 mg of MnTPPCl in a 3 mL of chloroform show the good sensing response with alcohol vapor.

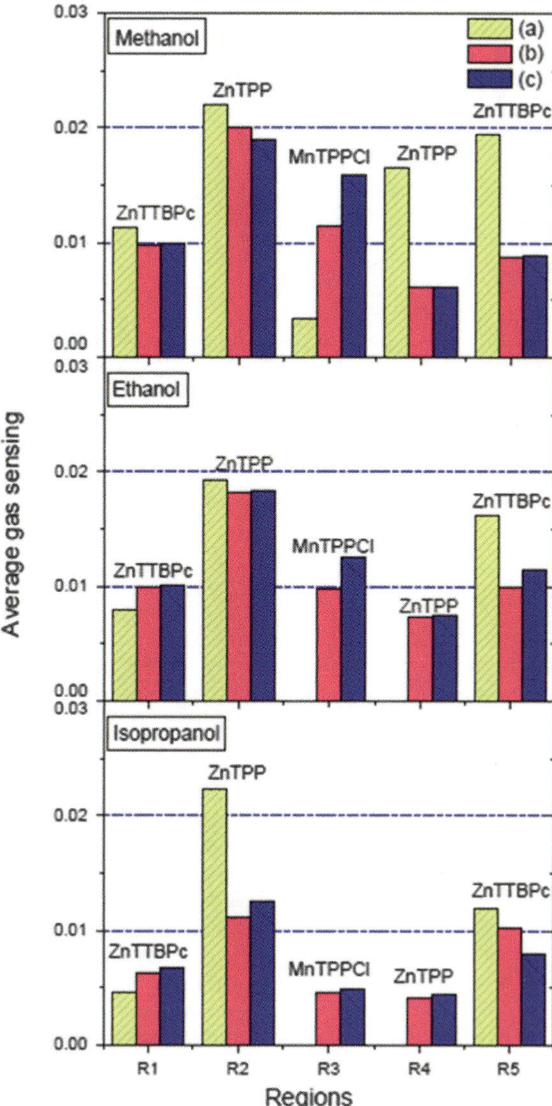

Fig. 11 The average gas sensing of (**a**) ZnTTBPc, ZnTPP, and MnTPPCl, (**b**) hybrid layer of ZnTTBPc: ZnTPP:MnTPPCl (12:5:10) and (**c**) hybrid layer of ZnTTBPc:ZnTPP:MnTPPCl (12:5:15) spin-coated thin film

6 Odor Classification Based on PCA

The PCA is a multivariate data analysis method that has been widely used as a pattern recognition algorithm in most e-nose systems to discriminate different VOCs, e.g., alcohols, toluene, methylethylketon, soft drinks, and alcoholic beverages [33–35].

Figure 12 shows PCA plots in the two dimensions as related to the sensing signal from the hybrid layer of ZnTTBPc/ZnTPP/MnTPPCl spin-coated thin film based on

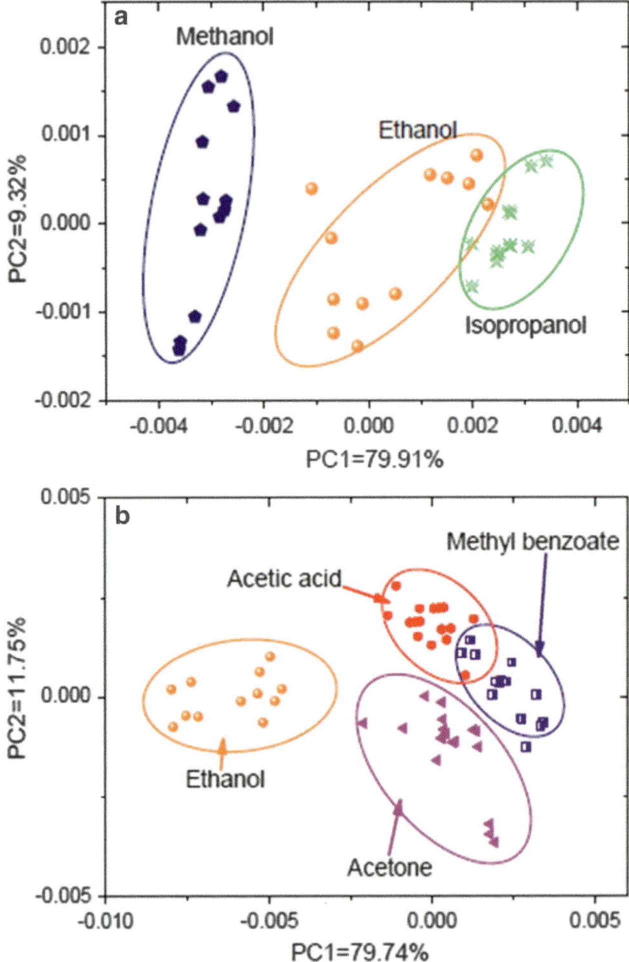

Fig. 12 PCA two-dimensional score plot related to the five arrays of hybrid gas sensor corresponding to (**a**) three types of alcohol vapor and (**b**) VOCs vapor

the optical absorption technique [36]. The results show that these hybrid materials can classify three types of alcohol, namely, methanol, ethanol, and isopropanol (Fig. 12a). These optical sensors separate the ethanol vapor from the other VOCs, methyl benzoate, acetone, and acetic acid (Fig. 12b). Furthermore, the classification of the different volume of ethanol in water was tested with the hybrid gas sensor. The ratios between ethanol and water were varied by changing the volume of ethanol and in water. The PCA calculation can separate the different volume of ethanol and in water from 0% to 100% (see Fig. 13).

The separation between 0% and 100% ethanol is shown clearly in PC1 axis direction. The PC1 axis, which is the highest variance value of principal coordinates

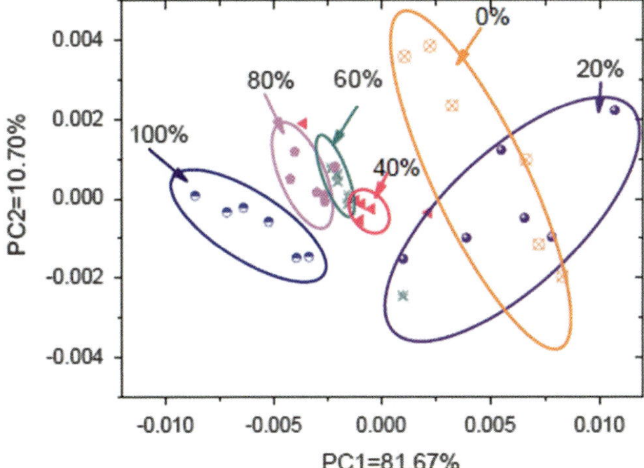

Fig. 13 PCA two-dimensional score plot related to the five arrays of hybrid gas sensor corresponding to the concentration of ethanol in water

(81.67%), represents the most important information of the data set [37]. The data set of 100% ethanol is located at the left-hand side of Fig. 13 while the data set of 0% ethanol (or 100% water) is located at the right-hand side of Fig. 13. It can be concluded that PCA is an effective statistical method in the electronic nose system to classify the group of odor. This method presents a high efficiency and a low demand on computing power in the system. Consequently, these results confirm that the optical electronic nose system based on the three types of organic compounds should be highly effective for discriminating VOCs and applicable to qualitative measurements of foods and beverages.

Acknowledgments Mahidol University and National Nanotechnology Center (Project No. P-12-01157) are acknowledged for supports.

References

1. Manno D, Micocci G, Serra A et al (1999) Gas sensing properties of meso, meso-buta-1,3-diyne-bridged Cu(II) octaethylporphyrin dimer Langmuir-Blodgett films. Sens Actuators B Chem 57:179–182
2. Montmeat P, Madonia S, Pasquinet E et al (2005) Metalloporphyrins as sensing material for quartz-crystal microbalance nitroaromatics sensors. J Sens IEEE 5:174–179
3. Uttiya S, Pratontep S, Bhanthumnavin W et al (2008) Volatile organic compound sensor arrays based on zinc phthalocyanine and zinc porphyrin thin films. In: Proceedings of the 2nd IEEE-NEC 2008, pp 618–623
4. Pearce TC, Schiffman SS, Nagle HT et al (2003) Handbook of machine olfaction electronic nose technology. Wiley-VCH Verlag GmbH & Co. KGaA, Weinheim

5. Campbell M (1997) Sensor system for environmental monitoring, vol 1: sensor technologies. Blackie Academic & Professional, London
6. Wongchoosuk C, Wisitsoraat A, Tuantranont A et al (2010) Portable electronic nose based on carbon nanotube-SnO_2 gas sensors and its application for detection of methanol contamination in whiskeys. Sens Actuators B 147:392–399
7. Lorwongtragool P, Wisitsoraat A, Kerdcharoen T (2011) An electronic nose for amine detection based on polymer/SWNT-COOH nanocomposite. J Nanosci Nanotechnol 11:10454–10459
8. Ding H, Erokhin V, Ram MK et al (2000) A physical insight into the gas-sensing properties of copper II tetra-tert-butyl-5,10,15,20-tetraazaporphyrin Langmuir-Blodgett films. Thin Solid Films 379:279–286
9. Akrajas SMM, Yahaya M (2002) Enriching the selectivity of metalloporphyrins chemical sensors by means of optical technique. Sens Actuators B Chem 85:191–196
10. Spadavecchia J, Ciccarella G, Siciliano P et al (2004) Spin-coated thin films of metal porphyrin-phthalocyanine blend for an optochemical sensor of alcohol vapour. Sens Actuators B Chem 100:88–93
11. Muthukumar P, John SA (2011) Highly sensitive detection of HCl gas using a thin film of meso-tetra(4-pyridyl)porphyrin coated glass slide by optochemical method. Sens Actuators B 159:238–244
12. Spadavecchia J, Ciccarella G, Rella R et al (2003) Metallophthalocyanines thin films in array configuration for electronic optical nose applications. Sens Actuators B 96:489–497
13. Smith KM (1975) Porphyrin and metalloporphyrins. Elsevier Scientific Pub. Co., Newyork
14. El-Nahass MM, Zeyada HM, Aziz MS et al (2005) Optical absorption of tetraphenylporphyrin thin films in UV–vis–NIR region. Spectrochim Acta A 61:3026–3031
15. Moser FH, Thomas AL (1963) Phthalocyanine compounds. Reinhold, London
16. Kladsomboon S, Pratontep S, Puntheeranurak T et al (2011) An artificial nose based on M-porphyrin (M = Mg, Zn) thin film and optical spectroscopy. J Nanosci Nanotechnol 11:10589–10594
17. Richardson TH, Dooling CM, Worsfold O et al (2002) Gas sensing properties of porphyrins assemblies prepared using ultra-fast LB deposition. Colloids Surf A 198–200:843–857
18. Li XY, Shen SY, Zhou QF (1998) The gas response behavior of spin-coated phthalocyanine films to NO_2. Thin Solid Films 324:274–276
19. Arshad S, Salleh MM, Yahaya M (2007) Detection of volatile organic compounds using titanium dioxide coated with dye-porphyrins thin films in bulk acoustic system. Solid State Sci Technol 15:175–181
20. Salleh MM, Umar AA, Yahaya M (2002) Optical sensing of capsicum aroma using four porphyrins derivatives thin films. Thin Solid Films 417:162–165
21. Amico AD, Natale CD, Paolesse R et al (2000) Metallo porphyrins as basic material for volatile sensitive sensors. Sens Actuators B 65:209–215
22. Luo T, Zhang W, Gan F (1992) Structural change of Langmuir-Blodgett film of tetra-neopentoxy phthalocyanine zinc during heat treatment. Opt Mater 1:267–270
23. Liu CHJ, Lu WC (2007) Optical amine sensor based on metallophthalocyanine. J Chin Inst Chem Eng 38:483–488
24. Gu Z, Yin J, Liang P et al (2004) Temperature characteristics of optical parameters of phthalocyanine LB films and spin-coated film. Opt Mater 27:1618–1622
25. Kladsomboon S, Pratontep S, Uttiya S et al (2008) Alcohol gas sensors based on magnesium tetraphenyl porphyrins. In: Proceedings of the 2nd IEEE-NEC 2008, pp 952–955
26. Nardis S, Monti D, Natale CD et al (2004) Preparation and characterization of cobalt porphyrins modified tin dioxide film for sensor applications. Sens Actuators B 103:339–343
27. Kato K, Saito Y, Ohdaira Y et al (2006) Evaluation of nanostructure and properties of aluminum phthalocyanine chloride thin films due to ethanol-vapor treatment. Thin Solid Films 499:174–178

28. Ho KC, Chen CM, Liao JY (2005) Enhancing chemiresistor-type NO gas-sensing properties using ethanol-treated lead phthalocyanine thin films. Sens Actuators B 108:418–426
29. Kladsomboon S, Pratontep S, Puntheeranurak T et al (2009) Investigation of thermal and methanol-vapor treatments for MgTPP as an optical gas sensor. In: Proceedings of the 4th IEEE-NEM 2009, pp 843–847
30. Poveda LA, Ferro VR, Garcia de la Vega JM et al (2000) Molecular modeling of highly peripheral substituted Mg- and Zn-porphyrins. Phys Chem Chem Phys 2:4147–4156
31. Dunbar ADF, Richardson TH, McNaughton AJ et al (2006) Investigation of free base, Mg, Sn, and Zn substituted porphyrin LB films as gas sensors for organic analytes. J Phys Chem 110:16646–16651
32. Zhang X, Zhang Y, Jiang J (2005) Infrared spectra of metal-free, N,N-dideuterio, and magnesium porphyrins: density functional calculations. Spectrochim Acta A 61:2576–2583
33. Fernandez MJ, Fontecha JL, Sayago I et al (2007) Discrimination of volatile compounds through an electronic nose based on ZnO SAW sensors. Sens Actuators B 127:277–283
34. Zhang C, Suslick KS (2007) Colorimetric sensor array for soft drink analysis. J Agric Food Chem 55:237–242
35. Zhang C, Bailey DP, Suslick KS (2006) Colorimetric sensor arrays for the analysis of beers: a feasibility study. J Agric Food Chem 54:4925–4931
36. Kladsomboon S, Pratontep S, Kerdcharoen T (2010) Optical electronic nose based on porphyrin and phthalocyanine thin films. In: Proceedings of the ECTI 2010, pp 565–568
37. Brereton RG (2003) Chemometrics data analysis for the laboratory and chemical plant. Wiley, England

Nanotechnology to Improve Detection Sensitivity for Electrochemical Microdevices

Masatoshi Yokokawa, Daisuke Itoh, and Hiroaki Suzuki

Abstract With the increase in demands and applications, techniques to create electrochemical microdevices have made a remarkable progress over the last four decades. Key components of the electrochemical devices are electrodes that are easily fabricated by microfabrication techniques. Because of this, miniaturization, batch-fabrication, and integration with other components can easily be realized. This is a contrast to devices based on other detection principles. Miniaturization of the devices also brings with it additional advantages such as very small consumption of sample and reagent solutions, rapid mixing, and parallel processing. On the other hand, however, a challenging issue we often encounter is that it becomes increasingly difficult to maintain the performance that has been achieved by conventional electrochemical devices used in laboratories. To cope with the problem, nanotechnology provides good solutions. Numerous papers have been published to demonstrate the effectiveness of nanotechnology. Therefore, it is impossible to cover all the contents. However, a convincing conclusion is that nanotechnology really has surprising effects on sensing performance. With the wealth of knowledge of nanotechnology, their application to microfabricated devices will be the subject of the next stage. In this chapter, nanotechnologies applicable to the improvement of the performance of existing microfabricated electrochemical devices will be introduced. Although various techniques have been developed for single independent electrodes, those that may be difficult to apply to microfabricated devices are excluded. On the other hand, those that are applicable to nanoelectrodes are included.

Keywords Electrochemical detection, Nanoelectrode, Array, Ensemble, Electrochemiluminescence, Nanofabrication

---

M. Yokokawa, D. Itoh and H. Suzuki (✉)
Graduate School of Pure and Applied Sciences, University of Tsukuba, 1-1-1 Tennodai, Tsukuba, Ibaraki 305-8573, Japan
e-mail: hsuzuki@ims.tsukuba.ac.jp

A. Tuantranont (ed.), *Applications of Nanomaterials in Sensors and Diagnostics*, Springer Series on Chemical Sensors and Biosensors (2013) 14: 257–280
DOI 10.1007/5346_2012_39,
Published online: 8 December 2012

Contents

1 Microelectrodes and Nanoelectrodes

1.1 Properties of Independent Micro-/Nano-electrodes

To realize highly sensitive detection in a solution of a very small volume, microelectrodes are effective [1, 2]. Typical dimensions of reported microelectrodes are from tens of μm down to sub-μm. Actually, however, there is no clear-cut boundary, and a microelectrode should refer to an electrode whose characteristic dimension is comparable with or smaller than the diffusion layer thickness [3]. The thickness varies as time elapses, which also affects electrode behavior. Here, let us suppose that a sufficiently large overpotential is applied to a microdisk electrode to oxidize or reduce a compound. The diffusion of the molecules to the electrode is quite different from that observed with electrodes usually used in laboratories. With a planar electrode of the mm order for example, the analyte molecules are depleted at the electrode surface and diffuses uniformly from the bulk of the solution to the electrode surface except for the edges of the electrode. On the other hand, with the microelectrode, analyte molecules move towards the electrode along the concentration gradient in the diffusion layer (Fig. 1).

The enhanced diffusion brings with it some advantages. These include a current increase, increase in the signal-to-noise ratio, fast establishment of the steady-state signal, and reduction of the influence of solution resistance [3]. As the size of the electrode decreases, the double-layer capacitance also decreases. This results in the reduction of charging current, which is beneficial in conducting a voltammetric analysis by scanning electrode potential.

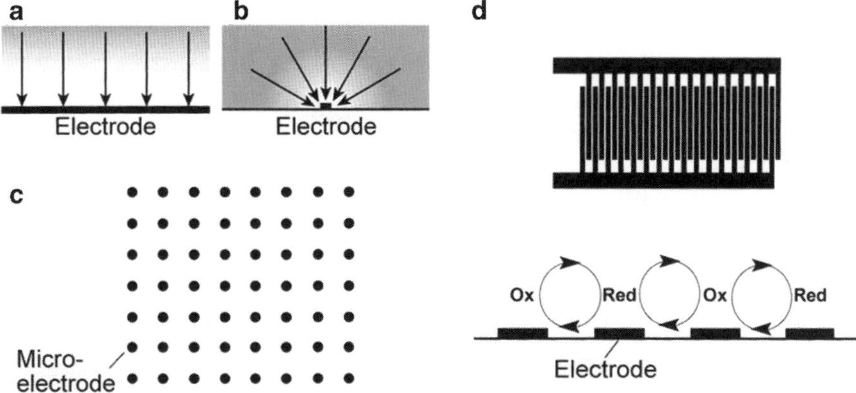

Fig. 1 Diffusion of analyte molecules to a planar electrode of the mm order (**a**) and to a microelectrode (**b**). (**c**) Microelectrode array. (**d**) Interdigitated microelectrodes and redox cycling

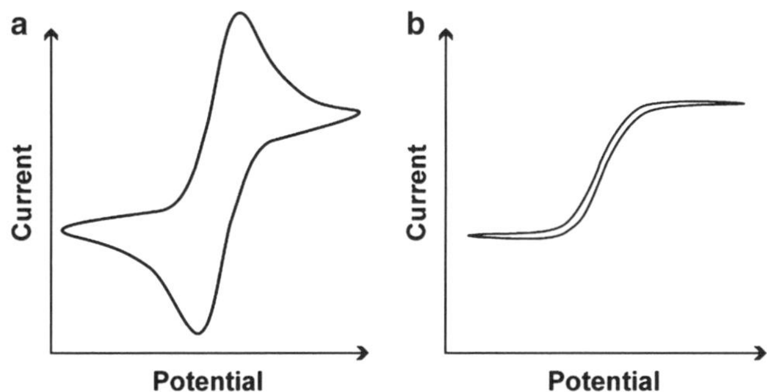

Fig. 2 Cyclic voltammograms observed with an electrode of the mm order (**a**) and with a microelectrode (**b**)

In cyclic voltammograms observed with an electrode of the mm order, for example, redox peaks are observed at scan rates usually used in electrochemical analyses (Fig. 2a). The peaks appear because the rate of mass transport is smaller than that of electrolysis at higher overpotentials, and the molecules that are subject to the electrode reaction continue to decrease in the vicinity of the electrode. On the other hand, with a microelectrode, a sigmoidal curve is observed under a typical condition (Fig. 2b). It should be noted that the shape of the voltammogram also depends on the scan rate of potential. With a very large scan rate, even the microelectrodes show peak-shaped voltammograms. On the other hand, at a very low scan rate, or in a long time scale, even larger electrodes show sigmoidal curves [1].

As mentioned, the microelectrodes have fascinating and excellent properties. However, a problem is its very small current. To solve this problem, microelectrode arrays have been used. In the following discussions, a group of electrodes with controlled shape, dimensions, and an ordered spacing is called an array, whereas a group of electrodes with a random spacing will be called an ensemble. Some geometrical variations are found for the arrays. One is an array of circular electrodes. The other is a row of thin strips of electrodes. The diffusion profile of a microelectrode array shown in Fig. 1c is influenced by the relative interelectrode spacing d/r, where d is the center to center distance and r is the radius of the electrode. With large d/r, the diffusion of the electroactive species to each electrode remains independent of all others. In this case, the limiting current is expressed as the sum of current generated on all the electrodes in the array. On the other hand, with the decrease in d/r, the diffusion layers overlap, and the diffusion profile approaches that of a uniform diffusion observed with a macroscopic electrode with the same geometric area. Although the diffusion is apparently the same as a macroscopic planar electrode, it should be noted that there is a significant difference between the microelectrode array and a macroscopic planar electrode. Because the background and capacitive currents change in proportional to the active area, they are reduced significantly with the microelectrodes. Consequently, substantial improvement in signal-to-noise (SN) ratio and Faradaic-to-capacitive current ratios is achieved with the microelectrode array.

1.2 Redox Cycling

Although the micro/nanoelectrodes themselves are very effective to improve the detection performance, a more fascinating feature called redox cycling can be realized by using two groups of microelectrodes [4]. An orthodox configuration is an interdigitated electrode array (IDA) that consists of a pair of fingers called a generator and a collector. They are held at appropriate different potentials (Fig. 1d). An electroactive analyte undergoes oxidation or reduction on the generator then diffuses to the collector where the analyte undergoes a reverse reaction. The species then diffuses back to the generator, and the cycle is repeated. This results in significant amplification of signal.

The array of micro/nanoelectrodes is not necessarily planar. Figure 3 shows a circular microcavity containing an addressable recessed microdisk gold electrode formed at the bottom and a tubular nanoband gold electrode on the vertical microcavity wall [5]. The microcavity consisted of layers (Fig. 3) consisting of gold layers with a chromium adhesion layer and polyimide insulating layers. The electrodes were separated from each other by 4 μm. The electrode structure was used to cause redox cycling. With this method, electrodes and spacing of the nm order can easily be realized. On the other hand, increasing the number of electrodes will be a tough work.

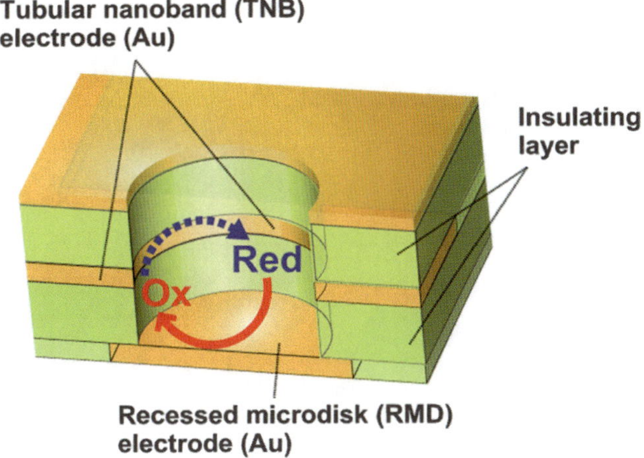

Fig. 3 Redox cycling in a microcavity with a recessed microdisk electrode and tubular nanoband electrodes

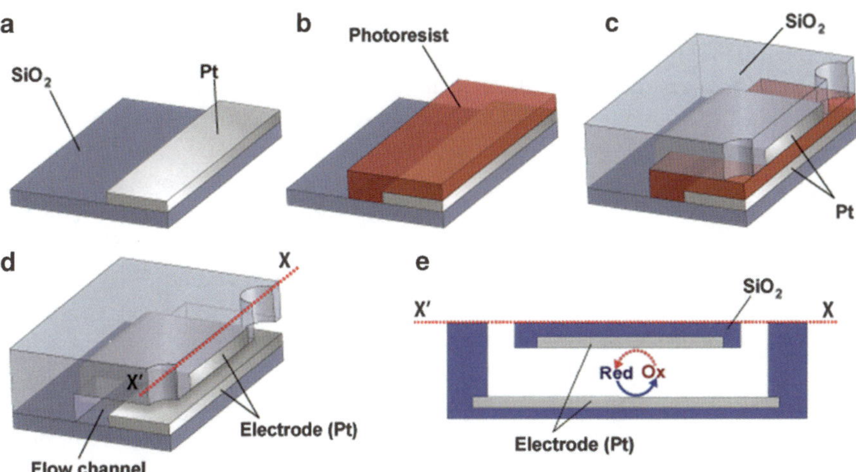

Fig. 4 Fabrication of a nanocavity with a pair of electrodes located at the bottom and ceiling (**a–d**). (**e**) Redox cycling caused between the electrodes

Another approach that has a potential to achieve significantly higher amplification is to use a nanocavity [6–9]. A pair of electrodes is located at the bottom and ceiling of a cavity with a height of the nm order. In this device, the electrodes can be simple planar electrodes. Redox cycling occurs between the electrodes facing each other in the nanocavity (Fig. 4).

Although the device uses a nanocavity, the method of formation of the structures is based on the conventional photolithography and micromachining [6–9]. First, the bottom electrode is formed on a substrate (Fig. 4a). Then, a pattern of a sacrificial

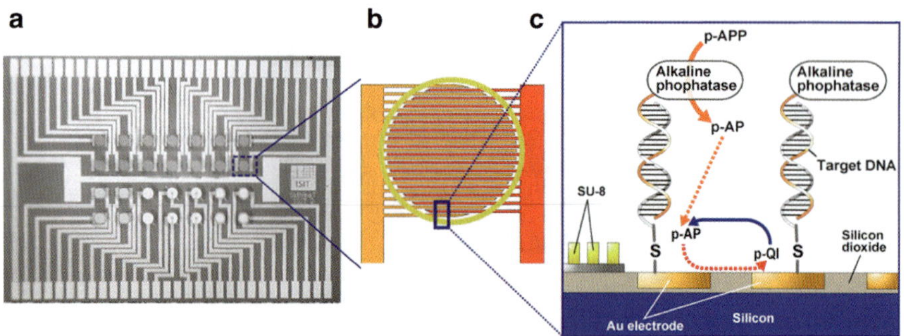

Fig. 5 Chip with multiple detection sites. (**a**) Top view of the chip. (**b**) Magnified view of an IDA electrodes. (**c**) Detection of a target DNA. The DNA binds to a probe DNA linked to the IDA electrodes. Then, the target DNA is enzyme-labeled. A reaction product of the enzymatic reaction p-aminophenole (p-AP) is detected by redox cycling. With kind permission from Springer Science +Business Media: [14]

layer such as photoresist, chromium, or amorphous silicon is formed on the electrode pattern (Fig. 4b). After the top electrode is formed on the sacrificial layer, an insulator layer is formed on the entire structure. Through-holes are formed in the insulating layer to remove the sacrificial layer and to inject a solution to be analyzed (Fig. 4c). Finally, by etching the sacrificial layer away by immersing the structure in an appropriate etchant, the electrode structure separated by a distance of the nm order is formed (Fig. 4d, e). When in use, the nanocavity is filled with an analyte solution and appropriate potentials are applied to the electrodes. Because of the short distance between the electrodes, nearly 100% of the generated product is cycled between the electrodes. With this approach, redox cycling was significantly enhanced, which realized the detection of a few hundred molecules [7].

Redox cycling not only brings with it the amplification of current but also brings additional advantages. In amperometric detection, the interference by electroactive compounds such as L-ascorbic acid, uric acid, and dissolved oxygen poses a problem. However, when the reaction for the analyte is reversible and that for the interferent is irreversible, redox cycling occurs only for the analyte, which suppresses the influence of the interferent relatively.

Redox cycling caused on IDAs has been used for the detection of proteins and DNAs. To couple the molecular recognition of probe molecules to electrochemical detection, the probe molecules are modified with an appropriate enzyme that produces electroactive molecules. One of such enzymes is alkaline phosphatase. This enzyme converts p-aminophenyl phosphate (PAPP) into electroactive p-aminophenol (PAP). PAP is oxidized into quinoneimine, which is also electroactive. In some reported cases, IDAs of the sub-μm order have been used to cause redox cycling for this purpose. The technique has been used for the detection of bacteriophages [10, 11], DNA [12], and RNA [12, 13]. An advantage of electrochemical devices is the integration of components. In some of the above-mentioned cases, simultaneous detection of different target molecules on an array of detection sites has been demonstrated (Fig. 5a) [14].

2 Activation of Electrodes Using Nanostructures

2.1 Activation of Electrode Reactions

Faradaic currents that originate from kinetically controlled electrochemical reactions depend on the real surface area of the working electrode rather than the geometric area. Therefore, by increasing the surface area by some orders of magnitude, the Faradaic current of a sluggish reaction can be enhanced. To this end, nanotechnology offers an excellent solution.

Hydrogen peroxide is often detected in biosensors that use enzymes categorized as oxidases. For biosensing, improvement of sensitivity is often required. To activate the surface of a platinum electrode and improve detection sensitivity, a relatively easy and effective method is to use platinum black. The platinum black is formed by electrodepositing platinum from a solution containing chloroplatinic acid. The platinum black has been used for the hydrogen electrode used for the reference electrode. Without the platinum black, the reversible potential of the hydrogen electrode is not expected. For H_2O_2, a significant improvement in the sensitivity has been observed [15]. Another effective structure is mesoporous platinum [16]. The structure is formed by depositing platinum using the three-dimensional structure of lyotropic liquid crystal phases. Voltammograms obtained with an ordinary planar platinum electrode do not exhibit well-defined plateaus for the oxidation and reduction of H_2O_2, which suggests that the response is under mixed kinetic and diffusion control [16]. Furthermore, linearity of the calibration plot is poor with the ordinary platinum electrode particularly at higher concentrations. This problem has also been solved with the mesoporous platinum and the lower detection limit has also been reduced compared with the conventional platinum electrode.

Glucose is one of the critical target analytes for biosensors. Although numerous papers have been published with regard to enzymatic biosensors, an inherent problem in biosensors in general is long-term stability. To solve this problem, non-enzymatic glucose sensors with nanostructures have been proposed. As demonstrated for the detection of hydrogen peroxide, porous structure of platinum is also effective to enhance direct oxidation of glucose [17]. Ordered porous platinum structure can be formed by depositing platinum through cdse-cds crystalline template (Fig. 6a–c) and dissolving the template in hydrofluoric acid (Fig. 6d). A significant increase in Faradaic current originating from the oxidation of glucose has been observed. The response of the electrode showed unique pH dependence with a maximum around pH 9.

Nanostructures can be formed in a variety of other methods. Ordered array of cylindrical platinum mesopores can be obtained by electrodepositing platinum within the aqueous domains of the liquid crystalline phases of oligoethylene oxide nonionic surfactants and removing the surfactant by rinsing with a large volume of deionized water [18]. Highly ordered platinum nanotube arrays have also been used for the direct oxidation of glucose [19]. The platinum nanotube arrays

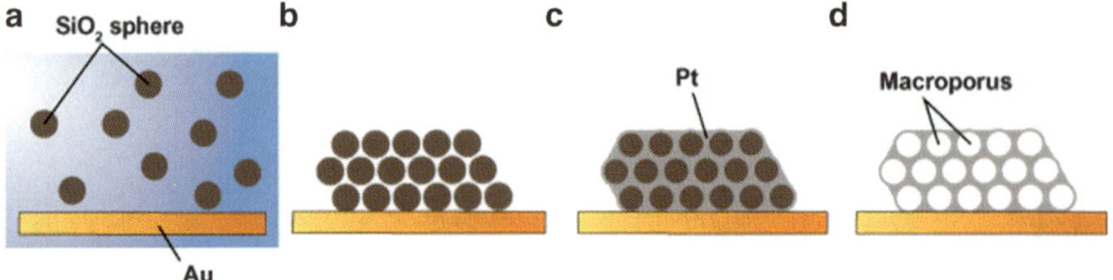

Fig. 6 Fabrication of the platinum microporous structure. (**a–b**) Deposition of SiO_2 spheres on a gold surface. (**c**) Electrodeposition of platinum into the interspace of the SiO_2 template. (**d**) Microporous platinum structure obtained after removing SiO_2 spheres by wet etching

were fabricated by electrodeposition of platinum in the pores of porous anodic alumina template. In these electrodes, the oxidation of glucose was enhanced with the increase in roughness factor. Furthermore, decrease in the influence of interferents such as L-ascorbic acid, uric acid, and p-acetamidophenol has been reported. Also, the electrode showed stable responses in the presence of chloride ions, which can be a cause of poisoning for noble metal electrodes.

Carbon nanotubes (CNTs) have also been used to improve the sensing performance. They are classified into single- and multi-walled tubes. The diameter of the former is typically 0.4–3 nm, whereas that of the latter ranges between 2 and 100 nm. The single-wall CNTs (SWCNTs) consist of a single cylindrical graphene sheet capped with hemispherical ends. The multi-wall CNTs (MWCNTs) consist of several to tens of concentric cylinders. Depending on the chirality of wrapping, CNTs show metallic or semiconducting properties.

Electrodes with a CNT/Nafion coating showed significant enhancement of sensitivity towards hydrogen peroxide and catecholamines [20]. Furthermore, the influence of interferents was suppressed significantly. Hrapovic et al. used SWCNTs and platinum nanoparticles immobilized with Nafion [21]. When compared with electrodes with only the CNTs or platinum nanoparticles, significantly larger current was observed. Also, due to the perm-selective nature of Nafion, no detectable responses were observed with L-ascorbic acid and uric acid of physiological concentrations.

An ensemble of MWCNTs also enhances direct oxidation of glucose in alkaline media [22]. The influence of poisoning by chloride ions was not observed. However, the influence of L-ascorbic acid and uric acid could not be eliminated completely. Enhancement of sensitivity is not limited to glucose. With single- or multi-wall CNTs, significant enhancement of sensitivity has also been reported for oxygen [23], dopamine [24, 25], epinephrine [25], L-ascorbic acid [25], and NADH [26].

The electrochemistry of CNTs is a little complicated, and it will not be appropriate to regard them as simple long electrodes of the nm order. Experimental facts accumulated up to now suggest that the open end of the MWCNT has a high electron transfer rate but the sidewall presents a low electron transfer rate [27, 28]. According to a recent report [29], however, the electrochemical activity of the parts of CNTs depends on the species to be analyzed and the existence of oxygen-containing surface functionalities. The accumulation of evidence is still necessary to reach the final conclusion.

Other than the CNTs, graphene sheets are beginning to be used for electrochemical devices, and have a potential to realize performance better than that of the CNT-based devices [28]. However, because there are technical difficulties to handle them at present compared with CNTs, the graphene-based devices are not addressed here.

2.2 Promotion of Direct Electron Transfer

Nanostructures have an effect to promote direct electron transfer between biomolecules and an electrode. In many enzymes, redox centers are located in the core of proteins. Therefore, direct exchange of electrons is often difficult. To promote electron transfer, mediators or promoters have been used. However, enzymes with capability of direct electron exchange facilitate fabrication of biosensors. To this end, CNTs have been used because of the remarkable electrocatalytic properties. With CNTs, there is a possibility to place them close to the redox centers of the proteins. This is actually the case. In many of the reported cases, a layer of single-wall or multiwall CNTs is formed on a base electrode by casting a CNT solution and the redox proteins are just placed on it. CNT electrodes have shown superior performance in promoting direct electron transfer with glucose oxidase [30], cytochrome *c* [31, 32], horseradish peroxidase [33, 34], hemoglobin [35], myoglobin [34], and microperoxidase [36], which was not observed with only the base electrode.

As for the realization of direct electron transfer, various nanomaterials other than CNTs have been tried, and numerous papers have been published. When improving sensing performance using this approach, previous trials should be checked by focusing on specific cases.

2.3 Activation of Electrochemiluminescence

Electrochemiluminescence (ECL) is generated by converting electrochemical energy into radiative energy [37, 38]. Advantages of the ECL detection when compared with fluorometry are excellent sensitivity and selectivity, broad dynamic range, spatial controllability, low cost, and compatibility with separation techniques. In particular, the most unique feature of ECL is that it can be initiated and controlled by applying a potential to an electrode. Several different mechanisms of ECL have been proposed: (1) annihilation ECL, (2) co-reactant ECL, and (3) cathodic luminescence. Among them, most of the ECL systems have been developed based on the co-reactant ECL. The co-reactant refers to a species to produce reactive intermediates, which react with a luminophore to form excited

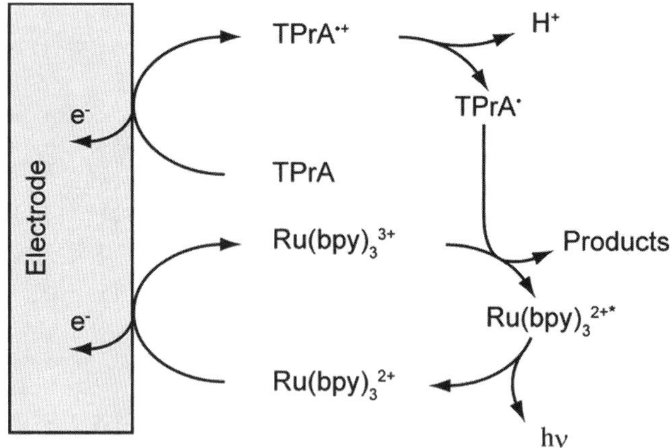

Fig. 7 Reaction mechanism of the $Ru(bpy)_3^{2+}$/TPrA ECL system

states of the luminophore. As an example, a widely used mechanism of tris(2,2′-bipyridyl)ruthenium (II) ($Ru(bpy)_3^{2+}$) and tri-*n*-propylamine (TPrA) system is described in Fig. 7.

Here, $Ru(bpy)_3^{2+}$ and TPrA are a luminophore and a co-reactant, respectively. First, $Ru(bpy)_3^{2+}$ is oxidized on the electrode. TPrA is also oxidized to produce a strong reductant TPrA•. Then, TPrA• and $Ru(bpy)_3^{3+}$ react to generate an excited state ($Ru(bpy)_3^{2+*}$) capable of emitting light. When the excited state returns to the ground state, luminescence whose emission peak is at 620 nm is emitted. Luminophores return to their initial state and can be used repeatedly, which is a marked contrast to luminophores used for chemiluminescence. This multiple excitation cycle amplifies the signal. On the other hand, the background signal is minimal because the stimulation mechanism is decoupled from light. The emitted light is detected using a commercial photomultiplier tube. The intensity of ECL depends on the applied potential and on the concentration of $Ru(bpy)_3^{2+}$ and TPrA.

To analyze nucleic acids, hybridization with DNA probes modified with ECL luminophores has been used. Zhang et al. reported a unique approach to detect a target single-strand DNA using a thiolated hairpin DNA tagged with $Ru(bpy)_3^{2+}$ assembled on a gold electrode (Fig. 8a) [39]. The hairpin DNA and the target DNA hybridize and form a rigid linear double-strand DNA (dsDNA), separating Ru $(bpy)_3^{2+}$ from the electrode. This results in the decrease of the ECL intensity, which can be used for DNA sensing. The lower detection limit was 90 pM. Intercalation of luminophore to dsDNA has also been used to enhance ECL. $[Ru(bpy)_2dppz]^{2+}$ (bpy = 2,2′-bipyridine; dppz = dipyrido[3,2-a:2′,3′-c]phenazine) itself emits negligible ECL. However, when it is intercalated into DNA, the ECL intensity increases by a factor of ~1,000 [40]. With a DNA aptamer against ATP, the ECL intensity decreases upon binding of ATP. With this technique, the lower

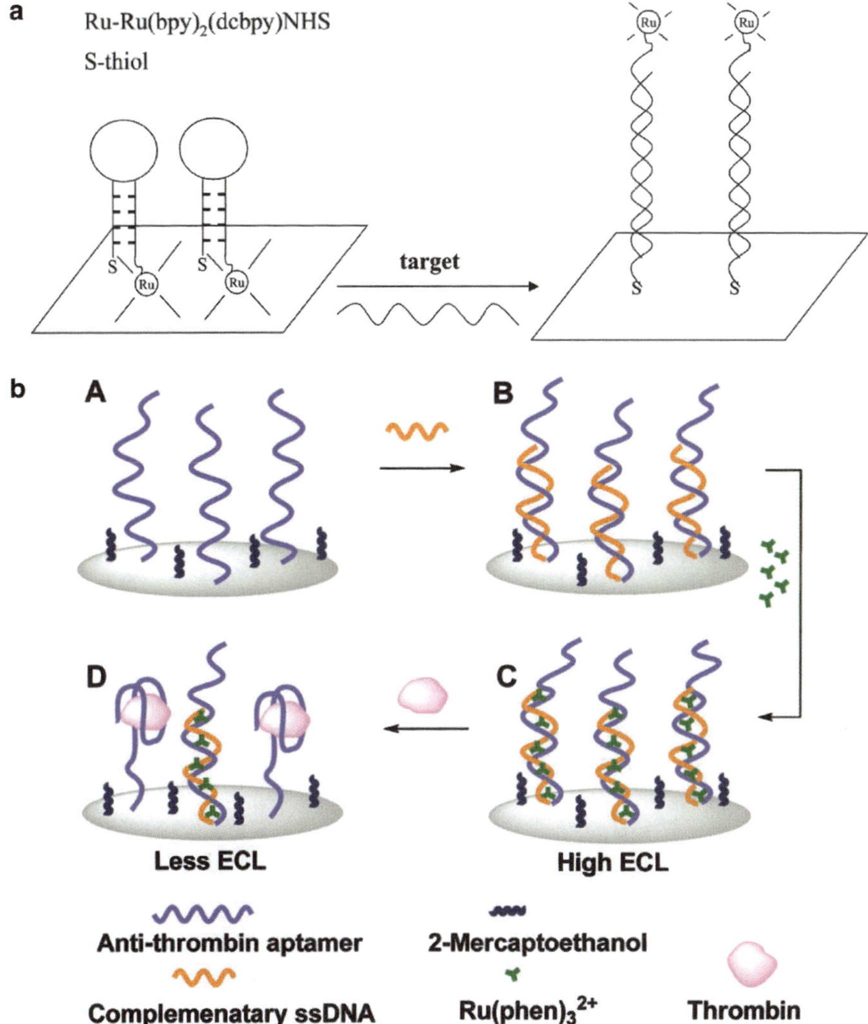

Fig. 8 ECL-based immunoassay. (**a**) Detection of DNA hybridization using hairpin-DNA probes. Reprinted with permission of [39]. Copyright 2008 American Chemical Society. (**b**) Detection of thrombin using label-free ECL aptasensors. (**A**) Attachment of the anti-thrombin thiolated aptamer to an electrode. (**B**) Formation of dsDNA with its cDNA. (**C**) Intercalation of $Ru(phen)_3^{2+}$ into the dsDNA. (**D**) Dissociation of dsDNA and release of $Ru(phen)_3^{2+}$ accompanying the binding of thrombin to its aptmer, resulting in the decrease of ECL. Reprinted with permission of [41]. Copyright 2009 American Chemical Society

detection limit was 100 nM [40]. The idea of using a DNA intercalator as an ECL fluorophore was further developed by Yin et al. (Fig. 8b) [41]. Their label-free ECL aptasensor was constructed based on the intercalation of $Ru(phen)_3^{2+}$ (phen = 1,10-phenanthroline) into dsDNA formed with an aptamer and its complementary DNA.

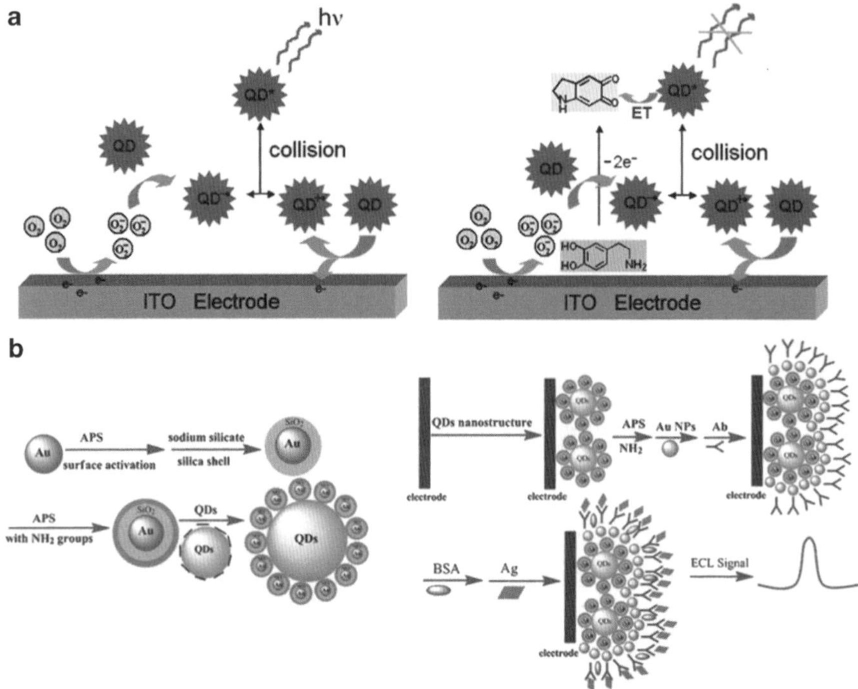

Fig. 9 Quantum-dot-based ECL. (**a**) ECL from QDs (*left*) and quenching (*right*). Reprinted with permission of [42]. Copyright 2007 American Chemical Society. (**b**) Synthesis of gold/silica/CdSe-CdS nanostructures (*left*) and fabrication of the ECL immunosensor (*right*). [44] – Reproduced by permission of The Royal Society of Chemistry

After the target molecule hybridized with its aptamer, the dsDNA dissociated and the intercalated $Ru(phen)_3^{2+}$ was released. The decrease in the ECL signal before and after the target molecule binding was used to quantify the concentration of the target molecule. For thrombin, the lower detection limit of 20 fM has been reported.

Unlike the other ECL methods, most of the quantum dots (QDs) ECL biosensors are developed based on quenching, inhibition, or enhancement of ECL. Quenching is caused by energy transfer between the excited QD and the analyte (or by-products generated from the analyte) when they are in close proximity. Liu et al. demonstrated an ECL quenching process of CdTe QDs and a new method to quantify catechol derivatives, which are ECL quenchers, as analytes (Fig. 9a) [42]. In the method, QD was excited by superoxide anion electrochemically generated at an electrode surface. In the presence of catechol derivatives, such as dopamine or L-adrenalin, energy transfer from the excited QDs to the catechol derivatives occurs, resulting in a significant decrease of ECL emission. The lower detection limit of dopamine was 50 nM.

One of the advantages of QDs is the easiness of functionalization. Yuan et al. reported signal amplification for ECL immunoassay [43]. In their system, cdse-cds nanoparticles were used as cores, which were covalently bound with QDs via polymer chains. The nanoparticles were further modified with antibodies to specifically bind target molecules. With the probe, the ECL measurement achieved 10 times as high sensitivity as that using unmodified QDs. The lower detection limit was a few pg mL^{-1}. As the other superstructure, a gold/silica/CdSe-CdS QD superstructure, formed by coating CdSe-CdS QDs with gold nanoparticles coated with SiO_2, emits ECL that is 17-fold higher than that from pure CdSe-CdS QDs (Fig. 9b) [44]. The structure was used for the detection of carcinoembryonic antigen, and a lower detection limit of 64 fg mL^{-1} has been achieved.

Detection sensitivity of ECL is significantly improved by using a SWCNT forest electrode. In the detection of a cancer marker, prostate specific antigen, that used nanoparticle labels containing $Ru(bpy)_3^{2+}$, 34-fold better sensitivity and 10-fold lower detection limit have been achieved compared with cases that used a pyrolytic graphite electrode [45]. CNT is also effective to enhance ECL from QDs. In the detection of H_2O_2, significant enhancement of ECL has been observed by using a CdS/CNT composite compared with the case without CNT [46].

In the detection of ECL using $Ru(bpy)_3^{2+}$, it is anticipated that the sensitivity is enhanced if $Ru(bpy)_3^{2+}$ ions are concentrated in the vicinity of an electrode. To this end, an effective approach is to use an ion-exchange polymer such as Nafion. With only this structure, the increase in the intensity of ECL was actually observed. However, by incorporating CNT further into this membrane, a 60-fold increase in the ECL intensity has been observed [47], which has been explained by an open structure realized accompanying the incorporation of CNT. Otherwise, co-reactants of $Ru(bpy)_3^{2+}$ can be concentrated on an electrode. In the enzymatic reaction of acetylcholinesterase, thiocholine is produced from acetylcholine. Thiocholine forms a monolayer on a gold electrode by gold-thiol bonding, which works as a concentrated layer of co-reactants for $Ru(bpy)_3^{2+}$ [48]. With antibodies labeled with the enzyme, anti tumor necrosis factor-α (TNF-α) of the sub-pM order has been detected by ELISA [48].

3 Fabrication of Nanoelectrode Arrays and Ensembles

The nanoelectrode structures discussed in previous sections are often similar. Therefore, it will be beneficial to describe the fabrication in a separate section to understand how these structures are formed. In this section, representative techniques to form nanoelectrode arrays and ensembles will be described. These techniques could be used independently or combined with other techniques to improve the detection performance.

3.1 Formation of Nanoelectrode Arrays by Electron-Beam Lithography or Focused Ion Beam Milling

IDAs can be formed by electron beam lithography [49]. By reducing the dimensions of the electrodes to the nm order, improvement of performance is expected. To realize it, techniques that are different from those used for microscale devices are used.

For the fabrication of electrodes of the μm order, photolithography is an appropriate choice. However, fabrication of electrodes with smaller dimensions must rely on other techniques. For electrodes of the sub-μm scale, electron-beam lithography has been used. Interdigitated electrodes of the sub-μm order have also been fabricated by deep UV lithography [50].

As a simple method, an array of nanoelectrodes can be formed by forming a layer of resist on a metal layer and by forming an array of nanoholes by the electron-beam lithography [51, 52]. Focused ion beam can also be used to form a nanoband electrode array. A platinum electrode pattern is formed by conventional photolithography and is passivated with a silicon nitride layer. Nano-scale openings are formed by milling the silicon nitride insulating layer to the bottom so that the underlying platinum is exposed (Fig. 10a) [53]. The exposed platinum areas work as the nanoelectrode array. Nanopore array electrodes have also been fabricated by this technique (Fig. 10b, c) [55].

By using additional techniques, the electrode size can be reduced further. Anisotropic etching is a technique often used for silicon bulk micromachining. This technique uses the difference in etching rates for various crystallographic orientations. The resulting structure can be used to shrink the patterns formed by electron-beam lithography (Fig. 11) [54]. After square patterns are formed in a silicon dioxide protecting layer formed on a very thin (100)-oriented silicon layer, the exposed areas of the silicon layer are anisotropically etched (Fig. 11a, b). This results in inverted pyramid through-holes (Fig. 11c, d). The dimensions of the through-holes formed on the other side of the silicon layer are much smaller than those formed in the silicon dioxide layer formed by electron-beam lithography. Nanoelectrodes with lateral dimensions of 15 nm have been obtained (Fig. 11e, f).

Conical microelectrodes have been fabricated by anisotropic etching of silicon (Fig. 10d) [56]. After the structure was formed, layers of insulators and platinum were formed. The platinum layer was exposed only at the tips by removing the outermost insulating layer after applying a photoresist.

Problems in the above-mentioned techniques are that they are high cost and low throughput. To solve this problem partially, a novel technique called nanoimprint lithography has been proposed (Fig. 10e) [57, 58]. In this method, a mold with nanostructures is first formed using the electron-beam lithography. This mold is then pressed into a thin film of thermoplastic polymer such as poly(methyl methacrylate) (PMMA) formed on a substrate that has been heated above its glass transition temperature. After separating the mold, the polymer residues in the compressed areas are removed by reactive ion etching. The patterns have been used to form interdigitated electrodes by a lift-off process [59, 60] or a nanodisc

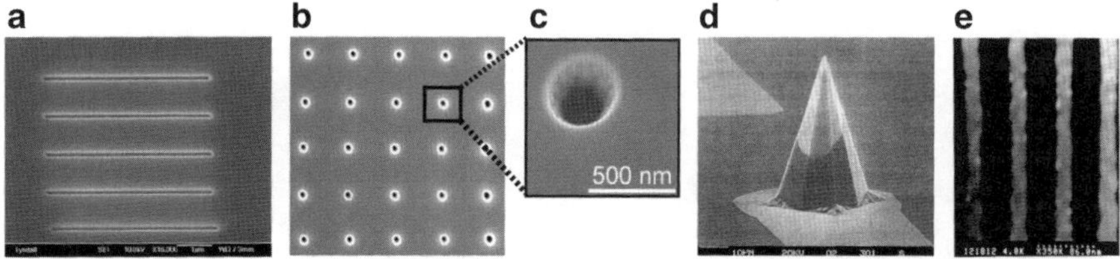

Fig. 10 SEM images of nanoelectrodes. (**a**) Nanoband electrode array fabricated by focused ion beam milling. Reprinted from [53], Copyright 2007, with permission from Elsevier. (**b**) Nanopore electrode array. (**c**) Magnified image of a nanopore. Adapted with permission from [55]. Copyright 2007 American Chemical Society. (**d**) Conical microelectrode with the platinum area exposed only at the tips. Reprinted from [56], Copyright 2000, with permission from Elsevier. (**e**) Metal lines (30 nm width and 70 nm pitch) fabricated by imprint lithography. Reprinted with permission from [58]. Copyright 1996, American Vacuum Society

electrode array by forming an insulating layer with an array of holes on a gold layer [52].

3.2 Formation of Nanoelectrode Arrays by Photolithography

In forming planar nanoelectrode arrays of controlled patterns, the electron-beam lithography is most widely used. Then, can't we form such nanoelectrode arrays if we do not have the very expensive instrument? The answer is yes. Some unique methods have been proposed based on patterning by ordinary photolithography.

An example is illustrated in Fig. 12 [61]. First, a nickel layer is deposited on a substrate, and photoresist patterns are formed. Then, the nickel layer is removed by electrochemically dissolving it. A point here is to overetch the nickel layer and form a horizontal trench under the photoresist layer. By depositing a metal such as gold, platinum, and palladium, and removing the photoresist and the nickel layer, nanoelectrode arrays can be obtained. The patterns of the nanoelectrodes can be designed in a desired manner and a long nanowire of the cm order can easily be obtained.

As already shown in Fig. 3, nanoelectrodes can also be formed on the cross-section of a sandwich structure. Single-band electrodes have been fabricated by depositing a metal electrode layer and an insulating layer and by exposing the electrode on one side [62–65]. Platinum and gold electrodes with widths of the nm order have been reported. Contrary to the width, the other dimension can be very long with this method. The cross section can be straight, circular, or a comb-like structure. The number of nanoelectrodes can be increased by stacking metal and insulator layers alternately [66].

In relation to this technique, a technique named "nanoskiving" has been proposed [67]. As in the previous cases, thin-film metal structures are embedded in epoxy. The epoxy matrix is then sectioned using an ultramicrotome. A section is placed on a substrate and epoxy is removed by oxygen plasma etching. Nanowires fabricated by this technique have been used for electroanalysis [68].

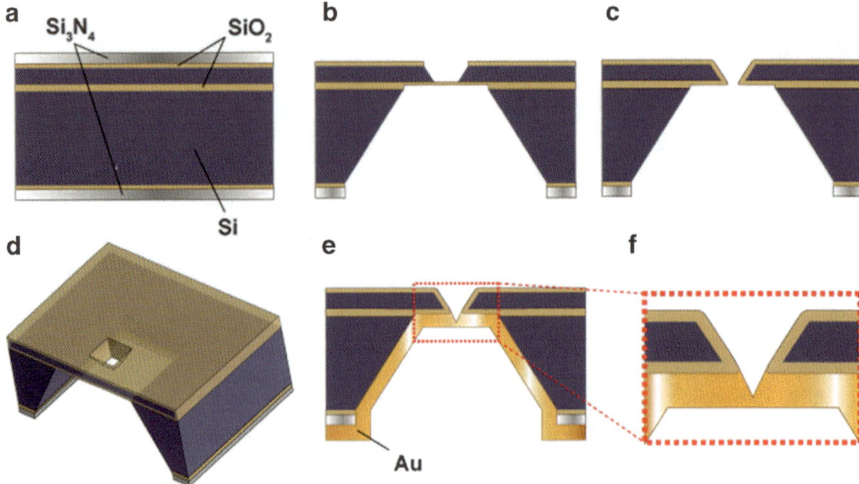

Fig. 11 Cross-sectional and three-dimensional illustrations of the fabrication process of a nanoelectrode using electron-beam lithography. (**a**) Si substrate covered with a SiO_2 layer and a Si_3N_4 layer. (**b**) Formation of a square pattern in the SiO_2 layer and anisotropic etching of the exposed Si layer. (**c**) Formation of pyramid-shaped holes through the wafer. (**d**) Three-dimensional view of the structure shown in (**c**). (**e**) Deposition of gold on one side of the device. (**e**) Magnified view of the nanoelectrode. Adapted with permission from [54]. Copyright 2005 American Chemical Society

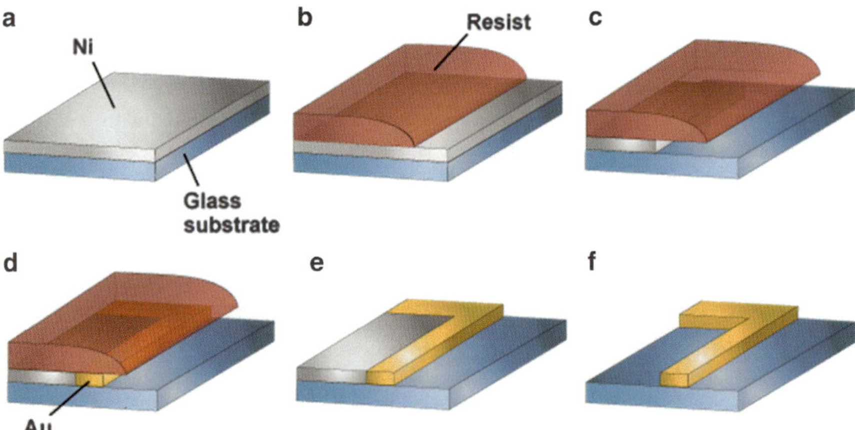

Fig. 12 Fabrication of a gold nanowire. (**a**) Deposition of a sacrificial Ni layer onto a glass substrate. (**b**) Formation of a photoresist pattern. (**c**) Overetching of the Ni layer. (**d**) Deposition of a gold layer into the trench between the photoresist and the glass substrate. (**e**) Photoresist removed. (**f**) Ni removed. Adapted by permission from Macmillan Publishers Ltd: [61], copyright 2006

3.3 Formation of a Layer of Randomly Oriented CNTs

CNTs activate electrode reactions. In taking advantage of this, the simplest method is to form a layer of randomly oriented CNTs by casting a solution containing CNTs. Here, a problem is unavailability of appropriate solvents [69]. CNTs can be made water-soluble by adsorbing surfactant molecules on the surface of CNTs [69–71]. CNTs are first dispersed in a solution of a surfactant whose concentration is higher than the critical micellar concentration, which is the concentration at which surfaces are saturated with surfactant and the surfactant molecules start self-aggregating into micelles. Then, the solution is sonicated. The solution is used to cast CNTs onto an appropriate substrate and form the layer. A concern with this method may be the existence of the surfactant, which may influence the physical properties of CNTs and induce unwanted chemical reactions. However, when the CNT layer is used for sensing, the surfactant can be removed by washing the layer with distilled water [69].

Solubility of CNT has also been improved by wrapping CNT in polymeric chains. Molecules of high molecular weight thread themselves onto or wrap themselves around the surfaces of CNTs and disrupt van der Waals interactions that cause CNTs to aggregate into bundles. For this purpose, poly (metaphenylne-vinylene) [72] or Nafion [20] has been used.

Solubilization of CNTs is a critical theme in many research fields. Therefore, many other methods have been developed and the effort continues even now [73].

3.4 Formation of CNT Nanoelectrode Ensembles

If CNTs are formed in a more controlled manner, nanoelectrode ensembles with appropriate interelectrode spacing could be realized by directly growing CNTs on a substrate. Low-density nanoelectrode arrays of CNTs have been fabricated by depositing Ni seeding nanoparticles first by electrochemical deposition and growing CNTs by plasma-enhanced chemical vapor deposition (Fig. 13) [74]. The CNTs were embedded in an epoxy layer. After polishing the surface, an array of tips of the CNT was obtained. Ni or NiFe alloy seeding spots can also be patterned by electron-beam lithography. Vertically aligned fibers with controlled interelectrode spacing have been fabricated [75–77].

3.5 Formation of Nanoelectrode Ensembles Using a Template

The porous structure of aluminum anodic oxide films has been used as a template to form various nanostructures [78]. Gold nanoelectrode ensembles were fabricated using this technique [79]. In forming the ensembles, a microporous oxide layer was first grown on an aluminum substrate. After the oxide layer was removed from the

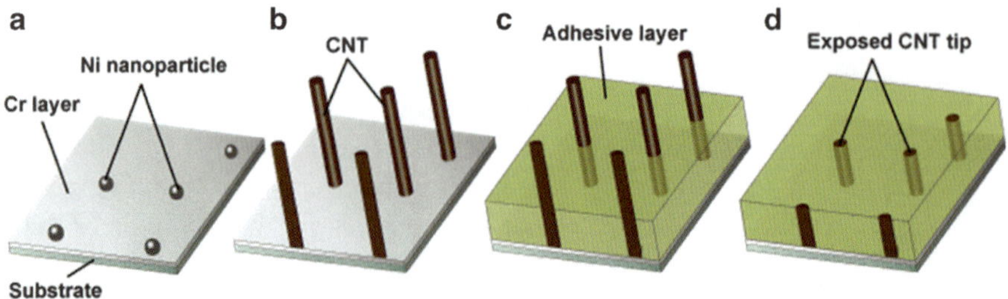

Fig. 13 Fabrication of a nanoelectrode array using vertically grown low-density CNTs. (**a**) Electrochemical deposition of Ni nanoparticles. (**b**) Growth of CNTs by plasma-enhanced chemical vapor deposition on the Ni particles. (**c**) Coating the surface with an adhesive. (**d**) Polishing of the adhesive layer. Adapted with permission from [74]. Copyright 2003 American Chemical Society

substrate to use it as a template, a gold layer was formed by vacuum-depositing gold onto the open ends of the pores. The side of the substrate covered with gold was attached to a glassy carbon electrode, and the other side of the porous oxide film was etched. As a result, the gold nanoelectrode ensemble was exposed.

Porous polycarbonate membranes have also been used as templates to form disk electrode ensembles of platinum [80] and gold [81]. The membranes were formed by an irradiation/chemical etch technique. In forming the ensemble of platinum electrodes, the membrane was fixed on a platinum electrode and platinum was deposited in the pores. After the pores were stuffed with platinum and the membrane surface was covered with overgrown platinum, the membrane surface was exposed again by removing the excess platinum [80]. The gold disk electrode ensembles were fabricated by depositing gold on the walls of nanopores by electroless deposition and following the same procedure [81]. With this technique, three-dimensional nanoelectrode ensembles can also be formed by partially or totally removing carbonate by dissolving it in an appropriate solvent such as dichloromethane [81] or by oxygen plasma etching [82].

3.6 *Other Techniques*

Nanoelectrode ensembles can be formed by opening up nanoholes in an insulating layer formed on an electrode. Defects in a self-assembled monolayer formed on a gold electrode work as nanoelectrode ensembles [83–85].

Like many other techniques to fabricate nanoelectrode ensembles, a problem is the control of pore size and distribution over the electrode surface. To solve this problem, the insulating film was formed with a highly ordered self-assembling block copolymer film [86]. The film is formed by spin-coating a polystylene (PS)/ PMMA diblock copolymer solution onto a gold electrode. The dried film is then annealed in the presence of a strong electric field to orient the PMMA perpendicularly to the electrode surface. Exposure to UV radiation simultaneously cross-links the PS and degrades the PMMA. The PMMA is finally dissolved in glacial acetic acid to form the pores.

Nanoelectrode ensembles can also be formed on a planar electrode even without an insulating layer. The overpotential to oxidize or reduce an electroactive analyte depends on the electrode material. Therefore, if nanoscale deposits of a metal that are active to an analyte are formed on a planar electrode of a different metal that is inactive to the analyte, the deposits work as a nanoelectrode ensemble. Platinum black particles were deposited on a gold electrode to form a nanoelectrode ensemble of platinum black. A significant enhancement of sensitivity to H_2O_2 was observed compared with the planar gold electrode used as the base electrode [15].

4 Concluding Remarks

In this chapter, techniques to improve the detection sensitivity of electrochemical microdevices were reviewed. Here, a convincing conclusion is that nanotechnology really has an effect for this purpose. Reflecting the fact and growing expectations, various nanomaterials such as nanotubes, nanowires, and nanoparticles have been used very actively. As a result, numerous papers have been published and the tendency continues even now [87–90]. At present, the application of nanotechnology to electrochemical microdevices is limited. However, with the wealth of knowledge of nanotechnology, devices whose performance is comparable with or better than those of macroscopic counterparts used for ordinary electroanalysis will surely be realized.

References

1. Wightman RM (1981) Microvoltammetric electrodes. Anal Chem 53(9):1125A–1134A
2. Niwa O (1995) Electroanalysis with interdigitated array microelectrodes. Electroanalysis 7(7):606–613
3. Štulík K, Amatore C, Holub K, Mareček V, Kutner W (2000) Microelectrodes. Definitions, characterization, and applications. Pure Appl Chem 72(8):1483–1492
4. Aoki A, Matsue T, Uchida I (1990) Electrochemical response at microarray electrodes in flowing streams and determination of catecholamines. Anal Chem 62:2206–2210
5. Vandaveer IV WR, Woodward DJ, Fritsch I (2003) Redox cycling measurements of a model compound and dopamine in ultrasmall volumes with a self-contained microcavity device. Electrochim Acta 48:3341–3348
6. Zevenbergen MAG, Krapf D, Zuiddam MR, Lemay SG (2007) Mesoscopic concentration fluctuations in a fluidic nanocavity detected by redox cycling. Nano Lett 7(2):384–388
7. Wolfrum B, Zevenbergen M, Lemay S (2008) Nanofluidic redox cycling amplification for the selective detection of catechol. Anal Chem 80:972–977
8. Zevenbergen MAG, Wolfrum BL, Goluch ED, Singh PS, Lemay SG (2009) Fast electron-transfer kinetics probed in nanofluidic channels. J Am Chem Soc 131:11471–11477

9. Kätelhön E, Hofmann B, Lemay SG, Zevenbergen MAG, Offenhäusser A, Wolfrum B (2010) Nanocavity redox cycling sensors for the detection of dopamine fluctuations in microfluidic gradients. Anal Chem 82:8502–8509
10. Łoś M, Łoś JM, Blohm L, Spillner E, Grunwald T, Albers J, Hintsche R, Węgrzyn G (2005) Rapid detection of viruses using electrical biochips and anti-virion sera. Lett Appl Microbiol 40:479–485
11. Bange A, Tu J, Zhu X, Ahn C, Halsall HB, Heineman WR (2007) Electrochemical detection of MS2 phage using a bead-based immunoassay and a nanoIDA. Electroanalysis 19 (21):2202–2207
12. Nebling E, Grunwald T, Albers J, Schäfer P, Hintsche R (2004) Electrical detection of viral DNA using ultramicroelectrode arrays. Anal Chem 76:689–696
13. Elsholz B, Wörl R, Blohm L, Albers J, Feucht H, Grunwald T, Jürgen B, Schweder T, Hintsche R (2006) Automated detection and quantitation of bacterial RNA by using electrical microarrays. Anal Chem 78:4794–4802
14. Albers J, Grunwald T, Nebling E, Piechotta G, Hintsche R (2003) Electrical biochip technology – a tool for microarrays and continuous monitoring. Anal Bioanal Chem 377:521–527
15. Niwa O, Horiuchi T, Morita M, Huang T, Kissinger PT (1996) Determination of acetylcholine and choline with platinum-black ultramicroarray electrodes using liquid chromatography with a post-column enzyme reactor. Anal Chim Acta 318:167–173
16. Evans SAG, Elliott JM, Andrews LM, Bartlett PN, Doyle PJ, Denuault G (2002) Detection of hydrogen peroxide at mesoporous platinum microelectrodes. Anal Chem 74:1322–1326
17. Song Y-Y, Zhang D, Gao W, Xia X-H (2005) Nonenzymatic glucose detection by using a three-dimensionally ordered, macroporous platinum template. Chem Eur J 11:2177–2182
18. Park S, Chung TD, Kim HC (2003) Nonenzymatic glucose detection using mesoporous platinum. Anal Chem 75:3046–3049
19. Yuan J, Wang K, Xia X (2005) Highly ordered platinum-nanotubule arrays for amperometric glucose sensing. Adv Funct Mater 15(5):803–809
20. Wang J, Musameh M, Lin Y (2003) Solubilization of carbon nanotubes by nafion toward the preparation of amperometric biosensors. J Am Chem Soc 125:2408–2409
21. Hrapovic S, Liu Y, Male KB, Luong JHT (2004) Electrochemical biosensing platforms using platinum nanoparticles and carbon nanotubes. Anal Chem 76:1083–1088
22. Ye J-S, Wen Y, Zhang WD, Gan LM, Xu GQ, Sheu F-S (2004) Nonenzymatic glucose detection using multi-walled carbon nanotube electrodes. Electrochem Commun 6:66–70
23. Britto PJ, Santhanam KSV, Rubio A, Alonso JA, Ajayan PM (1999) Improved charge transfer at carbon nanotube electrodes. Adv Mater 11(2):154–157
24. Britto PJ, Santhanam KSV, Ajayan PM (1996) Carbon nanotube electrode for oxidation of dopamine. Bioelectrochem Bioenerg 41:121–125
25. Luo H, Shi Z, Li N, Gu Z, Zhuang Q (2001) Investigation of the electrochemical and electrocatalytic behavior of single-wall carbon nanotube film on a glassy carbon electrode. Anal Chem 73:915–920
26. Musameh M, Wang J, Merkoci A, Lin Y (2002) Low-potential stable NADH detection at carbon-nanotube-modified glassy carbon electrodes. Electrochem Commun 4:743–746
27. Dumitrescu I, Unwin PR, Macpherson JV (2009) Electrochemistry at carbon nanotubes: perspective and issues. Chem Commun 6886–6901
28. Yang W, Ratinac KR, Ringer SP, Thordarson P, Gooding JJ, Braet F (2010) Carbon nanomaterials in biosensors: should you use nanotubes or graphene? Angew Chem Int Ed 49:2114–2138
29. Gong K, Chakrabarti S, Dai L (2008) Electrochemistry at carbon nanotube electrodes: is the nanotube tip more active than the sidewall? Angew Chem Int Ed 47:5446–5450
30. Cai C, Chen J (2004) Direct electron transfer of glucose oxidase promoted by carbon nanotubes. Anal Biochem 332:75–83
31. Wang G, Xu J-J, Chen H-Y (2002) Interfacing cytochrome *c* to electrodes with a DNA - carbon nanotube composite film. Electrochem Commun 4:506–509

32. Wang J, Li M, Shi Z, Li N, Gu Z (2002) Direct electrochemistry of cytochorome *c* at a glassy carbon electrode modified with single-wall carbon nanotubes. Anal Chem 74:1993–1997
33. Zhao Y-D, Zhang W-D, Chen H, Luo Q-M, Li SFY (2002) Direct electrochemistry of horseradish peroxidase at carbon nanotube powder microelectrode. Sens Actuators B 87:168–172
34. Yu X, Chattopadhyay D, Galeska I, Papadimitrakopoulos F, Rusling JF (2003) Peroxidase activity of enzymes bound to the ends of single-wall carbon nanotube forest electrodes. Electrochem Commun 5:408–411
35. Cai C, Chen J (2004) Direct electron transfer and bioelectrocatalysis of hemoglobin at a carbon nanotube electrode. Anal Biochem 325:285–292
36. Gooding JJ, Wibowo R, Liu J, Yang W, Losic D, Orbons S, Mearns FJ, Shapter JG, Hibbert DB (2003) Protein electrochemistry using aligned carbon nanotube arrays. J Am Chem Soc 125:9006–9007
37. Richter MM (2004) Electrochemiluminescence (ECL). Chem Rev 104:3003–3036
38. Miao W (2008) Electrogenerated chemiluminescence and its biorelated applications. Chem Rev 108:2506–2553
39. Zhang J, Qi H, Li Y, Yang J, Gao Q, Zhang C (2008) Electrogenerated chemiluminescence DNA biosensor based on hairpin DNA probe labeled with ruthenium complex. Anal Chem 80:2888–2894
40. Hu L, Bian Z, Li H, Han S, Yuan Y, Gao L, Xu G (2009) $[Ru(bpy)_2dppz]^{2+}$ electrochemiluminescence switch and its applications for DNA interaction study and label-free ATP aptasensor. Anal Chem 81:9807–9811
41. Yin X-B, Xin Y-Y, Zhao Y (2009) Label-free electrochemiluminescent aptasensor with attomolar mass detection limits based on a $Ru(phen)_3^{2+}$-double-strand DNA composite film electrode. Anal Chem 81:9299–9305
42. Liu X, Jiang H, Lei J, Ju H (2007) Anodic electrochemiluminescence of CdTe quantum dots and its energy transfer for detection of catechol derivatives. Anal Chem 79(21):8055–8060
43. Yuan L, Hua X, Wu Y, Pan X, Liu S (2011) Polymer-functionalized cdse-cds nanosphere labels for ultrasensitive detection of tumor necrosis factor-alpha. Anal Chem 83:6800–6809
44. Jie G-F, Liu P, Zhang S-S (2010) Highly enhanced electrochemiluminescence of novel gold/silica/CdSe-CdS nanostructures for ultrasensitive immunoassay of protein tumor marker. Chem Commun 46:1323–1325
45. Sardesai N, Pan S, Rusling J (2009) Electrochemiluminescent immunosensor for detection of protein cancer biomarkers using carbon nanotube forests and $[Ru\text{-}(bpy)_3]^{2+}$-doped silica nanoparticles. Chem Commun 4968–4970
46. Ding S-N, Xu J-J, Chen H-Y (2006) Enhanced solid-state electrochemiluminescence of CdS nanocrystals composited with carbon nanotubes in H_2O_2 solution. Chem Commun 3631–3633
47. Guo Z, Dong S (2004) Electrogenerated chemiluminescence from $Ru(bpy)_3^{2+}$ ion-exchanged in carbon nanotube/perfluorosulfonated ionomer composite films. Anal Chem 76:2683–2688
48. Kurita R, Arai K, Nakamoto K, Kato D, Niwa O (2010) Development of electrogenerated chemiluminescence-based enzyme linked immunosorbent assay for sub-pM detection. Anal Chem 82:1692–1697
49. Ueno K, Hayashida M, Ye J-Y, Misawa H (2005) Fabrication and electrochemical characterization of interdigitated nanoelectrode arrays. Electrochem Commun 7:161–165
50. van Gerwen P, Laureyn W, Laureys W, Huyberechts G, de Beeck MO, Baert K, Suls J, Sansen W, Jacobs P, Hermans L, Mertens R (1998) Nanoscaled interdigitated electrode arrays for biochemical sensors. Sens Actuators B 49:73–80
51. Hepel T, Osteryoung J (1986) Electrochemical characterization of electrodes with submicrometer dimensions. J Electrochem Soc 133(4):752–757
52. Sandison ME, Cooper JM (2006) Nanofabrication of electrode arrays by electron-beam and nanoimprint lithographies. Lab Chip 6:1020–1025
53. Lanyon YH, Arrigan DWM (2007) Recessed nanoband electrodes fabricated by focused ion beam milling. Sens Actuators B 121:341–347

54. Lemay SG, van den Broek DM, Storm AJ, Krapf D, Smeets RMM, Heering HA, Dekker C (2005) Lithographycally fabricated nanopore-based electrodes for electrochemistry. Anal Chem 77:1911–1915
55. Lanyon YH, Marzi GD, Watson YE, Quinn AJ, Gleeson JP, Redmond G, Arrigan DWM (2007) Fabrication of nanopore array electrodes by focused ion beam milling. Anal Chem 79:3048–3055
56. Thiébaud P, Beuret C, de Rooij NF, Koudelka-Hep M (2000) Microfabrication of Pt-tip microelectrodes. Sens Actuators B 70:51–56
57. Chou SY, Krauss PR, Renstrom PJ (1996) Imprint lithography with 25-nanometer resolution. Science 272:85–87
58. Chou SY, Krauss PR, Renstrom PJ (1996) Nanoimprint lithography. J Vac Sci Technol B 14 (6):4129–4133
59. Montelius L, Heidari B, Graczyk M, Maximov I, Sarwe E-L, Ling TGI (2000) Nanoimprint- and UV-lithography: Mix&Match process for fabrication of interdigitated nanobiosensors. Microelectron Eng 53:521–524
60. Beck M, Persson F, Carlberg P, Graczyk M, Maximov I, Ling TGI, Montelius L (2004) Nanoelectrochemical transducers for (bio-) chemical sensor applications fabricated by nanoimprint lithography. Microelectron Eng 73–74:837–842
61. Menke EJ, Thompson MA, Xiang C, Yang LC, Penner RM (2006) Lithographically patterned nanowire electrodeposition. Nat Mater 5:914–919
62. Nagale MP, Fritsch I (1998) Individually addressable, submicrometer band electrode arrays. 1. Fabrication from multilayered materials. Anal Chem 70:2902–2907
63. Morris RB, Franta DJ, White HS (1987) Electrochemistry at Pt band electrodes of width approaching molecular dimensions. Breakdown of transport equations at very small electrodes. J Phys Chem 91:3559–3564
64. Samuelsson M, Armgarth M, Nylander C (1991) Microstep electrodes: band ultramicroelectrodes fabricated by photolithography and reactive ion etching. Anal Chem 63:931–936
65. Caston SL, McCarley RL (2002) Characteristics of nanoscopic Au band electrodes. J Electroanal Chem 529:124–134
66. Odell DM, Bowyer WJ (1990) Fabrication of band microelectrode arrays from metal foil and heat-sealing fluoropolymer film. Anal Chem 62:1619–1623
67. Xu Q, Rioux RM, Whitesides GM (2007) Fabrication of complex metallic nanostructures by nanoskiving. ACS Nano 1(3):215–227
68. Dawson K, Strutwolf J, Rodgers KP, Herzog G, Arrigan DWM, Quinn AJ, O'Riordan A (2011) Single nanoskived nanowidres for electrochemical applications. Anal Chem 83:5535–5540
69. Bonard J-M, Stora T, Salvetat J-P, Maier F, Stöckli T, Duschl C, Forró L, de Heer WA, Châtelain A (1997) Purification and size-selection of carbon nanotubes. Adv Mater 9 (10):827–831
70. O'Connel MJ, Bachilo SM, Huffman CG, Moore VC, Strano MS, Haroz EH, Rialon KL, Boul PJ, Noon WH, Kittrell C, Ma J, Hauge RH, Weisman RB, Smalley RE (2002) Band gap fluorescence from individual single-walled carbon nanotubes. Science 297:593–596
71. Richard C, Balavoine F, Schultz P, Ebbesen TW, Mioskowski C (2003) Supramolecular self-assembly of lipid derivatives on carbon nanotubes. Science 300:775–778
72. Star A, Stoddart JF, Steuerman D, Diehl M, Boukai A, Wong EW, Yang X, Chung S-W, Choi H, Heath JR (2001) Preparation and properties of polymer-wrapped single-walled carbon nanotubes. Angew Chem Int Ed 40(9):1721–1725
73. Merkoçi A, Pumera M, Llopis X, Pérez B, del Valle M, Alegret S (2005) New materials for electrochemical sensing VI: carbon nanotubes. Trends Anal Chem 24(9):826–838
74. Tu Y, Lin Y, Ren ZF (2003) Nanoelectrode arrays based on low site density aligned carbon nanotubes. Nano Lett 3(1):107–109

75. Ren ZF, Huang ZP, Wang DZ, Wen JG, Xu JW, Wang JH, Calvet LE, Chen J, Klemic JF, Reed MA (1999) Growth of a single freestanding multiwall carbon nanotube on each nanonickel dot. Appl Phys Lett 75(8):1086–1088
76. Guillorn MA, McKnight TE, Melechko A, Merkulov VI, Britt PF, Austin DW, Lowndes DH, Simpson ML (2002) Individually addressable vertically aligned carbon nanofiber-based electrochemical probes. J Appl Phys 91(6):3824–3828
77. Koehne J, Li J, Cassell AM, Chen H, Ye Q, Ng HT, Han J, Meyyappan M (2004) The fabrication and electrochemical characterization of carbon nanotube nanoelectrode arrays. J Mater Chem 14:676–684
78. Masuda H, Fukuda K (1995) Ordered metal nanohole arrays made by a two-step replication of honeycomb structures of anodic alumina. Science 268:1466–1468
79. Uosaki K, Okazaki K, Kita H, Takahashi H (1990) Preparative method for fabricating a microelectrode ensemble: electrochemical response of microporous alminum anodic oxide film modified gold electrode. Anal Chem 62:652–656
80. Penner RM, Martin CR (1987) Prepartion and electrochemical characterization of ultramicroelectrode ensembles. Anal Chem 59:2625–2630
81. Menon VP, Martin CR (1995) Fabrication and evaluation of nanoelectrode ensembles. Anal Chem 67:1920–1928
82. Yu S, Li N, Wharton J, Martin CR (2003) Nano wheat fields prepared by plasma-etching gold nanowire-containing membranes. Nano Lett 3(6):815–818
83. Sabatani E, Rubinstein I (1987) Organized self-assembling monolayers on electrodes. 2. Monolayer-based ultramicroelectrodes for the study of very rapid electrode kinetics. J Phys Chem 91:6663–6669
84. Chailapakul O, Sun L, Xu C, Crooks RM (1993) Interactions between organized, surface-confined monolayers and vapor-phase probe molecules. 7. Comparison of self-assembling *n*-alkanethiol monolayers deposited on gold from liquid and vapor phases. J Am Chem Soc 115:12459–12467
85. Che G, Cabrera CR (1996) Molecular recognition based on (3-mercaptopropyl) trimethoxysilane modified gold electrodes. J Electroanal Chem 417:155–161
86. Jeoung E, Galow TH, Schotter J, Bal M, Ursache A, Tuominen MT, Stafford CM, Russell TP, Rotello VM (2001) Fabrication and characterization of nanoelectrode arrays formed via block copolymer self-assembly. Langmuir 17:6396–6398
87. Arrigan DWM (2004) Nanoelectrodes, nanoelectrode arrays and their applications. Analyst 129:1157–1165
88. Pumera M, Sánchez S, Ichinose I, Tang J (2007) Electrochemical nanobiosensors. Sens Actuators B 123:1195–1205
89. Noy A (2011) Bionanoelectronics. Adv Mater 23:807–820
90. Rassaei L, Singh PS, Lemay SG (2011) Lithography-based nanoelectrochemistry. Anal Chem 83:3974–3980

Index

A. Tuantranont (ed.), *Applications of Nanomaterials in Sensors and Diagnostics*, Springer Series on Chemical Sensors and Biosensors (2013) 14: 281–286
DOI 10.1007/5346_2013,

Printed by Printforce, the Netherlands